Europa – The Ocean Moon
Search for an Alien Biosphere

Cover images: The pole-to-pole color mosaic of the trailing hemisphere of Europa was assembled by Moses Milazzo at the Planetary Image Research Laboratory, of the Lunar and Planetary Laboratory, University of Arizona, from images taken by NASA's *Galileo* mission. The cover also includes a previously unpublished portion of the highest-resolution image of Europa made by *Galileo*, and an oblique airliner-type view of the distant horizon, through diamond-shaped frames, with a foreshortened cycloidal ridge.

Richard Greenberg

Europa – The Ocean Moon

Search for an Alien Biosphere

Published in association with
Praxis Publishing
Chichester, UK

Professor Richard Greenberg
Department of Planetary Sciences and Lunar and Planetary Laboratory
University of Arizona
Tucson
Arizona
USA

SPRINGER–PRAXIS BOOKS IN GEOPHYSICAL SCIENCES
SUBJECT *ADVISORY EDITORS*: Dr Philippe Blondel, C.Geol., F.G.S., Ph.D., M.Sc., Senior Scientist, Department of Physics, University of Bath, Bath, UK; John Mason, M.Sc., B.Sc., Ph.D.

ISBN 3-540-22450-5 Springer-Verlag Berlin Heidelberg New York

Springer is part of Springer-Science + Business Media (springeronline.com)

Bibliographic information published by Die Deutsche Bibliothek

Die Deutsche Bibliothek lists this publication in the Deutsche Nationalbibliografie; detailed bibliographic data are available from the Internet at http://dnb.ddb.de

Library of Congress Control Number: 2004114620

Apart from any fair dealing for the purposes of research or private study, or criticism or review, as permitted under the Copyright, Designs and Patents Act 1988, this publication may only be reproduced, stored or transmitted, in any form or by any means, with the prior permission in writing of the publishers, or in the case of reprographic reproduction in accordance with the terms of licences issued by the Copyright Licensing Agency. Enquiries concerning reproduction outside those terms should be sent to the publishers.

© Praxis Publishing Ltd, Chichester, UK, 2005
Printed in Germany

The use of general descriptive names, registered names, trademarks, etc. in this publication does not imply, even in the absence of a specific statement, that such names are exempt from the relevant protective laws and regulations and therefore free for general use.

Cover design: Jim Wilkie
Project Management: Originator Publishing Services, Gt Yarmouth, Norfolk, UK

Printed on acid-free paper

Contents

List of figures ix

Preface xiii

PART ONE   DISCOVERING EUROPA 1

1 Water World 3
   1.1   Life on a water world 3
   1.2   Is this for real? 4
   1.3   Tides 6

2 Touring the surface 7
   2.1   The global picture 7
   2.2   Zoom in to the regional scale 14
   2.3   Zooming closer: surface morphology 18
   2.4   Ridges 20
   2.5   Chaotic terrain 23

3 Politics and intellect: Converting images into ideas and knowledge 29
   3.1   Politics on board 31
   3.2   Methods of the geologists 32
   3.3   The rule of canon law 34
   3.4   *Galileo* in the 20th century 36
   3.5   Technological obsolescence 38

PART TWO TIDES 45

4 Tides and resonance 47
    4.1 Act locally, think globally 47
    4.2 Tidal distortion—the primary component 49
    4.3 *Galileo* data, the Laplace resonance, and orbital eccentricity. . . . 51
    4.4 The effect of orbital eccentricity—the variable component of the tides 54
    4.5 Effects of tides 57

5 Tides and rotation 61
    5.1 Synchronous rotation from the primary tidal component 61
    5.2 Non-synchronous rotation from the diurnal tide 63
    5.3 Rotational effects on Europa 67

6 Tides and stress 71
    6.1 Tidal stress due to non-synchronous rotation 74
    6.2 Tidal stress due to diurnal variation 80
    6.3 Tidal stress: non-synchronous and diurnal stress combined 83

7 Tidal heating 85

8 Tides and orbital evolution 91
    8.1 Orbital theory 91
    8.2 Politics takes control 97

PART THREE UNDERSTANDING EUROPA 101

9 Global cracking and non-synchronous rotation 103
    9.1 Lineaments formed by cracking 103
    9.2 The tectonic record of non-synchronous rotation 105
    9.3 How fast does Europa rotate? 112
    9.4 Large-scale tectonic patterns—summary 116

10 Building ridges 117
    10.1 Other ridge formation models 122
    10.2 Downwarping, marginal cracking, multi-ridge complexes, and dark margins 127
    10.3 Cracking through to the ocean 131

11 Dilation of cracks 133

12 Strike–slip 145
    12.1 Displacement at Astypalaea 146

	12.2	Tidal walking	149
	12.3	Predicting strike–slip	151
	12.4	Surveying strike–slip on Europa	157
	12.5	Particularly-striking examples	162
		12.5.1 The greatest displacement champion	162
		12.5.2 A time sequence of strike–slip	167
		12.5.3 A long, bent, equatorial cycloid in RegMap 01	168
	12.6	Polar wander	173
	12.7	Strike–slip summary	178

13 Return to Astypalaea 181

14 Cycloids 191

15 Rotation revisited 207
  15.1 Cycloid constraints on the rotation rate 207
  15.2 Contradictions with previous work 210
  15.3 Back to Udaeus–Minos 211

16 Chaos 219
  16.1 Characteristic appearance 219
  16.2 Three hypotheses for formation of chaos 227
  16.3 Our survey 231
  16.4 Melt-through 238
  16.5 Volcanism, not 243
  16.6 Heat for melt-through 247

17 Crust convergence 251
  17.1 Balancing the surface area budget 251
  17.2 Surface corrugations 252
  17.3 Chaotic terrain as a surface area sink 254
  17.4 Convergence bands 255
  17.5 The Evil Twin of Agenor 258

18 The scars of impact 265
  18.1 Gauges of age and crust thickness 265
  18.2 Numbers of impact features: Implications for surface age 266
  18.3 Appearance of impact features: Implications for ice thickness 269

19 Pits and uplifts 285
  19.1 Undeniable (if you know what's good for you) facts 285
  19.2 The myth of pits, spots, and domes 287
       19.2.1 PSDs and lenticulae 291

		19.2.2	Are any PSDs pits or domes?	293
		19.2.3	Farewell to PSDs	295
	19.3	Survey of pits and uplifts		297
		19.3.1	Pit counts	298
		19.3.2	Uplift counts	302
	19.4	Formation of pits and uplifts		306
		19.4.1	Survey results vs. the PSD taxonomy	306
		19.4.2	What are these things?	307

PART FOUR   LIFE ON EARTH AND EUROPA 311

20 The bandwagon 313
  20.1 Strike–slip in thick ice 314
  20.2 Overburden flexure 316
  20.3 Melt-through bashing 318
  20.4 Convection models 320

21 The biosphere 323
  21.1 Dreams of life 323
  21.2 Thin ice on a water world 324
  21.3 Substances above and below 326
  21.4 Life in the crust 327
  21.5 Planetary protection 331
    21.5.1 The possibility of contamination 331
    21.5.2 Standards and risk 332
    21.5.3 Getting it right 334

22 The exploration to come 337
  22.1 Plans for future space missions 337
  22.2 Look in the ice 340
  22.3 Mothballed data 342
  22.4 Weird features: The exceptions that hold the keys 343
    22.4.1 The many-legged spider of Manannán 343
    22.4.2 Disruption in the Sickle 343
    22.4.3 Short, curved double ridges within Astypalaea 346
    22.4.4 Isolated tilted rafts 346
    22.4.5 Horsetail of Agenor 348
    22.4.6 Multiple-cusp cycloids 348
    22.4.7 Old-style bands 348
  22.5 Self-correcting science 352

References 355

Index 375

Figures

2.1	(a) Galileo images of the Medician Stars, now known as the Galilean satellites. (b) A more recent telescopic view of Jupiter with the Galilean satellites	8
2.2	Full disk centered near 290°W	color section
2.3	Full disk centered near 40°W	color section
2.4a	Full-disk mosaic from orbit G1, centered near the equator at longitude about 220°W	12
2.4b	Identification of important landmarks that appear in Figure 2.4a	13
2.5a	Color composite of the Udaeus–Minos region	color section
2.5b,c,d	Separate filtered images used to make 2.5a	15, 16
2.6	Region from Pwyll to Conamara	color section
2.7	An enlargement of a 300-km-wide portion of Figure 2.6 showing Conamara Chaos	17
2.8	The Conamara region in a mosaic of images at about 180 m/pixel	20
2.9	A very-high-resolution image of densely ridged terrain	22
2.10	Part of a high-resolution image sequence of Conamara Chaos	24
2.11	A blow-up of the top center portion of Figure 2.8	26
2.12	A blow-up of detail of Figure 2.8	27
2.13	An interesting theory of chaotic terrain	28
4.1	Lunar tides on Earth	48
4.2a	The tide on Europa shown schematically	56
4.2b	Change in tide during an orbit	color section
4.3	Interrelationships among the processes that govern Europa's geology and geophysics	58
5.1	As Europa rotates, surface features are carried around relative to the direction of Jupiter	62
5.2	A schematic representation of the total tidal potential	65
5.3	As the ice shell of Europa slips around the body, the poles may move, as they do on Earth when the Arctic ice cap shifts position	69
6.1	Map of stress induced due to a 1° rotation	77
6.2	Maps of stress induced due to diurnal variation of the tide	81

x **Figures**

6.3	Maps of stress due to diurnal variation of the tide, added to the stress accumulated during 1° of non-synchronous rotation...................	82
8.1	Conjunction of Io with Europa always occurs on exactly the opposite side of Jupiter from conjunction of Europa with Ganymede	92
9.1	USGS map of Europa..	104
9.2	Three sets of lineaments in a time sequence........................	108
9.3	Images from *Voyager* and *Galileo* showing the same region with the terminator nearby...	114
10.1a	A double-ridge sliced off revealing its interior, like a road cut on Earth.....	118
10.1b	The highest resolution image ever taken by *Galileo*..................	119
10.2	A model for ridge formation by diurnal working of a pre-existing crack	120
10.3	Another picture of ridge formation........................ color section
10.4	Ridge formation on Japanese TV......................... color section
10.5	The archetypical example of a triple-ridge, a class of feature that does not exist	124
10.6	The terrain a few kilometers beyond the area in Figure 10.5.............	124
10.7	The complex of double-ridges mapped as a "triple-ridge"...............	125
10.8	An example of surrounding terrain purported to extend up the flank of the ridge...	126
10.9	Near the south pole, a ridge has extensions of older ridges on its flank	127
10.10	A double-ridge cut off by formation of chaotic terrain	129
10.11	A mature ridge weighs down the lithosphere, forming parallel cracks along the flanks..	130
11.1a	A dilational ridge, recognized by a central groove with symmetrical ridges on both sides...	134
11.1b	Reconstruction of part (a)	135
11.2	Ridge formation during dilation builds multiple, symmetrical ridges	135
11.3a	The "Sickle", a typical dilational band	136
11.3b	Reconstruction of the Sickle band and of the curved band to its south	137
11.4	Hybrid bands	138
11.5	Reconstruction of an old band near the large impact feature Tyre.........	139
11.6	Reconstruction of a portion of the Wedges region	140
11.7a	A pair of parallel dilation bands, located east of the Sickle..............	141
11.7b	A schematic of the dilational displacement that created the arrangement of crustal plates in Figure 11.7a	141
11.7c	Reconstruction of the band complex shown in Figure 11.7a	142
12.1a	Astypalaea Linea in a *Voyager* image	146
12.1b	A sketch of the key tectonic features in Figure 12.1a	147
12.1c	Randy Tufts' reconstruction of Astypalaea Linea prior to the strike–slip offset	148
12.2	Tidal walking at Astypalaea............................. color section
12.3	Principal stresses on an element of the surface......................	150
12.4	Diurnal stress variation during one orbital period at Astypalaea fault	152
12.5	An idealized model of tidal walking	155
12.6	A rubber sheet responds similarly to Astypalaea	156
12.7	The diurnal variation in stress at a fault running east–west in the Wedges region	158
12.8	Theoretical predictions of the sense of strike–slip displacement	159
12.9	Locations of "Regional Mapping" image sets	161
12.10a	Strike–slip faults in the far north (E19 portion) of RegMap 01	163
12.10b	Strike–slip faults in the E15 portion of RegMap 01	164
12.10c	Strike–slip faults in the equatorial regions of RegMap 01	165

12.11a	The fault marked as E in Figure 12.10b	166
12.11b	Reconstruction of Fault E (Figure 12.11a)...............................	167
12.12	Reconstruction of RegMap 01 .... color section	
12.13	The convergence site at location F in Figure 12.12.....................	169
12.14a	The equatorial portion of RegMap 01.....................................	170
12.14b	A large, bent, cycloidal lineament	171
12.14c	Reconstruction of Figure 12.14b...	172
12.14d	The area of convergence, inferred from the reconstruction in Figure 12.14c ..	173
12.15	Polar wander schematic	175
13.1	A mosaic of the high-resolution images of Astypalaea	184
13.2	A schematic of the general geometric relationship between strike–slip displacement and a pull-apart zone..	185
13.3	A schematic of the opening of pull-aparts when strike–slip displacement occurs along a cycloid-shaped crack	185
13.4	Enlargement of part of the high-resolution mosaic of Astypalaea in Figure 13.1	186
13.5	Enlargement of another part of the high-resolution mosaic of Astypalaea ...	187
13.6	Enlargement of another part of the high-resolution mosaic of Astypalaea in Figure 13.1...	188
14.1	Cycloidal ridges that appear prominently in a *Voyager* image	192
14.2	Cycloidal ridges in the northern hemisphere.............................	194
14.3	Examples of the double-ridge morphology of many cycloids	195
14.4	(top) A cycloidal crack, which has not developed ridges	196
14.5	Propagation of a cycloid	198
14.6	Cycloids can take on irregular geometries	199
14.7	Theoretical patterns produced as cracks propagate from starting points spaced every 10° in latitude and longitude	201
14.8	A preliminary map of observed cycloidal lineaments.....................	202
15.1	Several sets of lineaments cross one another in the southern-hemisphere portion of RegMap 02..	212
15.2	The intersection of the triple-bands Udaeus and Minos	215
16.1a	Conamara Chaos imaged during *Galileo*'s E6 orbit.......................	220
16.1b	Reconstruction of Conamara's rafts	221
16.2	A mosaic of images spanning Conamara at 54 m/pixel	222
16.3	A mosaic of very-high-resolution images (9 m/pixel) of Conamara	224
16.4	Part of the high-resolution image set shown in Figure 16.3 in its original viewing geometry	225
16.5	A high-resolution image of a large area of chaotic terrain...............	226
16.6	The "Mini-Mitten", a small patch of chaotic terrain at two different resolutions	227
16.7	Small patches of chaotic terrain, at different resolutions	228
16.8	The Mitten is a chaos of size comparable with Conamara................	230
16.9	Degradation of chaotic terrain by tectonics, and disruption of tectonic terrain by chaos formation	232
16.10	Global map of chaotic terrain .... color section	
16.11	A sample of "modified" chaotic terrain....................................	235
16.12	Size distribution of chaos patches	237
16.13	A schematic of the melt-through model of chaos formation	239
16.14	(a) Production of lateral cracking (e.g., near the Mitten). (b) Edge rafts in place. (c) Rafts migrate away from the edge.	241
16.15	The melt-through model explains relation of chaos to ridges	243

Figures

16.16	The Dark Pool	244
16.17	Volcanic flow would create topography similar to melt-through.	245
16.18	Thrace and Thera	color section
16.19	Thrace's southwest margins	247
17.1	Agenor and Katreus Linea	256
17.2	High-resolution image of Agenor	257
17.3	The Evil Twin of Agenor	259
17.4	Locations of Agenor and its twin on a USGS global mosaic	259
17.5a	An enlargement of Agenor's Twin	260
17.5b	Reconstruction along Agenor's Evil Twin	261
18.1	Tyre Macula	color section
18.2	Typical chaotic terrain? Interior of Tyre	270
18.3	Callanish	272
18.4	Close-up of Callanish	272
18.5	Amergin Crater has an interior indistinguishable from chaotic terrain	274
18.6	Amergin's interior hides in plain sight in the similar terrain of nearby chaos	275
18.7	Crater Manannán	276
18.8	Pwyll Crater in low-resolution color	color section
18.9	Pwyll Crater	277
18.10	Cilix Crater	color section
18.11	Depths, diameters, and classification of impact features	281
18.12	Crater Tegid	282
19.1	The six type examples for PSDs	288
19.2	PSD-type examples in the Conamara region	289
19.3	Examples of pits	294
19.4	A Venn diagram for variously-defined sets of features	296
19.5	Two of the largest uplift features	299
19.6	The locations, sizes, and shapes of pits	300
19.7	Size histograms of pits	301
19.8	A pit sheared apart by a strike–slip fault	303
19.9	The locations, sizes, and shapes of uplifts	304
19.10	Size histograms of uplifts	305
20.1	Elevations near Crater Cilix	color section
21.1	Tidal flow though a working crack provides a potentially-habitable setting.	color section
22.1	The Red Herring	color section
22.2	The interior of Manannán, indistinguishable from chaotic terrain	344
22.3	A portion of the Sickle dilation band	345
22.4	Terrain in the large parallelogram pull-apart	347
22.5	Isolated fins	347
22.6	The east end of Agenor	349
22.7	A branched-cusp cycloid	350
22.8	Cracks, ridges, and bands in the southern-leading hemisphere	351

Preface

"Europa! I found it."
"That's Eureka, Maynard."

Maynard G. Krebs and Dobie Gillis

At its heart, this book is about people and about one of the many remarkable things they do with their evolved mental and physical facilities. Exploring the Solar System, even a small part of it like the neighborhood of Jupiter, involves the coordinated efforts of a great many people with diverse talents, expertise, and abilities. In such a large undertaking, the richness of the human pageant is bound to be as amazing as what we discover about other planets, which is saying a great deal because Jupiter's oceanic moon Europa is far more exciting and interesting than we could have anticipated.

Exploration of Europa began with Galileo's early telescopic discoveries, and Earth-based observations continued for hundreds of years as instruments continually improved. For over a quarter century, starting in 1977, I served on the Imaging Team for the *Galileo* mission to orbit Jupiter. During that time the *Voyager* spacecraft gave us our first good look at Europa's surface during its quick fly-by in 1979, and a series of disasters delayed and degraded the *Galileo* spacecraft, so that it only arrived in orbit around Jupiter, where it would study the satellites for several years, late in 1995.

During that time I had a unique vantage point, formally an insider as a member of the Imaging Team, but always with the detached perspective of an outsider. For a variety of reasons, touched upon throughout the book, I never felt welcomed by the team. One reason was my scientific background. While most of the team was experienced in studying planetary surfaces or atmospheres, and most had been involved with space missions before, I was an expert in celestial mechanics, the study of the motions of planets and satellites. For most of the quarter century, the rest of the team seemed to be wondering why I was there. Then, when it became clear that my field was the key to understanding what we saw at Europa and evident how significant those discoveries were, attempts to keep me marginalized were driven by transparent social, political, and financial motives.

As a story about something people do, this book inevitably has villains, and although I often took their actions personally, I now realize that they were only following their nature. Mostly the story has heroes, a huge number of people who worked together to make a robot in space perform miracles, even as bad luck kept throwing challenges in the way. In the book I describe the conservatism of the engineers and administrators of the project, but without their careful efforts and dedication the scientific dimension of this story could never have been told. These remarkable people got the robot to Jupiter and they made it work. The discoveries described in this book could not have been made without them. While this book criticizes aspects of the conservatism of the *Galileo* project, that criticism only applies where that conservatism was extended inappropriately to scientific analysis, discouraging creativity and risk-taking just where they were needed.

Europa turned out to be arguably the most exciting subject of the *Galileo* mission's discoveries. To my amazement a series of circumstances brought me from a sense of marginalization in the early 1990s to center stage in Europa studies.

First, the premise of my original 1977 proposal that got me on the Imaging Team proved to be more accurate than anyone could have known at that time: Most of what has shaped Europa involved tides, which could be understood from the perspective of celestial mechanics.

Second, as *Galileo* approached Jupiter, several remarkable scientists joined my research group. The core group during many of the key discoveries was my postdoctoral associate Paul Geissler and my students Greg Hoppa and Randy Tufts. Their contributions have been widely recognized. Michael Benson's *New Yorker* article on the end of the *Galileo* mission highlighted Paul, Randy, and Greg's discoveries. This book puts them in a more complete context. Other students and associates in my group who contributed in important ways to our Europa work were Dave O'Brien, Alyssa Sarid, Terry Hurford, Jeannie Riley, Martha Leake, Sarah Frey, Dan Durda, Brandon Preblich, Gwen Bart, and Susan Arthofer. I am deeply grateful for all that they taught me and for the pleasure and comfort of their friendship. Their collective work makes up much of the scientific narrative of this book. It is a monument to their creativity and a fitting memorial to Randy, who inspired us all.

Third, I happened to work at one of the premier centers of research in planetary science, the University of Arizona's Lunar and Planetary Laboratory and its academic arm the Department of Planetary Sciences. The leadership of Mike Drake and his predecessor Gene Levy have protected an environment where creativity, intellectualism, discovery, and innovation flourish. The faculty and staff of LPL make this a wonderful place to work and many of them (too many to name them all) contributed to the Europa story in a wide variety of ways. Selina Johnson managed the affairs of our research group. Jay Melosh helped me and my students understand various crucial geophysical issues. LPL's Planetary Image Research Laboratory

(PIRL), under the direction of Alfred McEwen, facilitated our immediate access to *Galileo* data as they arrived on Earth, and produced a range of image products that were essential for our investigations. Most of the images in this book were processed at PIRL, including important mosaics and color versions produced by students Cynthia Phillips and Moses Milazzo. Joe Plassman, Chris Schaller, and Zibi Turtle of the PIRL group helped in many ways.

We were ready with expertise and resources when the images came down from the *Galileo* spacecraft, but, ironically, without the mission's problems we never would have been able to play a central role. The antenna failure reduced the number of images to a set that even a small group could handle. Greg, Randy, Paul, and I came to know every image. Moreover, our political isolation left us free to follow the evidence without pressure to conform to the constraints that governed so many others. Future investigations by ourselves and others will test our work, but nothing will top the thrill and satisfaction of seeing the pieces of the puzzle come together as we came to understand and to love Europa.

The research by my group described in this book was supported by NASA through the *Galileo* project, the Planetary Geology and Geophysics program, and the Jupiter Data Analysis program and by NSF through its Life in Extreme Environments program. Several chapters are based on material that we published in *Icarus*, and the section on planetary protection is based on our article in *American Scientist*.

I wrote the book while on a sabbatical visit at Dickinson College in Carlisle, Pennsylvania. I am grateful for the hospitality and support of the Dickinson Physics and Astronomy faculty, especially my hosts Priscilla and Ken Laws and department chairs Hans Pfister and Robert Boyle.

For help with the preparation of the book, I thank Philippe Blondel of the University of Bath, UK, and scientific editor for Praxis, whose careful editing and advice improved the manuscript immensely, and Clive Horwood of Praxis for his confidence and encouragement. Leontine Greenberg helped me to structure and clarify crucial material. Greg Hoppa counseled me throughout the process, provided a detailed review, and helped me see things from a more optimistic perspective. He insisted on the happy ending, and he was right.

Richard Greenberg
Carlisle, Pennsylvania
June, 2004

Part One

Discovering Europa

1

Water World

1.1 LIFE ON A WATER WORLD

Brisk tidal water sweeps over creatures clawed into the ice, bearing a fleet of jellyfish and other floaters to the source of their nourishment. As the water reaches the limits of its flow, it picks up oxygen from the pores of the ice, oxygen formed by the breakdown of frozen H_2O and by tiny plants that breathe it out as they extract energy from the Sun. The floating creatures absorb the oxygen and graze on the plants for a few hours.

The water cools quickly, but before more than a thin layer can freeze, the ebbing tide drags the animals deep down through cracks in the ice to the warmer ocean below. Most of the creatures survive the trip, but some become frozen to the walls of the water channels, and others are grabbed and eaten by anchored creatures waiting for them to drift past. The daily cycle goes on, with plants, herbivores, and carnivores playing out their roles.

The scene may be reminiscent of the Arctic, but such an ecosystem might occupy a setting more exotic and (from a human perspective) more hostile than any on Earth: the icy crust of Jupiter's moon Europa. Europa is about as big as our own Moon, but it is a water world, and tides are strong and active there. As far as we can tell, physical conditions on Europa could support life, even complex ecosystems that might exploit a variety of niches and environmental changes over various scales of time and place.

An active crack in the ice crust that links the ocean to the surface might support the daily lifestyle described above, as tides squeeze liquid water up and down between the ocean and the surface. Generations come and go as the stable daily cycle repeats for thousands of years. But longer than that, the tidal rhythm will change, so that any particular crack must eventually freeze shut. Life will go on only if it can adapt to change. Some organisms could make their way to another more recently opened crack. Others, frozen into place, must be able to hibernate as some bacteria do on Earth. Their wait would be rewarded by new cracking or by a melting event, either of which will bring back the warm oceanic water.

1.2 IS THIS FOR REAL?

Nobody on Earth knows whether life exists on Europa or ever did. We have considerable information about the appearance of this little "planet"[1] from spacecraft reconnaissance[2] and from Earth-based astronomical observations, but to understand why it looks that way, we rely on our understanding of the kinds of physical processes that may operate there. Deciding which physical processes have resulted in the observed appearance of Europa is crucial to assessing the likelihood of life at or beneath the surface.

This scientific process is inevitably imperfect and uncertain, especially in a big-ticket scientific project, like a space mission. It starts with the complex politics of deciding what to observe and how to do it, a process that determines how resources, prestige, and influence are allocated in the scientific community. Then comes the acquisition of data, which is a relatively objective step. After that stage, the study reverts to the heavily subjective: Language is used to describe and digest what we observe, ideally in terms that retain its essential character without prejudicing subsequent interpretations. Too often, these descriptions reflect the prejudices and personal interests of the project spokesmen. Then theorists construct models, which means considering an artificial physical system simple enough that the physics can be understood, and assuming that the results will apply to the real, complex body under consideration. Occasionally, theorists have first-hand acquaintance with the data, but too often their models are simply inspired by how the data were described to them. If the predictions of the artificial model compare well with the subjective description of reality, it is accepted that those physical processes were actually responsible for the character of the planet.

These scientific procedures depend on human judgment. What appears completely obvious to one scientist may seem questionable (or worse) to another. In the popular mind, science may be viewed as a systematic and perfectly logic-controlled process, while in fact it is a human endeavor, built on human creativity and numerous components of subjective judgment. Whatever the common impression may be, this judgment-dependent approach is an important part of the scientific method, once the raw data are in hand and we begin to try to make sense of it. Good scientists try to be as objective as possible and to follow the evidence where it takes them.

Based on our research on Europa, several lines of evidence, leading from the basic appearance of the surface through detailed modeling of the controlling

---

[1] Although Europa is formally a moon, planetary scientists often lapse into calling a moon a "planet" once its detailed character is revealed to us, and especially if that character and the processes involved appear to be planet-like.
[2] *Voyager 2* encountered Europa in July 1979, and the *Galileo* orbiter did several times from 1996 to 2000.

processes, point to physical conditions that may plausibly support life. A tidal ecosystem exploiting habitable niches is consistent with what we have learned about the crust of Europa and the diverse ways that it changes over the daily cycle, over millennia, and over hundreds of millions of years.

This book is about the lines of evidence that have led us to this picture of a permeable ice crust overlying a liquid water ocean. The story is inevitably about the human enterprise of science, how we learn about a planet from data returned from a spacecraft. Most of the story is an intellectual one, logically combining theoretical modeling with observational evidence to infer what we can about Europa. But Big Science done in the context of a large space mission is governed by politics and money, as much or more than by the search for truth. That aspect of the human enterprise cannot be uncoupled from the intellectual one, because political power has been used aggressively to promote a very different party line about Europa, in which an ocean, if any, is isolated deep below a thick layer of ice. It may seem bizarre that political clout would be used to promote a scientifically weak position, but for those of us long involved in the human enterprise of science, the situation is both familiar and disturbing. This book explains the scientific lines of evidence regarding Europa, but it also must address the political process that has effectively promoted an otherwise poorly justified party line.

I start with an overview of the major physical processes, driven predominantly by tides, that govern Europa's surface, with a tour of the surface based on images from the *Galileo* spacecraft (Chapter 2), and with a discussion of the "scientific method" as it plays out in the real world (Chapter 3). Then, in Parts Two and Three, I lay out the evidence about Europa in a logical sequence, explaining what led us to believe there is an ocean under the ice and that the ice is permeable. In Part Four I discuss the prospects for life in the physical setting that we have inferred.

Europa is exciting because it is so active. The icy surface is continually reprocessed at such a great rate that most of the observable terrain, structures, and materials have probably been in place for less than 50 million years, about 1% of the age of the Solar System.[3] By comparison, the ancient features that we see on the surface of our own Moon formed early in the Solar System's history and have changed little since then. The rate of change on Europa is more closely comparable with that on Earth. For example, on Earth during the same 50 million years continents have been significantly rearranged, with North America moving away from Europe, creating the Atlantic Ocean, and India crashing into Asia. In the cosmically short time since dinosaurs became extinct on Earth, the surface of Europa has entirely turned over a couple of times.

During all that reprocessing, the surface ice has been bombarded by material from comets and asteroids, as well as substances from the swarm of Jupiter-orbiting particles. Just below the ice, the global, liquid water ocean has received substances released from the deep rocky interior.

The dynamic ice crust serves as the interface between the substances from the interior and those from exterior space. This barrier is solid enough to keep the

---

[3] The youth of the surface is inferred from the paucity of craters (see Chapter 18).

chemistry in disequilibrium, but porous enough to allow interaction over various spatial and temporal scales. In broad terms, this physical and chemical setting seems to have the potential to support life, and a tidally-driven ecology in the crust might be able to exploit it effectively.

1.3 TIDES

On Earth, life tends to prosper at the boundaries of different physical regimes. Consider the diversity of life in a tide pool, at the land/sea interface. The natural flow of tidal water mixes the chemistry, allowing some organisms to commute between zones that satisfy their diverse needs, while others sit at anchor exploiting the flow, just as in our imagined view of the Europan crustal habitat.

On Europa, too, the geological activity that may provide the setting for a biosphere is driven by tides. These tides are enormous in comparison with terrestrial tides. On Earth, tides are driven by the pull of the Moon, and to a comparable degree by the larger but far more distant Sun. Europan tides are driven by Jupiter, which is about as close to Europa as the Moon is to Earth, but Jupiter is 20,000 times more massive. Such tides are bound to have major effects.

In fact, tides affect Europa in several crucial ways, discussed in detail in Part III:

- Tides distort the global shape of Europa on a daily basis, generating periodic *global stresses* that crack and displace plates of icy crust, driving a rich history of ongoing tectonics and surface change.
- Tidal friction creates *heat*. In fact, it is the dominant internal heat source, warming Europa enough to keep most of its thick water mantle melted, maintaining the global ocean, and allowing the frequent local or regional melt of the ice crust.
- Tides generate a *torque* that governs Europa's rotation. Unlike our Moon, which rotates synchronously with its orbit in such a way that the same face is always toward the Earth, Europa may rotate a bit faster than its orbital motion, so that the face it presents toward Jupiter slowly changes. Such non-synchronous rotation adds important components to tidal stress, leaving its imprint on the tectonic record of surface cracks.
- Tides control the long-term *orbital evolution* of several of Jupiter's largest satellites. A resonance among these moons, including Europa's, keeps the orbits from becoming circular, so they remain eccentric ellipses. In turn, the magnitude of the tides is directly dependent on the eccentricity. As tidally-driven orbital changes modify Europa's eccentricity, all of the effects of Europan tides gradually change over tens of millions of years.

Orbital resonance, a global ocean, and dramatic tectonics in the icy crust, with a daily ebb and flow of liquid from the ocean to the surface, are all interdependent through the mechanism of tides. And this interplay appears to have created a physical setting with all the ingredients and conditions for local habitable niches and for a long-lived global biosphere on Europa.

2

Touring the surface

2.1 THE GLOBAL PICTURE

Let's take a look at Europa and begin to explore the observable basis for our interpretation. Long before images were available, spectra obtained by Earth-based telescopic observation revealed that the surface is composed predominantly of water ice. Looking with a naked eye would show a nearly uniform white sphere, 1,565 km in radius.

Don't try that yourself: Your naked eyeball would freeze and explode in the vacuum of space. Even spacecraft cameras would have a fairly short life if they stayed near Europa for more than a few days. The intense energetic charged particles in Jupiter's magnetosphere would fry their electronics. Fortunately, the cameras on the *Voyager* and *Galileo* space probes were hardy enough to survive their short fly-by encounters with Europa, and sensitive enough to reveal much more detail than even a well-protected human eye (Figure 2.1).

The space probes also helped narrow the possibilities for what lies below the sunlit surface. Subtle variations in their trajectories, caused by Europa's gravitational field, revealed the interior layering of different densities: a metallic core of radius 700 km, a silicate (rocky) mantle, and an outer layer as thick as ~150 km with the density of liquid or solid water. Right at the surface the water is frozen, because, without a significant insulating atmosphere, heat radiates rapidly into space. But below the surface much of the H_2O layer is probably liquid according to estimates of internal tidal heating. With such a thick layer of water, this Moon-sized body has a global ocean comparable in volume with the oceans of the much larger Earth.

The presence of large amounts of water surrounding a rocky interior had been predicted by models of the formation of the Galilean satellites. The satellites formed within a huge gaseous nebula that surrounded the young Jupiter, just as the planets formed in the nebula around the Sun. As solids condensed from the Jovian nebula they accreted into satellites. The sequence of condensation governed the bulk

8 Touring the surface   [Ch. 2

(a)
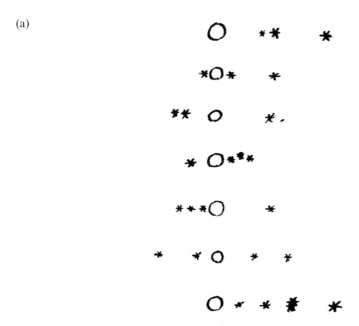

Figure 2.1. (a) Drawing by Galileo of the Medician Stars, now known as the Galilean satellites. This montage shows various drawings by Galileo showing the positions of the satellites (star shapes) relative to Jupiter (circle) at different times. From the Earth, the orbits are viewed edge-on, so the satellites appeared to the east and west of Jupiter, but Galileo inferred from their motion that they orbit the planet. (b) A more recent telescopic view of Jupiter with the Galilean satellites. You can track the motions of these moons with a standard pair of binoculars.

Image (b) by Michael Stegina and Adam Block of NOAO/AURA/NSF.

composition of the satellites, and internal heating separated the materials by density within each satellite. A thick layer of H_2O on Europa was expected long before *Voyager* or *Galileo* got there.

As soon as the possibility of an ocean was raised, speculation about life began. However, the presence of liquid water alone is not enough to support life. The character of the surface ice is equally critical, because connections through the ice, between the ocean and the surface, may be essential. Surface chemicals, especially oxygen, must reach the liquid water in order to support conventional forms of life, like those that predominate on Earth. This book will develop the several lines of evidence that point toward just those kinds of connections.

The 150-km layer of low-density material might not be all water. The density is also consistent with a portion of that layer consisting of hydrated silicates, such as clays, which are so rich in water that their densities are low. However, on geochemical grounds the likelihood of a substantial component of such minerals in the low density layer seems remote. Clays would probably occupy at most a small portion of this layer, mostly as a thin zone at the bottom of the ocean.

We know nothing about Europa's ocean floor, but that has not slowed speculation about undersea volcanism because, when the ocean was believed to be isolated below the ice crust, volcanoes seemed to be the only source of heat and chemicals that might offer hope for life on Europa. The volcanic models were not motivated by any observations of Europa, but rather by discoveries in the late 1970s of life supported by undersea volcanic vents on Earth and of dramatic volcanoes on Europa's sibling Io. For Europa, there is no observational evidence either for or against volcanism or any other seafloor structures.

We do have plenty of observations of the surface. Now, analysis of that evidence shows that the ice crust is thin enough to link the ocean to the surface. The possibility of life no longer depends on speculation about undersea volcanism.

We have excellent images of the surface of the ice because cameras on the *Voyager* and *Galileo* spacecraft had sensitivities and wavelength ranges that vastly exceeded those of the human eye.[1] Even on a global scale (with low resolution, >10 km per pixel), their images reveal much more than the uniform white water ice that would be visible to a human observer (Figure 2.2, see color section). At this resolution, orange–brown markings of still-unidentified substances[2] on the bright background come in two major categories: splotches ranging from tens to 100s of kilometers across and a global network of narrow lines. Even at this very low resolution, the lines and splotches provide a first indication of what prove (based on higher resolution and considerable analysis) to be the two major resurfacing processes on Europa: tectonics (the lines) and formation of chaotic terrain (the splotches). As we will see, each likely involves direct interaction of the ocean with the surface, by cracking or melting, respectively.

---

[1] The *Galileo* imaging camera had a spectral range of about 0.35 to 1.1 μm (extending into the near-IR), with 800 × 800 pixels spanning a field of view about 0.5° wide. The Near-Infrared Mapping Spectrometer (NIMS) covered the spectral range 0.7 to 5.2 μm with high spectral resolution, and spatial resolution of about 0.025°.
[2] The substances likely include hydrated salts and sulfur compounds, and even organic chemicals, as discussed in Chapter 21.

Figure 2.2 shows what is called the "trailing" hemisphere. The hemispheres are defined by their relationship to the orbital motion of Europa around Jupiter, just as our own Moon has a near side and a far side in relation to the Earth. Our Moon presents a constant face to Earth because its rotational period matches its orbital period. Europa's rotation is also synchronous with its orbit, at least close enough so that during the two decades covered by spacecraft images, the same hemisphere has continuously faced Jupiter. Later in this book, I discuss evidence that over the longer term Europa's rotation is actually non-synchronous, and that the whole surface slips around relative to the poles; but, at least during this period of human observations, the Jupiter-facing hemisphere has been nearly constant.

When Figure 2.2 was taken, Jupiter was toward the left, and Europa was moving in its orbit away from the camera. Thus, this view is of the side opposite the direction of orbital motion, hence the "trailing" hemisphere. North is at the top, and the equator crosses near the center of the image.

A major landmark in this hemisphere is the X formed by two globe-encircling dark lines that cross at a right angle slightly to the upper right of the center of the disk in Figure 2.2. This neighborhood is very important for several reasons. The two major types of geologic process are well represented here: The intersecting lines that form the X are major, global-scale tectonic features, and the dark patch in the lower crotch of the X, is Conamara Chaos, a well-known example of what has come to be known as chaotic terrain. Most importantly, this region was the subject of considerable targeted imaging, especially early in the *Galileo* mission, so we have unusually complete coverage of this area, with images with a variety of scales, resolution, filters, and lighting conditions. Indeed, Conamara itself came to be the archetype for chaos on Europa. The detailed images of this area have given us great insight into the character and dynamics of Europa's active surface. At the same time, over-reliance on this heavily studied site as a representative sample has produced some widely promoted generalizations about Europa that were not well justified, causing considerable confusion and misunderstandings. With appropriate attention to such pitfalls, imaging of the Conamara area provides a key to much of what is seen across Europa.

The X at Conamara also serves as a convenient landmark for relating imagery to global positions in latitude and longitude. The X lies just north of the equator and very close to the longitude of the center of the trailing hemisphere. That is to say it is nearly 90° east of the direction of Jupiter. The direction of Jupiter defines the prime meridian of Europa, where the longitude is zero, just as Greenwich Observatory defines the prime meridian of the terrestrial coordinate system. As a matter of convention, on Europa longitude is usually given in terms of degrees west of the prime meridian. Using that convention, the X is near longitude 270°W, rather than 90°E, although the two are equivalent.

Therefore, the X should be just north of the center of the trailing hemisphere. In Figure 2.2, it appears to be a bit too far to the right. In fact, the disk shown here is centered at about 290°W longitude, so it is not quite perfectly co-registered with the trailing hemisphere. The "sub-Jupiter point" (i.e., the point on the equator at longitude 0, from which Jupiter would appear directly overhead in the sky) lies to the far west (left) in this image.

The best single-frame color image of the entire sub-jovian region on a global scale is shown in Figure 2.3, see color section. The actual sub-Jupiter point is about half-way between the center of this full disk and the right-hand edge. The center of the disk in this view is located on the equator at about longitude 40°W (40° west of the sub-Jupiter point). In Figure 2.3, as well as in Figure 2.2, the north pole is at the top. The splotch and line patterns near the right (east) edge of the picture can be recognized as the same ones on the left in Figure 2.2.

A view of the full disk east of Figure 2.2 appears in Figure 2.4, which shows the disk centered on the equator at about longitude 220°W. Here, to the far left we can see a foreshortened view of the big X at Conamara, and we can identify the splotch pattern on the west side as the same pattern that is seen on the east side of Figure 2.2. The anti-jovian point, exactly opposite the direction of Jupiter, is 90° east of Conamara, toward the right side of Figure 2.4. This means that the center of this image is near the half-way point between the center of the trailing hemisphere toward the left, south of Conamara, and the center of the anti-jovian hemisphere toward the right.

The full disk in Figure 2.4 shows the usual splotches and lineaments. The lineaments that fill much of this view between the equator and about 30°S are known as "the Wedges", because many of these linear features are tapered in width from one end to the other. The wedge-shaped bands divide the surface into the distinctive rounded boxy shapes evident even on these low-resolution global views.

Near the top of this full-disk view lie several long, dark, roughly east–west lineaments. Where these lines cross one another, the intersection angles are oblique. The northernmost of the darker lineaments, near the top center of Figure 2.4 are Udaeus and Minos Linea. The region around the Udaeus–Minos intersection is important for several reasons: It contains these typical examples of global-scale (and even globe-encircling) lines; the tectonic features in this region played an important role in considerations of Europa's rotation; and (like the X at Conamara) these markings provide a point of reference for locating important sets of higher resolution images.

Moving from west to east around the equator, Figures 2.3, 2.2, and 2.4, in that order, provide a continuous global picture with the north pole always at the top. The extreme eastern edge of Figure 2.4 reaches just about all the way back around to the western edge of Figure 2.3 at longitude 130°W. Unfortunately, there is no overlap of global coverage there, so that the region (at the left of Figure 2.3 and right of Figure 2.4) is extremely foreshortened in both of those views. We do not have any global view to fill in these longitudes. We do have images at higher resolution of substantial portions of that region (from *Voyager* and *Galileo* regional imagery) and, as far as we can tell, there are no major surprises lurking there, at least none that would jump out on a global-scale, full-disk view. Wherever we look at the global scale, Europa is dominated by lines and splotches (i.e., by tectonics and chaos).

12 Touring the surface [Ch. 2

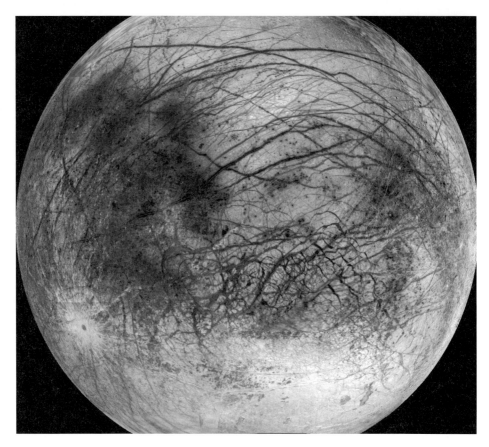

Figure 2.4a. Full-disk mosaic from orbit G1, centered near the equator at longitude about 220°W. The anti-jovian point is toward the right edge, about 2/3 of the way from the center.

Equally important as the ubiquitous splotches and lines is what is missing from these global views: craters. Unlike our own Moon, unlike the other icy moons of Jupiter, and indeed unlike most atmosphereless bodies in our solar system, Europa has hardly any craters. Apparently, external bombardment has had a minimal role in shaping the surface that we see today. Only a couple of impact features are evident at this scale. Crater Pwyll is about 1,000 km due south of Conamara. It is the dark spot (about 20 km wide) in the southern part of Figure 2.2 surrounded by an enormous system of bright rays of ejected ice that extend a thousand kilometers or more in every direction. Pwyll is also seen in the southwestern part of Figure 2.4. That Pwyll is an impact feature is evident from the ejecta rays. One ray even crosses the western side of Conamara Chaos. As a result of this whitening of its western side, Conamara, which is somewhat diamond-shaped, appears more triangular in the global view.

Another prominent impact structure is the dark circle called Callanish near the left side of Figure 2.2 (and also visible just at the eastern edge of Figure 2.3) just

Sec. 2.1] The global picture 13

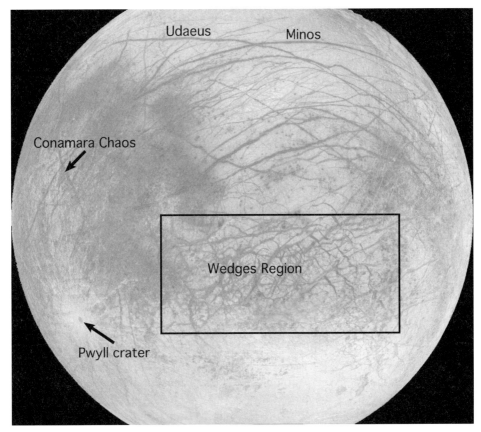

Figure 2.4b. Important landmarks in this hemisphere are the Udaeus–Minos intersection at the north, the "Wedges" region south of the equator, and Conamara Chaos and Pwyll Crater to the west (cf. Figure 2.2, see color section).

south of the equator. This large feature (about 60 km across) does not seem very different from the splotches of chaotic terrain, viewed at this scale. What hints at its impact origin at this scale is its round shape. Higher resolution images confirm its impact origin, but they also suggest some very interesting similarities to chaotic terrain: Both may represent breakthrough to the liquid ocean.

Despite a few such prominent examples, impact features are rare. Why are there so few craters? The answer is probably the same as for the Earth: The planet has undergone active geological resurfacing that has erased the record of nearly all impacts. The erasure must have occurred recently, because Europa is very susceptible to external bombardment: It has no atmosphere, it is bombarded by comets as well as by asteroidal material, and approaching trajectories are focused by Jupiter's gravity. The current surface must be less than about 50 million years old, or else it would show far more craters than it does.

Did some major event wipe the entire surface clean all at once about 50 million years ago? There is no evidence for that interpretation. Instead, the continual, gradual resurfacing manifested by chaotic terrain and tectonics provide a ready explanation for the youthful appearance of Europa. The crater-based age may simply represent the turnover time for gradual surface renewal.

Even in very-low-resolution global views, the main points of our story are introduced. Europa's appearance is dominated by features that represent the effects of resurfacing that has been rapid, recent, and probably ongoing. The global view shows the two major categories of the resurfacing processes: tectonics and chaos formation. As I show later, these processes not only modify the surface, but they generally provide, in diverse ways, access between the surface and the liquid ocean. The dark material that marks the sites of these processes is indistinguishable from site to site, independent of whether it lines tectonic features or marks chaotic terrain and its immediate surroundings. This similarity probably reflects a common feature of all these processes: They all involve interaction of the surface of the ice with the ocean. The dark markings may represent concentration of impurities by thermal effects due to the proximity of warm liquid, or they might simply consist of the most recently exposed substances from the ocean that lies just below the ice.

2.2 ZOOM IN TO THE REGIONAL SCALE

We can zoom in on surface features by examining regional-scale images, which are those taken with a resolution of roughly 1 km per pixel and covering regions several hundred to a thousand kilometers across. Figure 2.5 (color section, and pp. 15–16) shows a 1,000-km-wide portion of the Udaeus–Minos region at about 1.6 km per pixel.

Compare this relative close-up with the more distant view in the northern part of Figure 2.4. At this higher resolution, many of the global-scale dark lines resolve into double lines with a bright lane between them (Figure 2.5, see color section for part (a)). Counting the brighter zone between the dark lines, *Voyager*-mission geologists named these features "triple-bands", which is in retrospect an unfortunate term. These features are not really "triple", because the central bright lanes are not significantly brighter than the surface beyond the dark edges. The visible feature really consists of two dark lines, typically about 10 km wide. Moreover, on Europa the term "band" is more generally used to describe dilational lineaments (along which adjacent plates have pulled apart from one another). As we will see, the only thing that triple-bands and dilational bands have in common is that both initiated as tectonic faults.

In this region, at this scale, we also see a great many dark spots, typically about 10 km wide, and quite common in this region of Europa. During the early part of the *Galileo* mission, the International Astronomical Union, the organization charged with naming things in space, deliberated at length about what, or even whether, to name this class of feature. Eventually, they accepted my doctoral student Randy Tufts' proposal that they be called *lenticulae*, the dignified Latin for freckles. As we

Sec. 2.2]   Zoom in to the regional scale   15

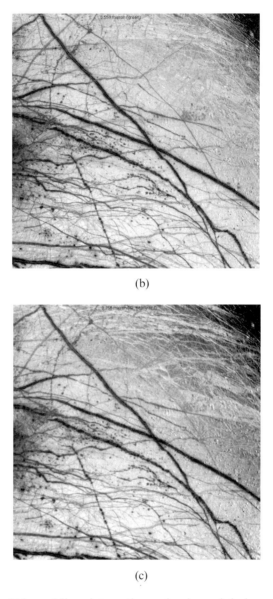

Figure 2.5b, c, d. The Udaeus–Minos intersection region imaged during orbit G1 at 1.6 km/pixel in various filters. Here we see separate images (b, c, d) taken through three filters, 0.559 μm (green), 0.756 μm (a wavelength slightly longer than red), and 0.986 μm ("near infrared"), respectively. The color composite of these images (Figure 2.5a), produced by Paul Geissler (LPL, Univ. Arizona), is in the color section. At this resolution, the global-scale dark lines have bright centers, so they were called "triple-bands", and the spots were dubbed *lenticulae*, Latin for freckles. The terminologies defined to describe appearances at this resolution have caused major misunderstandings when carelessly applied to high-resolution images.

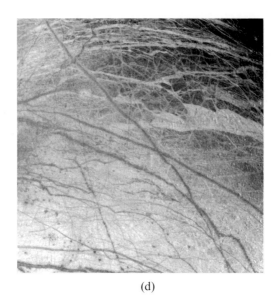

(d)

will see, most of these spots represent very small patches of chaotic terrain. The reason so many of them appear to be about 10 km wide is that such patches usually have a dark several-kilometer-wide halo around them. That typical halo size is independent of the size of the patch of chaos. Thus, at the resolution of Figure 2.5, 10 km is the minimum size of dark spot due to chaotic terrain; a patch of chaotic terrain smaller than that is still marked, with its extended halo, by a dark spot 10 km wide.

Unfortunately, the term "lenticulae" has been incorrectly applied to a poorly defined variety of morphological features that have been seen at high resolution, and the confusion has been used to promote incorrect generalizations about the true character of Europa's surface. This issue is discussed in considerable detail later in this book, especially in Chapter 19.

The important things to remember at this stage are that dark spots about 10 km wide are common on Europa (especially in certain regions like that shown in Figure 2.5); they mark most patches of chaotic terrain 10 km wide or smaller, which are usually surrounded by dark halos; and the word "lenticulae" was defined to describe these dark spots as they appear on low-resolution (~1 km/pixel as in Figure 2.5) images. The similar but larger splotches also prove at higher resolution to be chaotic terrain. Unfortunately, when those larger splotches were seen at low resolution beginning with *Voyager* images, they came to be called "mottled terrain". So a patch of "mottled terrain" is just a big lenticula, and it is identical to chaotic terrain. In this book I use the term chaotic terrain because it is based on better and more complete imagery than was available when the term "mottled terrain" was invoked to describe low-resolution pictures.

Regional imagery also allows us to zoom in on the broad area ranging from the big X at Conamara all the way south to the impact crater Pwyll (Figure 2.6, see color

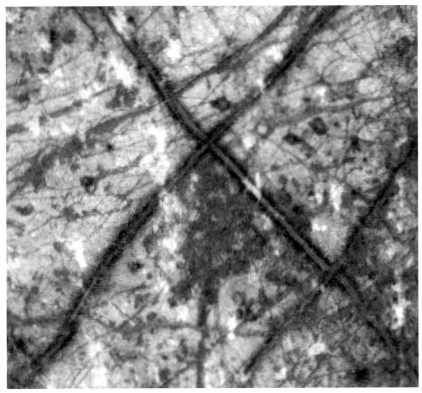

Figure 2.7. An enlargement of a 300-km-wide portion of the E4 image (Figure 2.6, see color section) showing Conamara Chaos (80 km wide) and the area around it.

section), about 1,000 km away. Compare this image with its context in the global images (just to the right of center in Figure 2.2 and to the far left in Figure 2.4). Note again that the global-scale lineaments (e.g., the lines of the X) divide into the double dark lines at this higher resolution. We can also see more clearly the ejecta pattern of material splashed out from the impact at Pwyll. It is now also much more apparent that the Conamara Chaos area is somewhat diamond-shaped, and nestled into the crotch of the big X, with a wisp of bright ejecta crossing its western side.

Conamara Chaos is zoomed even larger in Figure 2.7. At the center of the image the dark, diamond-shaped extent of Conamara is quite well defined, crossed at the left by the wisp of ejecta. Conamara itself is about 80 km across, and Figure 2.7 spans a region about 250 km across. Conamara is surrounded by numerous lenticulae. Comparing them with the size of Conamara, we can confirm that a typical diameter for these dark spots is about 10 km. Note too that, as in the Udaeus–Minos region (Figure 2.5), the width of the double dark lineaments ("triple-bands") is a similar scale, roughly 10 km.

The regional images shown here are parts of sets of images that were taken through various different filters by the *Galileo* camera. These images can be combined to produce color composites that demonstrate the variability of reflectivity with wavelength. Because the filter set does not correlate sufficiently with the wavelength ranges to which the human eye is sensitive, the color composite images do not show the colors that the eye might see. Of course, even if the filters had been selected to provide nearly "true" color, the contrast would need to be enhanced to show any features. It is important to remember that Europa is in fact quite white and bland, because its surface is overwhelmingly composed of water ice. The markings in all the global and regional images are only visible because the contrast has been enhanced, and color exaggerated.

What can we infer from the color information? The splotches and lineaments are generally an orangey-brown color. Since they seem to mark relatively recent geological activity involving oceanic exposure (specifically, tectonic cracks and chaos formation), the darkening and coloring agent may represent substances from the interior. The similar widths (~10 km) of the lenticulae around chaos patches and of the triple-bands around ridge complexes may represent the typical distances that such substances could spread from their exposure point. The colors themselves are not diagnostic of any particular substances, but spectra in the near-infrared range (1–10-μm wavelength) suggest they may contain hydrated salts, consistent with upwelling from the ocean. There are hints of sulfur compounds, which might help explain the color, and there is plausible speculation about organic chemicals as well. Between the dark markings, the bright ice in approximate color renditions varies from bluish to subtly pink, probably indicative of fine-scale structure, such as the average ice grain size, which affects the scattering of light.

In order to optimize the photometric value of these images (i.e., to enhance brightness differences and maximize the color information), they were all taken with direct sunlight illuminating the surface from nearly perpendicular to the field of view. What that illumination hides is the morphology (shape) of the surface. There are no shadows, no variations in brightness due to slope variations, none of the cues that indicate the structure of the surface. We cannot see any hills, ridges, pits, valleys, gorges, cliffs, or any of the other landforms that might appear on a planetary surface.

Partly, the lack of topographic features in these images is due to the viewing conditions. But it is also because there simply is not much topography on Europa. Hardly any features vary by more than a couple of hundred meters from the mean elevation, and slopes are very gradual everywhere. Europa is one of the smoothest bodies in the solar system. What we see in these images so far is the enhanced picture of subtle variations in surface reflectivity (or albedo) and color. We have seen markings corresponding to all of the important geological features and processes, but no direct evidence for them yet. For that, we need to look at higher resolution images.

2.3 ZOOMING CLOSER: SURFACE MORPHOLOGY

Galileo imagery includes coverage of about 10% of the surface at resolution of 200 m/pixel, a great improvement in revealing detail compared with the global or regional images. The improvement is not only due to the better resolution. Because these images were planned with a geological survey in mind, they were taken at times when the Sun angle was low enough relative to the surface that the illumination was favorable for revealing the bas-relief of surface structure. The high quality of these images exacerbates our frustration that they cover such a small fraction of the surface.[3] Nevertheless, the images we do have, combined with the broader context seen at lower resolution, suggests that we may have sampled most of the major types of terrains, structures, and processes.

The Conamara area that we saw in Figure 2.7 is shown at the higher resolution with the more oblique lighting in Figure 2.8. Remember, Conamara Chaos is about 80 km across, so this image shows an area about 180 km wide. With this illumination, the topography is evident. (Of course, when viewing an unfamiliar type of object, the brain can produce incorrect optical illusions. Sometimes, hills can look like holes or ridges can appear to be troughs. It may help if you bear in mind that in this image the lighting comes from the right, as it does in most of the images that show morphology in this book.)

Here, we begin to see the true character of the lines and splotches: The morphology of the large-scale lineaments is revealed, showing them to be complexes of ridges. The character of the chaotic terrain within Conamara Chaos is seen: There, blocks of older surface have been separated and displaced, like rafts of ice within a melted patch of lake ice on Earth. The bright ray of ejecta is still apparent running north–south across the western side of Conamara. Structural details of lenticulae are visible: Three dark spots north and northeast of Conamara in Figures 2.6 and 2.7 (and even in Figure 2.2) are resolved in Figure 2.8 into small patches of chaos each with a dark halo, detail that was not evident at lower resolution. Elsewhere, where there is no chaos, the terrain proves to be densely covered by mutually criss-crossing ridges ranging in width from about 1 km wide down to as narrow as the image can resolve. Uplift features are also visible, ranging in width from about 15 km on down. Many of these uplifts are up-bulged patches of chaos; others are irregularly-shaped, raised blocks of ridged terrain.[4]

In this brief introduction to the appearance of Europa's surface, I have already touched on many of the issues of Europan geology and geophysics that are discussed in detail in subsequent chapters. Europa is dominated by two classes of terrain, tectonic and chaotic, which correspond to the dominant processes that operate on Europa's icy crust. In the remainder of this chapter I focus on the appearance of chaotic terrain and of tectonic terrain, which is largely characterized by the presence of ridges.

---

[3] The high-gain antenna on the *Galileo* spacecraft failed to open, so we received only a tiny fraction of the anticipated imaging data originally expected. If it had opened properly, we would have the entire surface with this resolution and favorable lighting. The original picture budget is discussed in Chapter 3, along with a description of the spacecraft's road trips across America, which may have contributed to the mechanical failure of the antenna mechanism. The high-gain antenna failure and the consequent limited amount of image coverage is discussed in Chapter 12.
[4] The upward bulging of patches of chaos is discussed in Chapter 16, and other uplift features are described in Chapter 19.

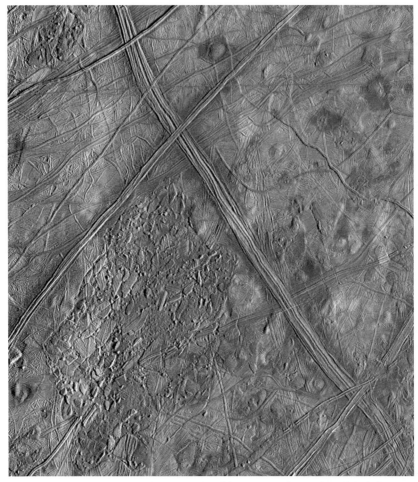

Figure 2.8. The Conamara region in a mosaic of images at about 180 m/pixel, here over an area about 180 km wide, shows morphological detail of the chaotic terrain. Conamara Chaos itself is about 80 km across and fills much of the lower left quadrant of this mosaic. Illumination is from the right. The structure of the global lineaments as complexes of double ridges is apparent. The darkening that dominates their appearance at low resolution is only a diffuse effect along the margins of these ridge systems. The lenticulae prove to be small patches of chaotic terrain with their similar dark margins.

2.4 RIDGES

The major ridge complexes that make up the global-scale lineaments (in this case the big X at Conamara) each consist of pairs of ridges that are roughly parallel, although crossing or intertwined in various places. The ridge complexes are fairly bright, at least as bright as any of the surrounding terrain. Surprisingly, the darkening that characterized these global-scale lineaments is hardly evident here at all.

How then does this structure relate to the appearance at lower resolution, the so-called "triple-bands"? If we look carefully at Figure 2.8, we see that along either side of the major ridge complexes the adjacent terrain is slightly darker than average. It is this subtle, diffuse darkening that shows as the dark lineaments on the contrast-enhanced global-scale images (e.g., Figure 2.2) or as the double dark components of the "triple-bands" at kilometer resolution (e.g., Figure 2.7). The ridge systems, which now are revealed to be structurally the most significant characteristic of the global-scale lineaments, lie between the faint double dark lines. In the lower resolution images they simply appear as the relatively bright center line of the triple-bands.

The dark coloration on the margins of the ridge complexes, which seemed so prominent on the global- and regional-scale images, proves to be barely recognizable in higher resolution images with oblique illumination (Figure 2.8). These darkened margins themselves do not directly mark any morphological structure, except that the surface seems to be somewhat smoother and lower, with a subdued topography, where it is darkest. The significance of the darkening is in the correlation of the coloring agents with ridge complexes: The ridges probably mark cracks along which oceanic substances have been able to reach the surface;[5] thus, these margins are the first of several examples of darkening (and perhaps thermal smoothing) associated with oceanic exposure.

The global- and regional-scale ridge complexes (like those in Figure 2.8) are composed of sets of double ridges. In fact, nearly all ridges on Europa come in pairs of identical components, each pair remarkably uniform along its length. Any case of a single ridge can nearly always be attributed to removal of its twin, due to overlapping by other intertwined ridges (as in some of the strands of the ridge complexes shown in Figure 2.8) or due to other resurfacing processes. The smallest ridge pairs would of course be unresolved in a given image if they are narrower than a single pixel, and thus may appear deceptively to be a single ridge.

The densely ridged terrain surrounding Conamara consists of criss-crossed double ridges. Figure 2.9 shows at very high resolution a sample of the densely ridged terrain near the northern edge of Figure 2.8. Here the resolution is about 21 m/pixel, one of the rare areas seen in such detail. With *Galileo*'s limited picture budget, only isolated sites have been observed so well, but they reveal the character of the widely distributed types of terrains. In Figure 2.9 we see rather typical densely-ridged terrain, where the previous surface has been covered repeatedly by ridge formation, each ridge crossing what was there before, until nothing is visible but ridges crossing ridges. Here, as everywhere on Europa, each ridge is part of a pair. The largest ridge pair crossing this field of view is about 2 km across and about 100 m high. This size is about as large as any ridge gets on Europa, a limit that may be controlled by the duration of the ridge-building process itself (Chapter 10). Most ridges seen here are much smaller. The older ones, which have been cross-cut many times, can only be seen in short surviving segments. Among the youngest features shown here are very fine cracks crossing and cutting older ridges. The double ridges

---

[5] The evidence for this statement is developed in Part Three of this book.

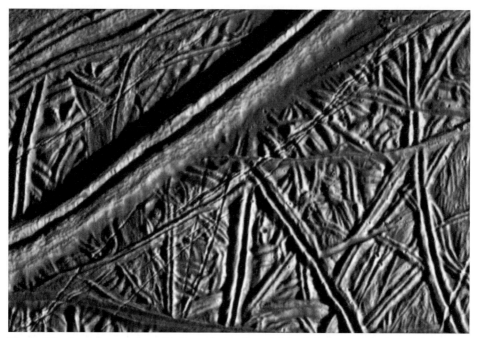

Figure 2.9. A portion of a very-high-resolution image (21 m/pixel) taken during orbit E6. The area shown is about 12 km across. This image sequence was designed to survey "bright plains", which proved to be densely ridged terrain. This area is located at the northern edge of the region shown in Figure 2.8. The common denominator of ridges on Europa is that they come in pairs.

may have formed along the borders of such cracks, while the recent cracks may not have had time for ridges to form.

In the literature, such densely ridged terrain is sometimes called "bright plains", a terminology that goes back to *Voyager* image interpretation, where the ridges were too fine to be seen at low resolution. Now, with higher resolution imagery (like Figures 2.8 and 2.9), where the true character of this terrain is apparent, that term should be obsolete, much as "mottled terrain" is an obsolete way to describe chaotic terrain. Both "bright plains" and "mottled terrain" are expressions that should have been abandoned once higher resolution images became available. Unfortunately, their use has continued, resulting in a confusing taxonomy.

In general, classifying a given type of terrain in various ways depending on the resolution of a particular image has been a major problem. In *Galileo* science analysis, this practice has caused confusion, misunderstandings, and incorrect generalizations and inferences.

Adding further to the confusion, and introducing a misleading implication, the geological mappers of the *Galileo* imaging team also called densely-ridged terrain "background plains", reflecting an assumption that this is the oldest type of terrain on Europa. That terminology suggests that the youth of Europa's surface is due to some sort of slate-cleaning event, which produced a background starting condition on which all subsequent geological processing acted. However, there is no evidence that resurfacing has ever been other than a gradual process of continual renewal by a fairly constant set of processes.

The common denominator for all ridge systems on Europa is that ridges come in pairs. Contrary to a too-common misconception, the double ridges do not correlate with the double dark lines of "triple-bands" that were seen at kilometer-scale resolution. That confusion may stem from an early *Galileo* press release montage (used in numerous review talks by *Galileo* spokesmen) that displayed triple-bands and examples of double ridges at similar size, without emphasizing the very different imaging circumstances and scale. Comparison of Figures 2.7 and 2.8 should disabuse one of that misconception: The ridges themselves are fairly bright and each is at most about 1 km wide. The dark margins are something else entirely. They are typically ~10 km wide, they are found outside the area of the ridges, and they are only associated with multi-ridge complexes (like those that make up the X in Conamara) or with a few of the largest ridges. The double dark lines are only pale diffuse markings, while the double ridges, as a class, are one of the most important and telling morphological features on Europa.

Ridges seem to correlate with tensile cracking of the icy crust, according to the following train of logic: The global and regional lineaments correlate reasonably well with theoretical tidal stress patterns. Because these lineaments seem to comprise complexes of double ridges, it is reasonably assumed that simple double ridges are similarly associated with cracks. Moreover, it has been generally assumed that the double nature is due to ridges running along each side of a crack.

Identification of ridges with cracks has been reinforced by investigations of their observable properties: their locations, orientations, geometries, formation sequences, and displacement of adjacent terrain. For example, as discussed in detail in Chapter 14, distinctive and ubiquitous lineament patterns in the shapes of cycloids (long chains of arcs joined together at cusps) usually comprise double ridges. These cycloidal ridge patterns follow from the diurnal variation of tidal stress during crack propagation. Cycloidal cracking also provides strong evidence for a liquid water ocean under the ice crust, because an ocean is required in order to give adequate tidal amplitude for these distinctive patterns to form. In this way and others, we will see how crack patterns marked by double ridges reveal the tectonic processes that drive much of the active resurfacing of Europa.

2.5 CHAOTIC TERRAIN

Inspection of Conamara Chaos itself at 200 m/pixel (Figure 2.8) shows the character of typical chaotic terrain. Details of this appearance are seen in a set of high-resolution images that span a belt across the southern half of Conamara, a portion of which is shown in Figure 2.10. Throughout Conamara the surface appearance suggests thermal disruption, leaving a lumpy matrix with somewhat displaced rafts, on whose surface fragments of the previous surface are clearly visible. Rafts

Figure 2.10. Part of a high-resolution (54-m/pixel) image sequence of Conamara Chaos taken during orbit E6. At the top and bottom, strips of the contiguous terrain are shown from the lower resolution image (Figure 2.8).

seen in considerable detail in Figure 2.10 can be seen in their broader context in Figure 2.8.

Typical of chaotic terrain, Conamara has the appearance of a site at which the crust had melted, allowing blocks of surface ice to float to slightly displaced locations before refreezing back into place.[6] Similar features are common in Arctic sea ice and even frozen lakes in terrestrial temperate regions, where the underlying liquid has been exposed.

Formation of chaotic terrain clearly represents the destruction of an earlier surface. In this case, that earlier surface was tectonic terrain. The surfaces of the rafts still display fragments of a terrain that was covered with ridges and cracks, essentially the same terrain that immediately surrounds Conamara. Like pieces of a picture puzzle, the rafts can be reassembled into fairly continuous areas, generally reconstructing a few of the major ridge systems that crossed the region. However, the entire destroyed surface cannot be reconstructed, because most of it has been broken into lumps too small to show their earlier surface or too melted to retain their original shape.

While formation of chaotic terrain has destroyed the previous tectonic terrain by breaking it up and melting much of it, we can also see in Figure 2.10 the revenge of tectonics. After Conamara formed and refroze, subsequent cracking has occurred. We can recognize cracks that formed after Conamara Chaos, because they cut through the lumpy matrix and when they reach rafts they either slice across them

---

[6] In Chapter 16 I discuss why such melt-through may be exactly what happened.

or wend their way among them. They contrast with those tectonic features that predate the formation of Conamara, which lie only on the rafts and do not extend into the matrix. One example of a post-Conamara crack runs diagonally across the lower left corner of Figure 2.10. Another toward the upper right snakes its way among rafts, along the northern edge of one of the larger rafts, and across the lumpy matrix. This example has already begun to form a double ridge.

In this way cracking and ridge-building have already begun to resurface Conamara. In other places on Europa, chaotic terrain has been covered by ridges to varying degrees, including many cases where the chaotic terrain is barely recognizable under the criss-crossing ridges.[7] Ridge formation seems to be as effective a resurfacing process as chaos formation. In Figure 2.9 we saw terrain where ridges had covered other ridges, and in Figure 2.10 we see where ridges have begun to cover chaos. Ridges cover what was there before; chaos formation destroys it.

Conamara is an unusually fresh example of chaotic terrain. For that reason it stood out so prominently in earlier imagery that it was selected for targeted high-resolution study, which led to its being widely cited as the archetype for chaotic terrain. If Conamara were really typical (rather than unusually fresh), the implication would be that chaotic terrain is rare and recent, a misconception that has been widely propagated in the literature. In fact, however, most chaotic terrain is much older than Conamara and modified by subsequent resurfacing. The older examples of chaos are simply harder to see, which led to the false impression generated early in the *Galileo* mission that chaos is a relatively recent phenomenon. On the contrary, like tectonic processes, formation of chaotic terrain has occurred throughout the geological history of Europa, as far back as the record goes.

Conamara is only typical as an example of very fresh chaos, not of Europan chaos in general. It must have formed recently relative to the rest of the surface. Given that the surface of Europa is less than 50 million years old, according to the lack of craters, Conamara itself probably formed within the past million years.

Another detail in Figure 2.10 worth noting is the presence of several tiny craters. Most obvious are two examples (each a couple of hundred meters across) lying on a raft near the far left of this picture. Based on relationships with other craters, these are probably part of the large population of tiny secondary craters, formed by ejecta thrown out by the rare larger impacts on Europa.

Returning to chaotic terrain, not only is Conamara not typical of chaos age, at 80 km wide it is not typical of the size of chaos patches either. The largest single patch that we have seen at 200 m/pixel is roughly circular and about 1,300 km across. Other even larger chaos regions are evident as the dark splotches in the global images (Figures 2.2–2.4). The distribution of sizes of patches of chaos is such that the smaller they are, the more there are, all the way down to patches so small they are barely recognizable. In 200-m-resolution images, it is hard to discern the matrix texture, tiny lumps, and rafts that identify chaotic terrain, if the patch is smaller than a few kilometers across. But where we have higher resolution, the

---

[7] Gradual erasure of chaotic terrain by tectonics, and limits of recognizability due to age and image resolution, are discussed in Chapter 16.

26 Touring the surface [Ch. 2

Figure 2.11. A blow-up of the top center portion of Figure 2.8, showing tectonic terrain, densely filled with double ridges, ridge complexes with dark margins that appear as "triple bands" at low resolution, and a small patch of chaotic terrain, surrounded by similar dark margins, which is a typical lenticula at low resolution. The area shown here is about 50 km across.

recognizability (and increasing numbers) of small patches of chaos extends down to proportionately smaller sizes.

A patch of chaos about 5 km across is shown at the top of Figure 2.11, which is an enlargement of the area just above the X in Figure 2.8. The X itself is shown at the lower left, and this small patch of chaos also appears as a lenticula in Figure 2.7. In this typical small patch of chaos, we see that the lumpiness of the texture is finer in proportion to the small size of the chaos area. There are no rafts large enough to reveal older terrain.

The patch of chaotic terrain in Figure 2.11 appears to be bulged upward. In some interpretations, features like this one are taken to be upwelling of magmas (slush or viscous ice) that rose and spread over the surface. However, such up-bowing would also follow naturally from exactly the same sort of fluid exposure as appears to have created Conamara and other chaotic terrain: After the melt-through from below, buoyancy would bulge up the surface during subsequent refreezing (Chapter 16).

The dark diffuse halo around the small patch of chaos in Figure 2.11 is

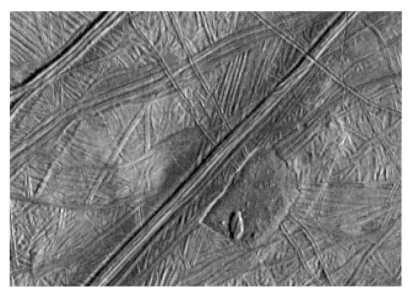

Figure 2.12. A blow-up of part of the same mosaic as shown in Figure 2.8. This area is to the northeast of the part shown in that figure. A small patch of typical chaotic terrain (~10 km across) lies surrounded by densely ridged tectonic terrain. It contains the usual lumpy matrix that characterizes chaos and, despite its small size, it contains a raft that displays a bit of the older ridged terrain. Patches of chaotic terrain are found at all sizes down to the limits of resolution in our images.

significant, because it explains the appearance of such features at low resolution, as well as the source of a crucial misconception: that the size of lenticula (low-resolution spots) can be used to infer the sizes of patches of chaotic terrain. This darkening shows up in Figure 2.7 as a lenticula, just as the similar darkening along the major ridge complex (that runs diagonally across Figure 2.11) shows as double dark lines (forming a so-called "triple-band") in Figure 2.7. The scale, the amount of darkening, the diffuse appearance, and the minimal effect on morphology are nearly identical whether this material is observed as a halo around a very small patch of chaos or as a pair of dark liners along a global ridge complex.

Once the relationships between appearance in global- to regional-scale images and appearance at this higher resolution is understood, we can interpret the true meaning of the archaic taxonomy developed for the earlier, low-resolution data: the splotches (or "mottled terrain") are chaotic terrain, "triple-bands" are major ridge complexes with diffuse dark borders, "bright plains" are densely ridged terrain, and "lenticulae" are small patches of chaotic terrain, enhanced in size by their diffuse dark halos.

Europa's surface is young, and it appears to be continually resurfaced by two dominant processes: tectonics forming cracks, ridges, and related features, and thermal processes that have created chaotic terrain. The area sampled in Figure 2.12 (just a bit further northwest than Figure 2.11 and visible to the upper left in

Sumo on Ice

Figure 2.13. Chaotic terrain is most likely due to melt-through from below, but the explanation suggested here is interesting. In any case, the appearance of chaotic terrain strongly suggests areas of exposure of the liquid water that is ordinarily below the ice.

© The New Yorker Collection 2000 Matthew Diffee from cartoonbank.com. All Rights Reserved.

Figure 2.8) encapsulates that fundamental character, showing densely-ridged terrain surrounding a small patch of chaos with all the appearance of a melt-through site, including a raft of displaced ridged terrain (Figure 2.13). Wherever it occurs, each of these dominant processes wipes out what was on the surface before. This resurfacing is rapid and recent. Each appears to involve the interaction of the liquid ocean with the surface. These processes and the conditions they produce may create and maintain a variety of habitable niches capable of supporting life. And as we will see, all of this activity is driven by tides.

3

Politics and intellect: Converting images into ideas and knowledge

> If what we were discussing were a point of law or of the humanities, in which neither true nor false exists, one might trust in subtlety of mind and readiness of tongue and the comparative expertness of writers, expecting him who excelled in these qualities to make his arguments the most plausible; and one might judge that to be correct. But in the physical sciences, where conclusions are true and necessary and have nothing to do with human preferences, one must take care not to place oneself in the defense of error.
>
> Galileo, 1632

Our introductory tour of Europa suggests that the important surface features, and the processes that formed them, all involved an ocean linked to the surface through a thin, permeable crust of ice in a variety of ways: narrow cracks allow ocean water to flow to the surface and form double ridges; cracks have often spread open, filling with fresh ocean water; from time to time, and place to place, local heating has thinned the crust, often to the point of exposing the ocean and creating chaotic terrain. A more comprehensive consideration of the available evidence, presented in subsequent chapters here, supports this model of a permeable ice crust.

This picture contrasts with a widely-promoted model in which the ocean, if any, is completely isolated from the surface. In that view, the character of the surface has nothing to do with the ocean. Instead, all the features are interpreted as being formed by processes within a thick layer of ice. The idea is based on the fact that, given enough time, even solid materials can flow. On Earth the movement of glaciers shows that solid ice flows easily. Even rock has deformed into seemingly improbable shapes over the aeons. On Europa, according to the isolated ocean model, the surface features have been attributed to upwellings within the ice, due to hypothetical solid state convection or low-density blobs or pockets of meltwater. The problem with such models is that there is no evidence that the hypothesized drivers exist, nor that the structures that we observe are what would be produced by such processes.

Nevertheless, the isolated ocean model was widely reported and accepted during much of the *Galileo* mission's time in orbit around Jupiter. Recognition of the direct role of the liquid ocean in shaping the surface has developed more slowly and recently.

Where did the isolated ocean model come from? To a large extent, it was the result of historical inertia. Until 1979, no one suspected that water on Europa would be anything but solid ice. Then, only a few days before *Voyager*'s arrival at Jupiter, Stan Peale (University of California at Santa Barbara) and his colleagues reported that the eccentricity of the orbits of the Galilean satellites would result in substantial heating by the friction of tides. The most optimistic estimates allow for enough ongoing heating in Europa to keep nearly all the water in a liquid state, except for a thin layer of ice at the surface. However, in the *Voyager* era of the 1980s, any liquid water so far from the Sun was a radical notion. To make a credible case for its plausibility, researchers needed to show that a liquid layer would be possible with even a minimal amount of tidal heating. Thus, the scientific literature was dominated by conservative estimates of heating rates, with correspondingly thick ice crusts.

The thickness of the ice layer on top of the ocean can be inferred from the rate of internal heating. We know the surface of the ice is kept cold (about 170°C below freezing) as heat radiates into space. If the internal heat flows out through the ice by thermal conduction, the thickness of the ice adjusts in accordance with the amount of heat it needs to transport. The faster heat is produced in the interior, the thinner the ice must be to carry away the excess. If the ice were too thick, extra heat would build up inside, melting the bottom of the ice until a balance was reached. In this equilibrium, the thickness of the ice is proportional to the rate of internal heat production.

For the highest plausible heating rates, each square meter must transport $\frac{1}{2}$ joule of energy per second (each 10-meter-square of surface would conduct out the heat equivalent to a 50-watt light bulb). In the steady state the ice would be one or two kilometers thick. More modest heating rates would imply ice thickness a few times greater. Compared with a frozen lake, or the Arctic ice sheet, kilometers of ice seem very thick. However, for a global solid crust overlying 150 km of liquid water in a place over five times as far from the warmth of the Sun as we are, an ice layer less than 10 km thick is thin indeed.

A much thicker crust could transport the heat almost as fast as that thin layer if the ice were convecting, instead of just conducting the heat. Remember, even solid ice can flow, albeit slowly. For convection within the solid ice crust, warm ice would flow up and cooler ice down, transporting heat like a conveyor belt. If the ice crust were 20 km thick or more and convecting, it could transport heat as quickly as a crust less than 10 km thick that is only conducting the heat. For a given heating rate, estimates of the ice thickness depend on whether it is convecting or not.

Theoretical considerations have been inconclusive, but it seems that convection may require that the ice have very special material properties, such as just the right grain size. We can only speculate whether the ice meets the very tight requirements. Some geophysicists think it is unlikely. What is more, if the ice is less than 15 or 20 km thick, the thermal instabilities needed to drive the conveyor belt of convection probably cannot develop. Thick ice is necessary, though not sufficient, for convection. If the crust is convecting, it must be considerably thicker than it would be if it were transporting heat by simple conduction.

3.1 POLITICS ON BOARD

These intellectual issues were central to considerations about Europa's ocean and icy crust during the years between the *Voyager* fly-bys of the Jupiter system in 1979, and the arrival of *Galileo* there. The most conservative estimates of tidal heating dominated the scientific literature from the early 1980s, not necessarily because the authors believed in the lower heating rates, but because the credibility of the idea of an ocean depended on a conservative estimate. As the possibility of a liquid layer under the ice began to be accepted, the more extreme idea that nearly all the water might be liquid remained a radical notion. The possibility that the ice layer might be thin enough for the ocean to be linked to the surface was considered by some scholars, but the conservative view that the ice must be thick was prevalent. That conservatism so dominated planetary geology that, even when in the early 1980s clear evidence was found in *Voyager* images for highly mobile crustal plates, which suggested that there was a low-viscosity layer just below the surface, publication of the discovery was blocked by peer review until the end of the decade.

As the first *Galileo* images arrived on Earth, the thick ice paradigm had remained in place. In itself, that intellectual climate would not have been stifling. Planetary scientists are agile thinkers and used to new ideas. The real problem was that the initial interpretation of *Galileo* images was dominated by academic politics and soon gelled into a party line.

The *Galileo* Imaging Team comprised a wide range of expertise in various aspects of planetary science, but it also included some of the more politically skillful, aggressive, and powerful members of the scientific community. While it was nominally a team, it resembled an arena full of professional gladiators, powerful infighters each giving priority to his own interests. This perverted definition of "team" was something I only gradually came to understand during my 26 years as a member.

These dozen scientists were employed by various separate academic (or quasi-academic) institutions, with which NASA contracted for their scientific expertise. Several team members had broad power bases that extended far beyond this single team. They had strong influence across NASA-supported space science in the arenas of funding decisions, policy, publications, and public relations. Their loyalties were to themselves and their universities (or research organizations), and their goals seem to have been to secure scientific recognition, research money, and influence. NASA got what it needed because the team members' interests and NASA's converged where it was important. Everyone needed the mission to succeed, at least to the extent that pictures would be taken and some interesting discoveries could be reported. But where their objectives diverged, each team member had his own agenda.

When NASA selected the team members at the start of the project, Michael Belton (Kitt Peak National Observatory) was designated team leader. Belton's work as an astronomer at Kitt Peak had not prepared him for the difficult management role he faced, but most people involved with the project think he did a good job. The camera got built, the pictures were taken, press releases were distributed, press conferences were held, scientific papers were published. Belton had to do all this while playing a weak hand relative to some of the heavy hitters on his team. A strong leader with vision might have done things differently, but Belton was not in a position where he could encourage or reward innovation. To keep the team from blowing up, and to retain some semblance of control, he had to give the most powerful players pretty much what they wanted.

During the two decades between the start of the *Galileo* mission and the spacecraft's imaging at Europa, it was arranged that those who planned each imaging sequence would get exclusive rights to do the initial interpretation of the data. And the heavy-hitting geologists made sure that they were assigned to do that planning.

The strategy paid these successful operators considerable dividends. For one thing, they received large amounts of money for their universities, because they made the case that they needed to hire staffs of graduate students, research associates, and technicians to do the work. Within their research universities, the overhead funds that come along with such large contracts are the currency for establishing a professor's salary, office size, and political clout. The other pay-off was that these individuals controlled the scientific results and interpretations that would be presented to the world in the name of the *Galileo* project. Through press conferences, major presentations at scientific conferences, and publications, they controlled how discoveries of the *Galileo* mission were presented to the public and to the broad scientific community. Their initial impressions of the images of Europa, and more frequently the quick-look qualitative impressions of their graduate students and young postdoctoral assistants, were preordained to become the canonical interpretation of the character of Europa.

Given how this canonical view was developed, it was probably bound to be wrong. It was based on initial qualitative impressions, because of contractual obligations to make quick definitive pronouncements. It was controlled by a small subset of the team, mostly geologists. The interpretative work was delegated to their inexperienced students. Other team members, with the complementary expertise necessary to do the job right, were locked out. All of this resulted in inexcusable errors in research methods and results.

3.2 METHODS OF THE GEOLOGISTS

Even at its best, the "scientific method" as actually practiced has little to do with the idealized version that schoolchildren are required to memorize. Moreover, there is considerable variation in practice from one field to another. The methods and habits of mind developed for geology are based on centuries of experience in obtaining and interpreting close and detailed observations of the surface of the Earth. But information about the surface of Europa came from images taken of an alien planet using a telescope flying through space. We can learn a great deal from such pictures, but if the traditional techniques of geology are used without attention to the unusual subject and novel sources of information, they can lead to trouble, and they did.

Consider the use of qualitative analogies, a standard technique in geology, where new discoveries are interpreted in terms of experience with similar features or structures. This approach has served the field well in exploring and understanding the Earth. But there is a danger in unduly applying terrestrial experience to a planet that may be completely different. The initial considerations of Europa were based on choosing the most similar type of geological feature on Earth. Inevitably, the surface was interpreted in terms of the types of processes that operate on a solid planet.

Another problem has been that geological tradition does not address the crucial issues regarding how data about Europa were obtained. In geology, there is usually a glut of data. Researchers walk all over their subject, chipping it with hammers, magnifying it under microscopes, even tasting it.[1] In order to reconstruct and interpret the processes that produced what is seen, traditional geology needs to cope with the problem of too much data. The issue there is how to recognize underlying patterns and generalities, and how to discriminate between those seemingly minor details that are critical constraints on the big picture and those that are simply local anomalies of no significant consequence.

However, with *Galileo* we have too little data, not too much. The problems with this kind of data are better known to astronomers than geologists: the remote sensing involved, the dependence on image data, the sparseness of the data, and the varying circumstances under which they were obtained all produce observational selection biases. This "bias" refers to unavoidable circumstances of the observations, not a failure of the scientific process. For astronomers, quantitative corrections for such types of biases are a standard part of data analysis. Bias corrections are understood in astronomy to be an essential precursor to any physical interpretation.

Unfortunately, the initial interpretations by the dominating geologists of the *Galileo* project, or their students, did not take such effects into account. In classifying features, differences in lighting and image resolution were confused with actual characteristics on the surface. The ease of recognizing certain types of features during a quick look was taken as a measure of their greater physical significance compared with features that were just as real, but harder to see. Generalizations were made on the basis of appearances in special selected locations. Anecdotal impressions became reported facts.

Ultimately, in the rush to publicize and publish the initial results of *Galileo* imaging, several factors came together, driving the geology to be described in terms of thick ice over a largely irrelevant ocean:

- The intellectual context in which a liquid ocean, let alone thin ice, remained uncertain.

---

[1] For example, to recognize umber, which sticks to the tongue.

- The tight political control of the initial quick-look interpretation of *Galileo* data that limited interdisciplinary consideration so that only a narrow, conservative perspective could dominate.
- The dependence on forced analogies with familiar features and processes on the solid Earth.
- The insufficiently quantitative analysis of the data.

Despite their shortcomings, these qualitative early impressions were prominently published and widely presented at scientific conferences by the chosen few, each wearing the mantle of authority of this major mission. Magazines for science professionals, especially *Science* and *Nature*, fast-tracked publication of the preliminary interpretations by the designated spokespersons. The same story was put out by the NASA and JPL publicity machines. The party line was intensified by active promotion and by frequent repetition and citation. Those interpretations, though based on only preliminary first impressions, were quickly accepted as canonical fact.

In fact, rather than being fast-tracked into the canon, such quick-look results should more appropriately have been treated with extra skepticism. This rule holds especially for Europa, where images show types of terrain that may reflect processes very different from familiar bodies, where data are complex and require quantitative assessment of observational selection biases, and where meaningful interpretation requires quantitative theoretical study, none of which was understood or available yet at the time of the early reports of the *Galileo* Imaging Team. Nevertheless, by early in the *Galileo* spacecraft's 7 years in the Jupiter system, the isolated ocean interpretation had become the canonical model of Europa.

3.3 THE RULE OF CANON LAW

Canonical doctrine has been known to stand in the way of scientific progress before. In fact, Europa played a central role in the all-time greatest example, the case that blasted open the door to modern scientific inquiry during the Renaissance.

For over a thousand years, the Ptolemaic theory of celestial motions had been a very successful model. Every moving body in the sky was deemed to be in orbit around the Earth. These orbits were approximately circular, but they also involved epicycles, *ad hoc* corrections in which the bodies followed small circles around points that orbited the Earth. With these epicycles (a concept that is still useful in modern celestial mechanics), this ancient model fit the observational data very well, and was an effective tool for prediction of future motion of stars, planets, and the Moon across the sky.

That theoretical model also had a huge amount of political traction. For one thing, it fit the observations very well. Beyond that, with the Earth at the center of everything, it was appealing to anyone contemplating his place in the universe. Most important, it was perfectly consistent with a literal reading of the Biblical description of the relationship of the heavens to Earth. It is no surprise that this world view was considered obligatory by the Western religious power structure.

The alternative model developed by Copernicus in the early 16th century did not seem to have much going for it. For one thing, it was no more accurate than the Ptolemaic paradigm. Having the planets orbit the Sun, rather than the Earth, did make things simpler. Fewer *ad hoc* epicycles were needed than in the Earth-centered model. But the intellectual community at the time had not yet decided whether simpler was better, although that broad question had been a philosophical issue for some time.

The merits of simplicity were overshadowed by a much bigger problem. Just as most modern U.S. planetary scientists get their research funding from NASA, and only if they stay close enough to the mainstream, Copernicus was funded by the Church, the same institution that promoted the Ptolemaic paradigm. No wonder that he was reluctant to stick his neck out publicly and publish his results. Of course, in those days his neck was what was in jeopardy. Nowadays, just the job and the funding would be cut.

That kind of career damage had happened to William of Ockham in the early 14th century. Ockham was kicked out of Oxford for his radical philosophical ideas, and was at odds with the papal court for most of his career. Ockham's most lasting and influential philosophical contribution is known as "Ockham's razor". The idea is that, of all possible theoretical models that could be constructed to fit observations, the simplest one, the one that shaves off non-essential details, is best. It is the one most likely to describe the essential underlying mechanism. Clearly, there is no guarantee that a simple model will be more correct than a complicated one. Sometimes a natural system may be so complicated that a too-simple model may miss what is really going on. But, throughout the scientific age, experience has shown the power of Ockham's principle. Ockham's razor underlies most theoretical modeling in modern science.

The medieval church hierarchy in the 14th century had been prescient in recognizing the threat of Ockham's ideas to canonical doctrine. Two hundred years later, Ockham's principle gave value to the strongest asset of the Copernican model: the fact that it was elegantly simple compared with the Ptolemaic model.

In modern physics and astronomy we take Ockham's razor for granted, so the strength of Copernicus's elegant model seems obvious. But during the 16th century acceptance of the Copernican model was slow. A widely-accepted older paradigm seemed adequate and a powerful establishment opposed change. Then, as now, being right did not necessarily count for much and was often a liability.

The observations that finally turned the tide of scientific opinion in favor of Copernicus surprisingly had nothing to do with the controversial orbits of planets around the Sun. They involved instead celestial bodies that were unknown to Copernicus. The four largest moons of Jupiter were discovered by Galileo almost 70 years after the death of Copernicus, and almost 400 years before the *Galileo* spacecraft imaged those same satellites. Galileo named these bodies in honor of his funding agency, the Medici family, but that was not enough to keep him out of trouble with the Church.

If these bodies were not part of Copernicus's model of planetary motion, how could the observations support the model? According to Galileo's 17th-century imaging data, the satellites were clearly and obviously orbiting Jupiter. He had shown by direct observation that not all astronomical bodies orbit the Earth. He had *not* shown directly that Copernicus's model was correct, but he did provide collateral evidence that it could be. That contribution was enough to change the prevailing mindset and allow people to think about Copernicus in a different way.

We now know these moons as the Galilean satellites Io, Europa, Ganymede, and Callisto. Once they were discovered, the Earth-centered canonical paradigm was overturned and the door was opened to acceptance of the Copernican world view. Beyond that, the Copernican description of planetary motions led science much further. It formed the observational basis for the development of Newtonian physics. Most profoundly, this demonstration of how simplicity can bring deep understanding of physical principles gave legitimacy to Ockham's principle. It formed a template for the interrelationship between observation and theory for all of modern science.

Nevertheless, the same forces that resisted the Copernican revolution are still with us, and still discourage breakaway ideas in science. Canonical models are challenging to shake, because of the strong, even ruthless, defense by the politically powerful or adept. There is a strong disincentive to resist the canon, especially for the young and untenured. Even setting politics and careerism aside, inertia favors an entrenched paradigm. Finally, Ockham's razor is often disregarded. For all of these reasons, the canonical thick ice, isolated ocean model of Europa still has powerful advocates and sycophantic adherents.

3.4 *GALILEO* IN THE 20TH CENTURY

Another *Galileo*, this time a robotic spacecraft, observed Europa in the late 20th century. Again, the images were viewed in the context of their times. There might have been an ocean, but the idea that it might be near the surface was a radical notion. The images of Europa were viewed and interpreted under an assumption that the ice was so thick that the ocean had little to do with what we see at the surface. Once these rushed geological pronouncements were made public, there were strong reasons to resist change. The powerful people who committed to this model could not afford to appear wrong. Even greater resistance came from those theorists who had believed and blindly accepted the initial authorized descriptions of the surface and had already developed artificial models to explain them. They could not afford to look gullible. And a flood of seemingly supportive work was published by young researchers whose job prospects and tenure decisions depended on adhering to the party line.[2]

The organizational structure and policies of NASA's space program and the *Galileo* mission in particular also worked against innovative thinking. *Galileo* was an extremely expensive project, so failure was not an option. It was not likely that Congress or the public would pay for a second try. As a result, there was a strong

---
[2] More on them in Chapter 20.

incentive to do things as they had been done before. Unfortunately, as I was to learn, this culture of resisting new ideas permeated all aspects of the mission.

In October 1977, having been selected to be a member of the "Solid-State Imaging" team,[3] I attended the kick-off meeting of the *Galileo* project at the Jet Propulsion Lab in Pasadena. A banner across the front of the room shouted "May the Force Be with Us", setting a date marker in memory: the first *Star Wars* movie had just come out. It was very exciting to be at what I thought was the frontier of science and technology.

I was wrong. In order to guarantee success, the project needed to stay as far back from the frontier as possible to be sure that the technology worked. When failure is not an option, neither is innovation. By the time *Galileo* was launched in 1989, its technology was behind the curve: CCD cameras and computers more capable than those on board had already become consumer commodities.

To be sure, in large part the flight technology was behind the times because of gross delays in the launch schedule. *Galileo* was originally to have been launched in 1982 using the space shuttle. Having already sold the shuttle idea to Congress, NASA needed to find some use for it. In order to make sure that the shuttle would be used, NASA dismantled the infrastructure for its *Titan-Centaur* rocket, its old reliable work horse for interplanetary launches. The space shuttle was not an especially desirable way to go, and at the time of the *Galileo* kick-off meeting no one knew how the space probe would get beyond Earth orbit once the shuttle got it there. Eventually, with considerable corporate lobbying in Congress, a modified version of the *Centaur* upper stage was selected to fly on the shuttle with *Galileo* to propel it onward toward Jupiter.

Selecting and developing the upper stage rocket delayed the mission substantially. *Galileo*'s voyage did not begin until 1985, and then it was only a road trip along Interstate 10 from southern California to Florida. As *Galileo* was resting at the beach in Florida at the beginning of 1985, waiting its turn for a shuttle launch, everything changed with the explosion of the shuttle *Challenger*.

In addition to the delays inherent in checking out and re-establishing the shuttle program, it became clear that a bomb as big as a modified *Centaur* rocket was not going to be allowed to ride along with humans on the shuttle. So, the next leg of *Galileo*'s journey was not toward Jupiter, but instead back over the I-10 highway to California.

---

[3] "Solid-State Imaging" (SSI) is technospeak for "digital camera". When the *Galileo* project got started it was known as the "Jupiter Orbiter with Probe" or JOP. Shortly afterward it was named *Galileo*. The obligatory three-letter acronym started appearing on memos as GLL. So, I found myself a member of the GLL SSI team. There is also a very arcane difference, important to some people, between the *Galileo* Mission and the *Galileo* Project. The Mission refers to the overall NASA program, which is managed at NASA headquarters in Washington, D.C. The *Galileo* Project is the development and operational activity that is managed by JPL under a contract with NASA. Usually, the word "Project" was used as a proper noun to refer to the top project management at JPL. When JPL engineers assigned to *Galileo* said things like, "Project has determined that we cannot do what you asked for", they meant their bosses told them not to do it. Finally, while the GLL camera was called SSI, the camera on the *Cassini* mission to Saturn is called ISS, or "Imaging Sub-System". Go figure.

As *Galileo* waited in California again for the next several years, civilization marched on: the shuttle program was reinstated; an "Inertial Upper Stage" rocket was developed to propel *Galileo* onward; and *Galileo*'s once-advanced onboard technologies aged while they became consumer commodities. Finally, *Galileo* hit the interstate again for a second road trip to Florida, heading for points beyond. It was launched from Cape Canaveral in late 1989 on the shuttle *Atlantis*. The Inertial Upper Stage was too weak to carry *Galileo* directly to Jupiter, so encounters with Venus and the Earth were added to give extra gravitational kicks, but that route doubled the travel time. The bundle of shopworn 1970s' technology reached the Jupiter system late in 1995 and didn't complete its mission until 2003, more than a quarter-century (and four more *Star Wars* episodes) after the kick-off meeting. As any superstitious person could tell you, that 1977 *Stars Wars* banner was a stupid idea.

3.5 TECHNOLOGICAL OBSOLESCENCE

Even if *Galileo* had gotten to Jupiter on time in 1985 its technology would have been old. The absolute requirement for success constrained engineers to avoid the forefront of technologies. During the first years of the project, as I listened to the reports and discussion at Imaging Team meetings, I noticed that selection of components generally seemed to require that they be available in commercial catalogs. While it was disappointing not to be as close to the forefront of technology as I had thought, it did make sense. Anything that flew on the spacecraft had to work.

More disappointing and inexcusable was the way this caution pervaded all aspects of mission planning, even extending to planning for science analysis procedures and technology that would be used here on planet Earth. While there was some rationale for using old reliable technology as much as possible for flight hardware in order to insure against failure, I could never have imagined that the same conservatism would apply to planning for ground-based activities that would take place many years into the future. What well-managed enterprise would specify technology purchases and functionality 5 years into the future on the basis of what was already available on the market?

It was not until 1980 that I began to recognize this aversion to the cutting edge of technology, and then only because the Hunt brothers of Texas had tried to corner the market on silver.

Photographic prints are dyed with silver. Mission planning in the late 1970s included consideration of the production and distribution of the pictures that the spacecraft would take when it got to the Jupiter system. For budgeting the picture production, planners estimated that 100,000 pictures would be taken, and each would need to be printed in a few different versions, with varying contrast, brightness, sharpening, and other enhancements. All these large-format pictures would need to be distributed to all the Imaging Team members and to various archives. Millions of photos would need to be printed. When the Hunts had bought up nearly half the world's silver supply, the *Galileo* darkroom budget soared to $12 million.

Several things about this plan impressed me. The first was that the spacecraft was expected to be able to get us many more than the 100,000 pictures in the budget, perhaps twice as many or more. The number of pictures would not be limited by the durability of the spacecraft, nor by the rate at which the digital images could be radioed back to Earth. Instead, the number was going to be limited by the printing costs. To me, this plan seemed like deciding not to take wedding photos because your computer printer was out of ink.

I was also impressed by the dollar budget. With 12 team members, the mission would be spending a fat round million dollars on photo printing for me. It was not obvious that this plan was how I would choose to spend my share of the resources. Finally, I could hardly imagine doing science with hundreds of thousands of large-format photographic prints. The mere physical manipulation seemed mind-boggling, let alone using them for research in any systematic or quantitative way.

Galileo images were going to be handled the way spacecraft images always had been. The bits of data would be radioed to Earth from the spacecraft. Each image was to be composed of 640,000 measurements of brightness in the camera's field of view, arranged in 800×800 pixels. The brightness values were encoded as 256 levels of gray (256 levels $= 8$ bits $= 1$ byte), so each image could be encoded as 640,000 bytes or 2/3 of a megabyte, or a few times less with clever data compression schemes. A central processing facility on Earth would spend \$12 million converting that information into photographic hard copies for delivery to the science teams. Once the hard copies would be printed, the bits would be stored on magnetic tape, where it would be inaccessible and vulnerable to loss and physical deterioration. Photographic hard copy was considered to be the final, permanent product.

As a medium, photographic hard copy has some advantages: the printed resolution can be very high; you can quickly look over many images spread across a table; no special viewing equipment is needed; and they can be tacked to a bulletin board. However, compared with digital data, there are overwhelming disadvantages. A photographic print cannot display the full range of brightness and contrast that is contained in the original digital data. Studying or archiving hard copy means that you are throwing out a good part of your expensive and valuable information. Moreover, further processing beyond a predictable, automated routine is difficult and expensive, and usually degrades the information content even more. Once hard copy is printed, further enhancements are difficult.

I had watched as *Voyager* images were processed back in 1979. A team of skilled craftsmen from the U.S. Geological Survey was detailed to JPL, with sharp knives and glue, to cut-and-paste photos into large mosaics. These people were very capable and hard-working, but the activity looked to me like something from the era of the Second World War, rather than the new age of information technology.

In the *Voyager* encounter with the satellites of Jupiter, scientific understanding was often delayed by the limitations of what can be seen in printed hard copy. For example, when the *Voyager* Imaging Team had received the photos that were taken during the brief encounters with the satellites of Jupiter, they were mystified by the appearance of Io, the innermost Galilean satellite. The photos showed a strange blotchy place that did not look like any planetary surface that they had seen before. In fact, the explanation was clear in the bits of information that had been radioed to Earth, but it was lost when the photos were printed. Fortunately, different versions of the same images were prepared for the optical navigation team. These engineers were given overexposed, high-contrast versions because they needed to see the stars in the field of view of the camera in order to navigate the spacecraft. Io appeared only as a white disk, but it was this overexposure that allowed an optical navigation engineer at JPL, Linda Morabito, not an Imaging Team scientist, to discover volcanism on Io when she spotted the gigantic volcanic plumes spraying into the dark sky. This information was in the original digital bits sent by *Voyager*, but the prints that showed Io's surface did not show the plumes, and the prints of the same image that showed the plumes did not show the surface. The photographic prints were degraded versions of the image data that *Voyager* had sent to Earth.

At the time that the *Galileo* Imaging Team was planning its photo budget, elsewhere computers were already being used to display and process images. The field of digital image processing was still in its infancy, but its potential was clear. The commercial sector, already active in Silicon Valley, was developing applications poised to revolutionize image processing in all sorts of fields, especially where the money was, in biomedicine and military defense. But, on the frontiers of planetary science, where revolutionary digital data were being received from Jupiter, they were being processed using 19th century technology. *Galileo*'s $12 million plan for the future was for more of the same.

I decided that I would prefer to let the Hunt brothers keep my share of the silver. I would be happy to forego any hard copy if the *Galileo* project would simply send me the bits of information when they arrived on Earth. All that I asked in exchange was that I might get part of the million dollar savings from my share of the darkroom expenses to buy image-processing computer equipment. The advantages seemed obvious, at least to me. The project would save money. I could buy a great computer system, and by waiting until I needed it I could buy an even better and cheaper one. I would not have to manage a huge library of photographic hard copy in my lab. And I would have the digital bits to look at in any way I wanted to. Moreover, I would be able to do precise quantitative measurements with greater ease and precision than with paper photos. It seemed to me that this approach might work for others on the team as well.

I gave a presentation of this idea at a team meeting in Hawaii in 1980 and I was aggressively attacked, and virtually hooted out of the room. The reasons seemed incredible to me at the time, and are even more incredible in retrospect. For example, one geologist could not understand how you could make measurements on a digital image. He was unable to envision modern graphical measuring tools on a computer. He thought that my proposal would require him to lay a wooden ruler against his computer monitor. For his ignorance, I was ridiculed. Some of the old-timers could not imagine changing their ways. Others could imagine it, but must have realized that it would mean relinquishing the power and authority and mystique that they had cultivated since the beginning of planetary exploration, by doing things the same old way. My idea was roundly mocked and rejected. I learned an important lesson: At the frontiers of planetary exploration, new ideas are not welcome.

My plan was at odds with another aspect of the *Galileo* mentality. The planning process had no way to accommodate the rate of change of technological advancement. The plan for 1985 had to be based on computers that could be found in a 1980 catalog. The same went for data storage media. My plan was attacked on the grounds that it would require rooms full of computer tapes to store the data at each of our home institutions. My prediction that a few laser disks would be able to hold the whole *Galileo* data set on a single bookshelf was roundly ridiculed. Actually, this was not a radical idea at all. Video disks were already on the commercial market and the CD format was in the works, but such notions were too risky for *Galileo*. In short, the project was as conservative in planning technology for my office, which was to be bought years in the future, as they were in planning for the technology that was to fly into space. If I had understood the politics and mentality of the *Galileo* project and Imaging Team, I would have kept my mouth shut. Being right did not do you any good in the *Galileo* project, unless you kept it to yourself. Sharing good ideas was definitely a mistake.

Of course, over the next few years as technology marched on, the project had to let go of its old ways. The whole world was adopting digital imaging; image processing could be done on home computers; and the whole *Galileo* data set could fit in a sun visor CD caddy. Eventually, the Imaging Team adopted my plan, and even used the name that I had proposed in that first humiliating presentation, the Home Institution Image Processing System (HIIPS). I have never heard my name mentioned in connection with HIIPS, although I heard a rumor that in private someone once overheard team leader Mike Belton acknowledge that it was my "prescience" that got it all started. He would never say it publicly, or to me. As team leader, he had little enough political capital without offending his most powerful team members.

The same stifling politics and operational conservatism affected the so-called team's way of doing science. For Europa the thick-ice, isolated-ocean paradigm was embraced because any other interpretation would have suggested a high-profile commitment to something new. Authorized speakers propagated a mantra of questionable facts and anecdotal evidence selected to bolster the model. The isolated-ocean model was widely disseminated as the authoritative *Galileo* mission dogma.[4]

At the same time as the isolated-ocean paradigm was taking on a life of its own, a less rushed, interdisciplinary review of the data began to show that what is seen on

---

[4] *Galileo* is not the only NASA mission where political clout enforced a party-line dogma. For example, several people involved with the *Magellan* misson have described what happened with considerations of resurfacing of Venus. As one told me, "For a long time, there could only be plate tectonics on Venus, and no one was allowed to publish articles to the contrary. And, suddenly, when *Magellan* data showed there really was no plate tectonics, global catastrophic resurfacing events became the new dogma, and again no dissent was allowed. You could draw an analogy between what happened with Europa and with *Magellan*." Another person told me that the party line was maintained by teams of enforcers who were organized and coordinated to attack any dissenting speaker during the question periods after his or her talks. Publications and grants can also be blocked by coordinated tactics. The same thing happened with the *Galileo* mission. In fact, some of the same political hustlers and enforcers involved with *Magellan* have established the thick ice dogma for Europa.

Europa is actually quite different from much of what had been reported earlier. Some key anecdotal evidence from the earlier quick look proved to have been premature. Observational selection effects were proving to have been significant, and quantitative corrections could be made. An interdisciplinary approach, in which the strengths of geological methodologies are enhanced by complementary ways of representing and interpreting the data, has now produced a very different big picture of the character of Europa (described in Parts 2 through 4 of this book), in which the ice is thin enough for the ocean to be linked to the surface.

At first, this permeable crust model faced considerable opposition. The isolated ocean model had become as firmly entrenched as the canonical *Galileo* result, and, like the Ptolemaic model of the heavens 500 years earlier, it was under the protection of powerful interests, resisting paradigm change. And, once again, the canonical paradigm was challenged by a model that has Ockham's razor on its side. As we will see, one of the great strengths of the new model is that nearly all of the major observed characteristics of Europa's surface can be readily explained with a single assumption: that the ice is thin enough for cracks and occasional melting to expose the ocean.

Part of the resistance to change may have come from the fact that this subject was initially relegated to the realm of geology. The use of broad, underlying conceptual models, like other aspects of scientific methodology, differs among various disciplines. In astronomy, theoretical models are constructed to connect the dots between sparse data; conceptual models must be rebuilt frequently as new information becomes available. In geology, such broad conceptual models are harder to construct, because so many pesky details get in the way of the big picture.

Geologists seem to have a harder time shifting paradigms. The field of geology may never live down the story of continental drift during the first two-thirds of the 20th century. Any layperson with a world map could see that the Americas had broken away from Europe and Africa. But geologists knew too many details that did not seem to fit the simple story. Acceptance of the global processes of plate tectonics and continental drift was resisted for decades. But once those broad principles were accepted, the details fell into place.

A cautious resistance to paradigm shifts is reasonable when a model has been serving well. But the isolated ocean model for Europa had become the canonical paradigm for all the wrong reasons. Now, however, a very different interpretation of Europa is emerging, in which intimate linkages with the ocean have continually reshaped and replaced Europa's surface. This result is based on quantitative investigations of the tidal processes that drive activity on Europa and quantitative analyses of observations, especially correcting for observational selection effects that may have skewed initial impressions of the surface character of the satellite.

In what follows, I describe the various lines of evidence and show how they have led to this emerging picture of Europa's history and physical structure. While we have seen the folly of accepting any scientific model as dogma, the evidence for this broad picture is compelling. If we are correct, the physical setting on Europa, with its ocean linked to the surface, provides potentially hospitable environmental niches that meet the requirements for survival, spread, and evolution of life. If there is life on Europa, the biosphere extends upward from the ocean to within centimeters of the surface. And everything is driven by tides, as I describe in the following chapters.

Part Two

Tides

4

Tides and resonance

4.1 ACT LOCALLY, THINK GLOBALLY

On Earth, twice each day, tides wash up over beaches, flow in and out of estuaries, and fill and empty harbors. In New Brunswick, Canada, at low tide water drips onto the beach from wet seaweed on cliffs 16 m above. Yet, there are no discernible tides in the Mediterranean, and sailors crossing the oceans see no effects at all. Diverse as these phenomena seem from a local or regional perspective, the fundamental process can be understood on a global scale.

The self-gravity of a planet tends to make it spherical. Each bit of mass pulls on each other bit, with their cumulative gravity pulling the material into a symmetrical ball. The strength of the material can support some deviations, but not much. And continual shaking and erosion allows gravity to pull mountains down.

But no planet is alone. Any neighboring bodies exert forces on each bit of its mass (e.g., our Moon acts on the Earth). The force of gravity depends strongly on distance, so the Moon pulls more strongly on the part of the Earth nearest to it, less strongly on the center of the Earth, and even less strongly on the mass in the Earth on the side farthest away. The Moon stretches the globe of the Earth (Figure 4.1), elongating it into an oval (more precisely, an ellipsoid). What matters for this stretching is not how hard gravity pulls, but rather how much harder it pulls on one part of the Earth compared with another. The amount of stretching is nearly the same on each side, so the two tidal bulges are equally high.

Things get more complicated because every other body in the universe is also pulling and stretching the Earth. Fortunately, only one other body makes a significant contribution, and that is the Sun. It acts to stretch the Earth toward itself, so this stretching is in a different direction and different amount than the tidal stretching due to the Moon. Even though the Sun is 3×10^7 times more massive than the Moon, it is 300 times farther away, and by coincidence these effects roughly balance, so it stretches the Earth by a similar amount. At any instant, the globe of the Earth is

48   Tides and resonance   [Ch. 4

Figure 4.1. The Moon pulls harder (long arrow) on the nearest side of the Earth, and less (short arrow) on the farthest side, so the Earth gets stretched. The Earth does the same thing to the Moon, but that stretching is not shown here. For the moment, just consider tides raised on the Earth. This sketch is schematic, and not to scale.

being stretched in two different directions by similar amounts. Each of these two tide raisers, the Sun and the Moon, tries to mold the Earth into an ellipsoid elongated toward itself. The sum of these two distortions is also an ellipsoid, whose elongation roughly tracks the direction of the Moon, but with large variations in magnitude and direction due to the solar component. The force field that tries to stretch the planet is described by the "tidal potential".

    The shape of the Earth continually responds to this effect. If the Sun and Moon stayed still in the sky, the Earth would stretch out into the ellipsoidal shape governed by the tidal potential. At its surface, the ocean layer would be pulled about 3 meters upward, at the ends of the stretched ellipsoid. Within the Earth, each of its spherical layers, the metallic core, the rocky mantle and crust, as well as the thin water layer on top, would also stretch into ellipsoids, all aligned in the same direction and all conforming to the combination of the gravitational potential of the Sun, the Moon, and the self-gravity of the Earth.

    But the Sun and the Moon do not stay still in the sky. The direction of the Sun circles around the Earth each year and the Moon orbits the Earth each month. The direction and magnitude of their combined tidal potential changes. What is more, the Earth spins all the way around each day. The material of the Earth, the nickel–iron of the core, the rock, the water, and the air, continually deforms so the planet can remold to the ever-changing tidal potential.

    It cannot keep up. The rock that makes up the bulk of the Earth cannot stretch or flow fast enough to change the shape and keep it lined up with the tidal potential. The alignment is always a bit behind where it should be. Moreover, the potential goes around so fast relative to the body of the planet that the stretching of the rock never gets nearly as elongated as it would be if it had time to respond. The rock is too rigid for the solid part of the planet to respond more than a small amount to the tidal potential. Only the fluid ocean (and the atmosphere, which has very little mass) can respond in a timely way. But the ocean layer is thin, only a few kilometers. In order to change the outer shape of the planet, the water flows to fill in each end of the stretched ellipsoid with an extra couple of meters of fluid. The oceans must keep rushing in changing directions as the tidal potential moves around. Pesky continents cover only a small fraction of the surface, but they get in the way. The water flows and sloshes in all sorts of complex ways as it rushes to keep up with the tidal potential. The flow gets trapped in places, funneled into random coastline configurations like Canada's Bay of Fundy. The water moves especially fast as it passes through the Bering Strait and across the shallow Bering Sea. The Mediterranean Sea is so landlocked, that there is not much flow at all; relatively little water can squeeze past Gibraltar.[1]

Local effects of tides vary tremendously, but those are details. Tidal flow is a global-scale phenomenon. It is the continual reshaping of the Earth as its fluid outer layer accommodates to the tidal potential of the Sun and Moon.

4.2 TIDAL DISTORTION—THE PRIMARY COMPONENT

On Europa tides are much simpler, and much larger. Only one tide raiser dominates, and that is Jupiter. It is so massive (30,000 times the mass of the Moon) and so close to Europa (only about 1.5 times as far as the Moon is from Earth), that its tide stretches the 3,300-km-wide globe of Europa by about 1 km. Compare that value with the 13,000-km-wide Earth being stretched by about 6 m.

Europa's tide is huge, but it does not change much. Europa's ocean, with its thin ice crust, has long been accommodated to that figure. It does not have to play the losing game of catch-up that the Earth's oceans do. Below the ocean, the rocky mantle and iron core of Europa also have had time to conform by slow viscous flow to the stretching effect of Jupiter.

The reason this tide is nearly constant is that Europa rotates nearly synchronously with its orbital motion. Its spin period is nearly the same as its orbital period, so it keeps one face toward Jupiter, much like the way the Earth's Moon always presents the same face toward the planet. An important difference is that the Moon is exactly synchronous, while Europa may rotate very slightly nonsynchronously. Europa's rotation is near enough to synchronous, however, that the tide is practically fixed. Therefore, its body and ocean are stretched close to the shape dictated by Jupiter's tidal potential.

The shape of Europa's tidal distortion is described by:

$$H = h_2 \{R_E(M_J/M_E)(R_E/a)^3\}(\tfrac{1}{4} + \tfrac{3}{4} \cos 2\theta) \tag{4.1}$$

where H is the height of the surface (relative to the radius of a sphere of equal volume) at a location an angular distance θ from the direction of Jupiter, R_E is the satellite's radius, M_J and M_E are the mass of Jupiter and Europa, respectively, and a is the distance from Jupiter. The coefficient h_2, which was introduced by the British mathematician and geophysicist A.E.H. Love early in the 20th century, is a factor that takes into account the effect of the physical properties of the material.

The "Love number" h_2 would have a value of 5/2 if the satellite were of uniform density and fully relaxed (like a fluid, without internal elastic stresses) to conform to the tidal potential. Actually, if the tidal potential were only due to the gravitational pull of Jupiter, such a relaxed Europa would have $h_2 = 1$. However, the tidal potential includes not only the direct tidal stretching by Jupiter, but also the

---
[1] Tides in the Mediterranean are only ~10 cm in height and can be hidden by the greater effects of wind and air pressure.

gravitational field due to the distortion of the satellite itself. This latter effect enhances the tides, which allows h_2 to be substantially greater than 1.

Whatever the value of h_2, Eq. (4.1) does give a shape that is consistent with the expected elongation, or stretching, of Europa expected from our physical understanding of tides. Going around the equator, according to Eq. (4.1), the height of the surface has two peaks, one in the direction of Jupiter ($\theta = 0°$) and one in the opposite direction ($\theta = 180°$), corresponding to the stretching of the globe by the tides.

Putting some numbers into Eq. (4.1), with Europa's mass being 4.68×10^{-5} times Jupiter's, Europa's radius of 1,560 km, and Europa's orbital radius of 422,000 km, we find that the quantity in brackets has a value of 800 m. With $h_2 = 5/2$, the height H of this fixed tide would be 2 km at the sub- and anti-Jovian points.

For a real moon, the value of h_2 must be much less than for the simplified uniform, incompressible model. The density of the satellite increases with depth, and the material is compressible. For Europa, we must take into account the dense metal core, the rocky mantle, and the thick outer layer with the density of water. Computation of the amplitude of tides for such a layered model is challenging. There is no simple formula available. We have to rely on results from proprietary computer codes, whose owners and developers tend to keep them close to their chests. Given the non-uniform density distribution, and assuming that most of the H_2O layer is liquid, h_2 is probably ~ 1.2, about half that for the uniform, low-rigidity, incompressible case. If the H_2O layer were completely frozen, with no liquid layer, the height of the tide would be much smaller, about 3% as high, with $h_2 \sim 0.04$. Most likely though, there is a substantial ocean, in which case, according to Eq. (4.1) with $h_2 \sim 1.2$, H is probably ~ 1 km at the sub- and anti-Jovian points.

In other words, tides on Europa are expected to be roughly a kilometer high. If a future spacecraft measures Europa's shape, the height of this tide will help constrain the possible interior structure.

If Europa were on a perfectly circular orbit and rotating synchronously, there would be no variation in this tide. The amplitude would not change, because the distance from Jupiter would be constant. The orientation of the elongation relative to the body of Europa would also be fixed, because with synchronous rotation the body retains a fixed orientation relative to the direction of Jupiter. As huge as this tide is, other than establishing the shape of Europa, it would have no interesting effects. The body would not be continually reshaped, so there would be no frictional heating and no stressing of the crust. The surface would have been relatively boring, showing little more than craters formed during the bombardment by small bodies over billions of years.

Early in its history, Europa's rotation was surely not synchronous. The direction and rate of its rotation would have been determined by the heavy bombardment by bodies accreting onto the growing moon. The dominant effects might have come late in the growth period from some of the larger impactors, whose collision geometries would determine the rotation. Whatever the details of the accretion process, it probably did not lead to synchronous rotation.

In the case of such non-synchronous rotation, the direction of the Jupiter-aligned tidal potential moves around relative to the body of the satellite. As a result, the orientation of the kilometer high tidal bulge must continually change. This continual deformation would have stressed the body, straining and cracking its surface, and friction would have generated huge amounts of internal heat.

During non-synchronous rotation, the amplitude of the tide remains approximately the same as for the fixed tide (Eq. 4.1), but because the distortion cannot respond instantaneously the direction of the distortion would slightly lag behind the changing tidal potential, so that it would be asymmetrical relative to Jupiter. This slight tidal-lag asymmetry would have an important effect: It would result in a tidal torque that would have rapidly (in only $\sim 10^5$ yr) reduced the satellite's spin rate toward synchroneity with its orbital period (about 3.6 days), as discussed in Chapter 5. During spin-down, the decrease in centrifugal flattening of the body would also stress and crack the surface (see Chapter 6).

Such interesting dynamical events, with their major tectonic and thermal effects, surely happened early in the history of Europa, but the short duration of the spin-down process suggests it would have happened fast and been over early. Any geological record of the early tectonic and thermal processes would have been blasted away long ago, and Europa would have settled to conform to a fixed tidal figure. In that case Europa, like the Earth's Moon, would be rotating synchronously with its surface dominated by impact scars.

4.3 GALILEO DATA, THE LAPLACE RESONANCE, AND ORBITAL ECCENTRICITY

That boring scenario did not occur. The evidence for something much more exciting comes from Galileo data. In fact, not from the *Galileo* spacecraft in the 20th and 21st centuries, but rather from the original observations of the moons of Jupiter in the early 1600s. Galileo made careful plots of the movements of the satellites, and the orbital periods were quite well defined. The periods of the inner three are approximately the following: $42\frac{1}{2}$ hours for Io, 85 hours for Europa, and 170 hours for Ganymede.

The ratios of these periods are striking. From these numbers, going back to the original observations by Galileo, it appears that they are locked into a ratio (or "commensurability") of $1:2:4$. In the time it takes Ganymede to go around Jupiter once, Europa goes around exactly twice and Io goes around four times. This ratio also means that the location in the orbit where Io overtakes Europa, where they are both in line with Jupiter (a configuration called "conjunction"), is always the same. After a conjunction, Io advances ahead of Europa, and next catches up with Europa at the same location, after Europa has made one whole orbit and Io has gone around twice. The conjunction of Europa with Ganymede also occurs at a fixed location, because of the $1:2$ ratio of their periods. Moreover, Galileo's plots of the orbital motion also showed that the conjunction of Europa with Io always occurs at exactly the opposite side of Jupiter from the conjunction of Europa with Ganymede.[2]

The scientific world at the time of Galileo found it sufficiently traumatic to deal with the simple issue that the satellites seemed to be orbiting Jupiter and not the Earth, in violation of canonical wisdom. It took almost two centuries before the strange and amazing commensurabilities among the orbital periods were understood. The explanation was part of the great body of research carried out by the French mathematician Pierre-Simon Laplace, and published early in the 19th century. Laplace successfully navigated difficult political times, including pressure to go along with some crackpot governmental astronomy related to calendars during the French Revolution. Fortunately, the Galilean satellites were no longer the political issue they had been in the 17th century.

Laplace showed that the periodic repetitions of the conjunctions of the Galilean satellites enhanced their mutual gravitational effects. Although each satellite's orbit was predominantly governed by Jupiter's gravity, the satellites' effects on one another can perturb the orbits in various ways. Because the commensurability of their orbital periods causes exactly the same geometries to repeat every few days, the gravitational forces are repeated periodically, creating a resonance. The cumulative effect is remarkable. It maintains the whole-number ratios of the periods, and accordingly maintains the alignments of the conjunctions.

Most important is that the resonance pumps the eccentricities of the orbits. Because of the resonant mutual interactions, none of the satellites could remain in a circular orbit, so they all follow elliptical trajectories. In fact, the conjunctions of the satellites are aligned with the major (i.e., long) axes of the orbits. For example, each time that a conjunction of Io and Europa occurs, Europa is at the apocenter[3] of its orbit (where it is farthest from Jupiter) and each conjunction of Europa with Ganymede is 180° away, at Europa's pericenter (where it is closest to Jupiter).

Remarkably though, more than 150 years after Laplace, as astronomical interest in the planets and their moons grew during the 20th century, a misunderstanding led to a widespread belief that the orbits of the Galilean satellites are circular. In nearly every textbook and authoritative tabulation of orbital parameters, the eccentricities of the Galilean satellites were listed as zero. For example, the eccentricity of Europa is listed as 0.000 in the book *Planetary Satellites*, part of an authoritative series on space science.

The misleading table entries had propagated from a peculiar tradition in celestial mechanics of recording only one component of the eccentricity. For a satellite that is not in resonance, the orbital eccentricity is determined by initial conditions. The initial position and velocity (e.g., at a time when early solar system forces like

---

[2] Here I have glossed over a subtlety that is not critical here, but which is addressed in Chapter 8. In fact, the longitudes of conjunction are locked to the orientations of the long ("major") axes of the elliptical orbits, not to a longitude that is fixed in space. This means that the exact 1 : 2 : 4 ratio of periods is actually relative to the major axes, which precess at a slow rate. The actual "mean motions" (their mean angular velocities) of Io, Europa, and Ganymede are 203.4890°/day, 101.3747°/day, and 50.3175°/day, respectively.

[3] "Apocenter" and "pericenter" are generic terms applying to distance from the central body in any orbit; for orbits around Jupiter, they are often called "apojove" and "perijove".

collisions and gas drag had ceased) determine the size and shape of the orbit. For a given distance from the planet, only if the velocity were tuned to exactly the right value would the orbit be circular. For a satellite in resonance, the orbital eccentricity gets more complicated. It is the sum of two components: the effect of the resonance, called the *forced* component, plus the effect of initial conditions, called the *free* component. No matter how you tuned the initial velocity you could never get rid of the forced component, as long as the satellite is in a resonance.

From a mathematical perspective, the free eccentricity in the resonant case is analogous to the eccentricity of the ordinary orbit. Each is the parameter determined by initial conditions. Partly for that reason, a tradition developed of tabulating only the free eccentricity.

There was another reason for not showing the total eccentricity. Orbital eccentricity can be thought of as a vector quantity. It has a magnitude (how off-center the orbit is) and a direction (which way it is off-center). The magnitudes of the forced and free components are fixed, but their directions change relative to one another. Sometimes they are aligned, so that the actual eccentricity is the sum of the free and forced components; Other times they are opposite so the actual orbit has an eccentricity that is the difference between the two components. Usually, the true eccentricity oscillates between those extremes. What was the poor table writer to do? Even if both components were tabulated, the actual eccentricity at any time would be something different and ever-changing. Given limits on available columns and given the need to keep it simple (few planetary scientists understood these distinctions), tabulations followed the mathematical preference: only the free eccentricity was listed.

For most satellites, the free eccentricity describes the actual orbital motion, because the forced components are negligible. However, for Io or Europa, the forced eccentricity is the dominant component and the free eccentricity is negligible. For many decades, astronomers and students were informed that the eccentricities of these satellites were zero or negligible. The true eccentric motion remained unappreciated.

One reason that no one was surprised to see the value "zero" listed is that tides were expected, in general, to circularize orbits. For most satellites, after the early rapid tidal despinning, if a satellite's orbit is eccentric, the strength, and even the direction, of the tidal potential changes throughout each orbital period. As Eq. (4.1) shows, the height of the tide varies inversely as the cube of the distance; so, even a modest eccentricity can result in a big difference between the tide at pericenter and at apocenter. It had been understood for decades that, if the delayed response of the material to such changes in the tidal potential is taken into account, the forces on the satellites will tend to circularize its orbit. No one was surprised to see values of zero in the eccentricity columns of the standard tables.

If a satellite is in resonance, the eccentricity-damping process is different. As I had demonstrated in a 1978 paper, processes that tend to damp eccentricities actually only damp down the free component. That is why Io and Europa, being so close to Jupiter and susceptible to tidal effects, had such small free eccentricities compared with the component of the eccentricities forced by the resonance. As a result, standard tables gave a value of zero, rather than a value that described the actual eccentricity of the satellites' trajectories.

In the mid- to late-1970s, I wrote a few articles about resonances among satellites, and gave several presentations about them at conferences. A few weeks before *Voyager* reached the Jupiter system, I got a phone call from my colleague Stan Peale, a Professor of Physics at the University of California in Santa Barbara. Stan is a few years older than me, and I had learned a great deal about tides and celestial mechanics from studying his work. This time Stan had questions for me. He had heard me speak about the Laplace resonance, and he wanted to confirm that the eccentricity of Io was 0.0041 and the eccentricity of Europa was 0.01. Had he understood me correctly when I had said to disregard the tabulated values of zero? I told him that was correct. "Do you know what this means?" he asked. "No, what?" I cluelessly responded. "It means that there must be an incredibly high rate of tidal heating in Io and probably in Europa as well." I became the first of many celestial mechanicians to slap their foreheads: Why didn't I think of that!

If a satellite's eccentricity is damped down to zero by tidal effects, the tidal bulges become fixed. Even if the bulges were big, they would have no interesting geophysical effects. The continuous readjustment to a changing tidal potential comes to a halt. There is no friction, no internal energy dissipation, no heating. But, if a resonance maintains an eccentric orbit, the tidal heating continues. And, if the tide is raised by the nearby largest planet in the solar system, as in the case of Io and Europa, a great deal of heat must be produced.

A few days before *Voyager* reached Jupiter in 1979, the magazine *Science* published the article by Peale et al. that predicted something dramatic would be found at Io and possibly at the other Galilean satellites. The rest is history: *Voyager* found Io to be actively spewing volcanic plumes into space, and the surfaces of the other resonant satellites, Europa and Ganymede, appeared far more active than frozen ice balls in space had any right to be.

A few years later, what had been learned about the Galilean satellites from *Voyager* was reviewed in a reference book *The Satellites of Jupiter*. In my chapter in that book, I covered the dynamics of the Laplace resonance and described the distinction between the free and forced eccentricities, including a table that showed their values, with the forced eccentricity of 0.0101 and the tiny free eccentricity of 0.000 09. The free eccentricity causes the total eccentricity to vary slightly, but it is always near 0.01. Nevertheless, the opening chapter of that same book contains a table summarizing orbital parameters of Jupiter's satellites: Europa's eccentricity is listed as 0.000. Established ideas are hard to change.

Europa's orbit is not circular. Its eccentricity is about 0.01. Everything interesting about Europa follows from the fact that the eccentricity is not zero.

4.4 THE EFFECT OF ORBITAL ECCENTRICITY—THE VARIABLE COMPONENT OF THE TIDES

Because Europa's orbit is eccentric, tides on its body change over the course of each 85-hour orbital period. The variation in the tide is responsible for the character of

Europa's surface, providing frictional heat for melting the ice, and stressing the crust to drive the tectonics. The period of tidal variation driven by orbital eccentricity is simply identical to the orbital period from pericenter to pericenter. Because the length of a day on Europa is very close to the duration of an orbit, we have come to call the tidal variation the *diurnal tide*.

That term is something of a misnomer, because the length of a Europan day depends on the rate of rotation relative to the Sun, not on Europa's orbital motion. If Europa rotates non-synchronously, we would not expect the length of the day to match the orbital period. Even if Europa's rotation is synchronous with its orbit, so that one hemisphere is locked toward the direction of Jupiter, the length of a day would not quite match the orbital period, because of Jupiter's motion relative to the Sun during that time. Nevertheless, we refer to the tidal variation that is generated by the orbital eccentricity as the diurnal tide because its period is very close to the length of a day on Europa and, moreover, it happens to be comparable with the length of a day on Earth.

Now, consider the variation of the tide due to an eccentric orbit. Even if rotation is synchronous keeping one face toward Jupiter, the magnitude and orientation of tidal distortion changes throughout each orbital period. The tide-raising gravitational effect is at a maximum at pericenter, when Europa is closest to Jupiter, and a minimum at apocenter, when it is farthest away. Moreover, the orientation of the tide advances ahead relative to the body of the satellite after pericenter and falls behind before pericenter (Figure 4.2, see also color section for Figure 4.2b).

Figure 4.2 represents schematically how the orbital eccentricity e drives tidal variation on Europa. Bear in mind the true proportions of the geometry: Europa's radius is about 1,560 km and it averages a distance of 671,000 km from Jupiter, whose radius is about 71,500 km. The average height of each tidal bulge on Europa is about 500 m. Europa's orbital eccentricity is 0.01, so the elliptical epicycle traced out by Jupiter relative to Europa (Figure 4.2a) has dimensions 13,400 km by 26,800 km and Europa's distance from Jupiter varies by less than 1%. That small variation in distance means that the orbit is nearly circular, but the variation is enough to drive dramatic effects on Europa.

The magnitude of the diurnal tidal variation (the change in the height of the tidal bulges shown in Figure 4.2) can be found by replacing the distance from Jupiter a with the pericenter distance $a(1 - e)$ in Eq. (4.1). The height of the tide is found to vary by $3e$ times the value 1 km given by Eq. (4.1) for the primary tide. Thus, for Europa with $e = 0.01$ the amplitude is 30 m (i.e., the tidal height at the sub- and anti-Jove points at pericenter (or apocenter) is 30 m higher (or lower) than the surface of the fixed tide given by Eq. (4.1)).

That result assumes that the value of h_2 is the same for the rapidly varying diurnal tide as it is for the nearly constant primary tide. Recall that the coefficient h_2 in Eq. (4.1) accounts for the cumulative effects of all of the material properties of the satellite in responding to the tidal potential.

Because the timescale for this change is so short, the physical properties of the materials involved (viscosity, as well as rigidity) may play a role in limiting the extent of tidal deformation. How can we estimate the value of h_2 for this dynamic situation? Computed results (from proprietary codes) are available for only a few interior models.

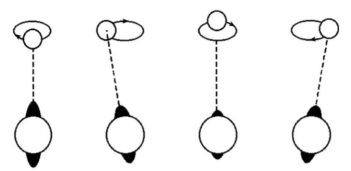

Figure 4.2a. The tide on Europa (at the bottom) is shown schematically at four points (from left to right) in its orbit. The black bulges represent Europa's stretched shape. Jupiter is the small circle (obviously the giant planet is not to scale) near the top. In reality, Europa orbits around Jupiter presenting one face toward Jupiter, but the frame of reference used here is locked to the body of Europa. In effect, Europa is held still in each time step, so that Jupiter always remains on the same side. The eccentricity of the orbit produces small variations in the distance and direction of Jupiter relative to the body of Europa. In this reference frame, Jupiter seems to follow a small oval trajectory. This path is an artifact of the particular Europa-centric reference frame, and is sometimes called an epicycle, harking back to the historic terminology of Ptolemaic astronomy. At the left, Europa is at the perijove of its orbit (closest to Jupiter), and its tidal elongation is greatest. A half-orbit later, Europa is at the apojove of its orbit (farthest from Jupiter), and its tidal elongation is smallest. In-between, the direction of Jupiter relative to the body of Europa is shifted slightly. The body of Europa must continually remold in order to conform to the shape driven by the tide. (See Figure 4.2b in the color section for a less abstract representation.)

They indicate however that, even with the rapid, continual change in the tidal potential over the course of each day on Europa, the outer surface can conform to the shape dictated by that tidal potential. Therefore, the value of the Love number h_2 for the diurnal tide should be reasonably close to the value ($\sim$1.2) that it had for the primary (nearly constant) component of the tide. So, as long as Europa's water is largely liquid, our estimate of 30 m for the amplitude of tidal variation is probably accurate.

The results computed by the various proprietary models show that tidal variation is fairly independent of whether the ice crust is very thin or several tens of kilometers thick, as long as most of the water under the ice is liquid. Only if the water is nearly all frozen does the tidal amplitude drop significantly, and then it can drop by an order of magnitude.

If and when spacecraft return to Europa, they may be equipped with altimeters capable of measuring the amplitude of this tidal variation, which would complement measurement of the gravitational effects of the tidal shape on the spacecraft's motions. In principle, these measurements could help constrain internal properties, such as the thickness of the ice. However, the fact that tidal amplitude is fairly insensitive to ice thickness, if the thickness is less than a few tens of kilometers, is somewhat discouraging, because determining the thickness of the ice is a crucial issue. Unfortunately, other uncertainties in the details of internal structure and materials probably will swamp out the effect of ice thickness on tidal variation. Measurements of tides will tell us whether there is an ocean, but we would need very precise measurements and improved interior models if there is to be any hope of using this approach to determine the thickness of the ice.

Over the course of each orbit, the position of Jupiter relative to Europa not only gets closer and farther, but it also advances and regresses in Europa-centric longitude as shown in Figure 4.2. At pericenter the tidal figure is the primary component (Eq. 4.1), simply augmented by about 3%. At 1/4 orbit before or after pericenter, Europa is at the mean distance from Jupiter, so the magnitude of the tide is the same as the primary component given by Eq. (4.1), but the orientation of tidal elongation is rotated by an angle $2e$ (about 1.5°). It is important to understand that the change in orientation of the elongation of Europa (as in Figure 4.2) does not represent rotation of the body, but rather a "remolding" of the figure of Europa in response to the changing direction of Jupiter.

4.5 EFFECTS OF TIDES

Were it not for the eccentricity-driven diurnal tides, Europa would be as inactive as Callisto, the farthest out from Jupiter of the Galilean satellites. Callisto, not part of an orbital resonance and thus with no diurnal tides, is simply a heavily cratered target for bombardment by every type of debris in the solar system. The three resonant satellites (Io, Europa, and Ganymede) all have much more interesting geology and geophysics because of the various effects of the tides.

Tides affect the satellites in four major ways. First, the continual remolding of the figure of the body entails friction and so generates heat. Second, the change in shape stresses the cold, brittle, elastic outer layer (usually called the "lithosphere"; see Chapter 6 for a discussion of this strange terminology) causing cracking and a variety of related tectonic processes. Third, a slight lag in the tidal response, inevitable for any real material body, causes an asymmetry in the alignment of the elongation with the tide raiser, Jupiter. The resulting gravitational torques can affect a satellite's rotation. Fourth, these torques, as well as torques involving tides raised on Jupiter by the satellites, have probably caused long-term variation in the orbits of the satellites.

The heating and stress generated by tides have direct effects on the observable surfaces of the satellites, while rotation and orbital change modify the tides and thus indirectly affect what we see on the surface. All of these processes are interdependent, as shown in Figure 4.3, where arrows show where one type of phenomenon affects another. Here, tides are linked to the four effects of heat, stress, orbits, and rotation, with stress corresponding to observable tectonics and heat corresponding to observable thermal features, like chaos.

Consider each of the four effects in turn. (1) Tidal distortion stresses the surface driving tectonics. (2) Tides can modify the satellites' orbits. In turn, the orbits also drive the tides. We have seen how the Laplace resonance and the forced eccentricities

58   Tides and resonance   [Ch. 4

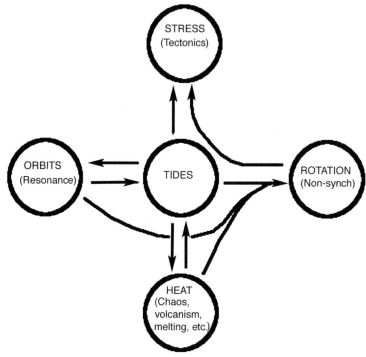

Figure 4.3. Interrelationships among the processes that govern Europa's geology and geophysics. Tides lie at the hub.

are necessary for tides to persist. (3) Tidal friction heats the satellites. In turn, the heating can modify the tides. For example, we have already discussed how the height of a tide depends strongly on whether part of the interior has been melted. (4) Tides may drive non-synchronous rotation, but only (as indicated by the lower curved lines in Figure 4.3) if the satellite is well heated and in an eccentric orbit; otherwise, tides alone would result in synchronous rotation, as they do for the Earth's Moon. Non-synchronous rotation in turn modifies tides in a way that does not affect orbits or heating very much, but that may make a profound addition to the way that tidal distortion stresses the surface (upper curved line in Figure 4.3).

   Not all of these effects are evidenced significantly on all of the Galilean satellites. On Io, because it is closest to Jupiter, tidal heating is so great that volcanism swamps out and hides most large-scale tectonic effects. It is not obvious whether any of the tectonic features observed there are related to tides. At the opposite extreme, far from Jupiter, poor pathetic Callisto has no significant tides because it is not part of the orbital resonance. Ganymede shows little indication of heating, because it is the farthest of the three resonant satellites from Jupiter, although a substantial portion of the surface is dominated by tectonics. Europa lies in the sweet spot between Io and Ganymede, where there is considerable tidal heating, but not so much that thermal effects hide the tide-dominated tectonics.

All four of the interacting tidal processes (Figure 4.3), rotation, stress, heating, and orbits, have been in play in creating the surface that we have observed. In the next four chapters, we consider each of these four processes in more detail, in preparation for a more comprehensive look at the features observed in *Galileo* images and their implications regarding the character of Europa.

5

Tides and rotation

Europa's rotation results from the influence of tides. We have seen that the tide consists of a primary component, which would be present whether or not the orbit were eccentric, and a variable or "diurnal" portion, which results from the rapid change in tidal potential due to orbital eccentricity. These components of the tide have opposite effects on rotation. The primary one tends to drive the spin rate toward synchronous rotation, while the diurnal part tends to make the rotation go a bit faster than synchronous. In the following sections, I explain each of these tendencies by reviewing the dynamics of tides in more detail. Readers less interested in the theory of tides might jump to Section 5.3 of this chapter, which summarizes how tides may have affected Europa's rotation, laying the foundation for understanding much of what we see on the surface of Europa.

5.1 SYNCHRONOUS ROTATION FROM THE PRIMARY TIDAL COMPONENT

Tides affect the rotation of a satellite or planet because tidal elongation is not necessarily aligned with the direction of the tide-raising body (contrary to the seeming alignment in Figures 4.1 and 4.2). In an idealized world, the symmetry of the system would yield an exact alignment, so no torque would be exerted. However, for real planets the positions and orientations are continually changing, so that the material that composes a planet must be continually rearranged or remolded to conform. We see part of that rearrangement of material on the Earth as the oceans rush around to accommodate the changing tidal potential of the Sun and Moon. Real materials take time to move, and friction slows them down. The response to tidal potential is always late, so tidal bulges are never perfectly aligned with the tide raiser.

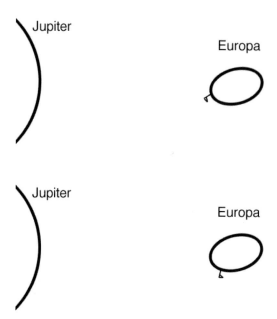

Figure 5.1. As Europa rotates (counterclockwise in this hypothetical schematic), surface features are carried around relative to the direction of Jupiter, as shown by the flag. The flag is at a point at the end of the tidal elongation in the top picture, and a short time later (bottom) it has rotated to a new location. Even as Europa rotates, the tidal elongation has a fixed orientation, relative to the direction of Jupiter. The orientation of the tide, angled slightly relative to the direction of Jupiter, is fixed in a steady state controlled by the tidal potential of Jupiter and the lagging response of the material of Europa.

In order to understand the effect of this lag, suppose Europa were in a circular orbit and rotating not synchronously but rather much faster than its orbit around Jupiter. Because the body of Europa cannot remold itself instantaneously, the lag means that tidal bulges are carried by the rotation slightly ahead of the tidal potential, so they are no longer aligned with the tide raiser. As rotation continues, the material tries to remold accordingly, but the response can never catch up. The tidal bulges reach a steady state of slight misalignment (Figure 5.1).

The amount of this misalignment, or lag angle, depends on how quickly the material of Europa can respond to the tide-raising potential. This response depends on the character of the materials and the structure of the satellite. If it were elastic material, the periodic behavior would resemble simple harmonic oscillations. It is conventional (e.g., in mechanical or electrical engineering) to represent frictional damping in a harmonic oscillator (like a spring) by a parameter Q. The larger the amount of damping, the smaller the parameter Q. A planet is much more complex than a simple harmonic oscillator. Much of the material responds more in a viscous manner than elastically; on the Earth, much of the response involves flowing water. Nevertheless, by analogy with the simple harmonic oscillator, the combined, total effective damping is usually represented by a parameter Q. In theoretical formulations the lag angle, which is always very small, is inversely proportional to Q; for Europa, the lag angle would be a tiny fraction of a degree.

Once the symmetry is broken by this slight lag, Jupiter can exert a torque on Europa. The bulge that is closest to Jupiter (Figure 5.1) gets the strongest gravitational pull, which exerts a torque in the direction opposite the rotation. As long as Europa continues to rotate relative to the direction of Jupiter, this torque will tend to slow the spin.

Conservation of angular momentum requires that Europa exert an equal and opposite torque on Jupiter. Assuming Europa's spin is in the same direction as the orbit, this torque adds energy to Jupiter's orbit relative to Europa or, equivalently, to Europa's orbit around Jupiter. This effect slowly increases the size of Europa's orbit, which in turn slowly decreases its orbital period. So, this tidal torque decreases both Europa's spin period and its orbital period. The spin rate changes much more quickly, because there is so much less angular momentum in Europa's spin than in its orbit (the distance from Jupiter being so much greater than the radius of Europa). The spin rate changes until the rotation period matches the orbital period.

This process brought the Moon into its current state of rotation, synchronous with its orbit around the Earth, so as to present a constant face toward us. The same process is slowing down the rotation of the Earth due to the tides raised by the Sun and Moon.

In each case, the strength of the torque depends on the lag angle, and thus ultimately on the details of frictional energy dissipation. Details can be important. For example, the spin of the Earth is being slowed down in this way, and a major portion of total energy loss is in the friction of ocean currents moving across the shallow Bering Sea. For Europa we can only guess at the mechanisms for energy dissipation and how important they are. We do have some idea of the typical behavior of geological materials from studies of the tidal distortion of the Moon and Earth, and from seismic studies of dissipation in rock. Almost certainly, the spin-down for Europa would have been very quick, probably requiring not much longer than 100,000 years.

Europa did slow down to near-synchronous rotation. The evidence comes from observations of the orientation relative to Jupiter, which did not change appreciably during the years between the *Voyager* encounter in 1979 and the *Galileo* mission in the late 1990s. Also, Earth-based observations, though they could not show the surface in detail, had been consistent with synchronous rotation, at least to the degree of precision that was possible.

5.2 NON-SYNCHRONOUS ROTATION FROM THE DIURNAL TIDE

If we introduce the effect of Europa's orbital eccentricity, the tidal amplitude and direction continually change as shown in Figure 4.2a. At any instant, the tide-raising potential (or the tidal shape to which the body tends to conform) can be thought of as the sum of the primary component, plus the additional diurnal variation that represents the effect of orbital eccentricity. We have already seen that diurnal change in the amplitude has a magnitude of about 3% of the primary one.

The total tidal potential is always aligned with the instantaneous direction of Jupiter (Figure 4.2a), so that, without any dissipation, Europa would continually remold toward conformance with that symmetrical, but ever-changing, shape. But, just as in the case of a circular orbit with non-synchronous rotation, the real material of Europa lags behind the tidal potential. It is difficult to envision the geometry of the lagging response of the material of Europa to this complicated continual remolding. However, it becomes much simpler if we follow an approach pioneered by George Darwin in the late 19th century, which decomposes the distortion into separate components, using Fourier analysis. An elegant application of this mathematical analysis was done quantitatively in a short, classic paper by the geophysicist Sir Harold Jeffreys, but the essence can be understood in physical terms as well.

In order to take into account the effect of the diurnal tide, we note that the tidal shape at any time is essentially a sinusoidal wave pattern (Eq. 4.1). In other words, if we consider a tidally-elongated body, as we go all the way around the equator the height relative to a sphere rises and falls twice. This shape can be considered to be a wave with two peaks: one facing the tide raiser, and one on the opposite side. We can break this wave down into several components. The first is the primary component, which is independent of e. This component represents a constant elongation oriented toward the average direction of Jupiter (i.e., toward the center of the orbital epicycle—Figure 4.2a). The primary component is the elongation of Europa that would be present if the orbit were circular instead of eccentric. For synchronous rotation, this primary component would be fixed relative to the body of the satellite, even as the diurnal tide varies relative to the body.

One component of diurnal variation is a standing wave that slightly elongates Europa at pericenter along the direction of the axis of the primary component of the tide and at apocenter along a direction perpendicular to the primary component (i.e., with bulges in the leading and trailing hemispheres). This standing wave adds to the amplitude of the primary component at pericenter, and subtracts from the amplitude at apocenter. It accounts for the variation of the amplitude of the total tidal bulge with the periodically-varying distance from Jupiter. It gives the maximum and minimum of the varying tide at pericenter and apocenter, respectively (as shown in Figure 4.2). As with any standing wave, this "amplitude-varying standing wave" is equivalent to the sum of two waves traveling around Europa in opposite directions.

The remaining portion of diurnal variation is another standing wave that alternately raises small bulges aligned 45° ahead of the orientation of the bulges of the primary component (peaking 1/4 orbital period after pericenter) and 45° behind the orientation of the primary component (peaking 1/4 orbital period after apocenter). This standing wave, when added to the primary component, alternately shifts the direction of the total tide slightly ahead and slightly behind the mean direction of Jupiter; it accounts for the variation in the direction of Jupiter due to Europa's periodically-changing velocity on its eccentric orbit (as shown in Figure 4.2). This "direction-varying standing wave", like the "amplitude-varying standing wave", is also composed of two waves traveling around Europa in opposite directions.

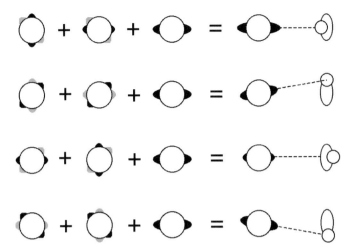

Figure 5.2. In this schematic representation, the total tidal potential is shown at the right, at the same four quarter-orbit time steps as in Figure 4.2a. The tidal potential responds to the changing direction and distance of Jupiter (the small circle moving at the far right as in Figure 4.2a). The total potential can be broken down into the primary (or fixed) component, locked to the average direction of Jupiter, plus two traveling components. The first wave, indicated by the dark bulges in the first column, travels retrograde relative to the orbit with amplitude $2e/3$ times the primary component, and the second one travels prograde with an amplitude seven times the first one. Both traveling waves go half-way around the satellite in each orbital period, as shown. The lag in the tidal response to the traveling waves can be represented by very small waves (gray bulges) following behind. (In this schematic, scales are not correct.)

In fact, the two traveling waves that compose the "direction-varying standing wave" are the same two traveling waves that compose the "amplitude-varying standing wave", except for different coefficients. Thus, even though each of the standing waves can be expressed as the sum of two traveling waves, the two standing waves can be expressed as the sum of only two traveling waves.

The breakdown of tidal potential into these components is summarized in Figure 5.2. The major components of diurnal tides are equivalent to the two traveling waves represented in the left two columns. Note that each wave moves around Europa relative to the direction of Jupiter at a rate such that the wave peaks move half-way around in each orbital period.

We know that the amplitude of these diurnal components of the tide must be proportional to the orbital eccentricity e, because variation in the height and the orientation of the total tide is proportional to e. In fact, the wave that moves in the prograde direction has an amplitude of $(14/3)e$ times the amplitude of the primary (non-diurnal) component. The wave that moves in the retrograde direction has an amplitude of $-(2/3)e$ times the amplitude of the primary (non-diurnal) component (the minus sign reflects the opposite phase as shown in Figure 5.2). With these values we get exactly the change in tidal height predicted by Eq. (4.1), with the distance a replaced by the distance in eccentric motion, which varies from $a(1-e)$ to $a(1+e)$.

Bear in mind that the amplitude of the diurnal components would be expected to be even smaller if Europa were completely solid, because a rigid body could not distort as much as a fluid one. This effect is built into the Love number parameter h_2 in Eq. (4.1). We have already discussed how numerical models have shown that, as long as Europa has a liquid water layer of roughly 100 km thick, diurnal tides will probably have a Love number as great as that for the primary tide. In the case of the primary tide, the Love number is large because there is so much time for even a solid body to relax to conformation with the tidal potential.

The traveling waves shown by the dark bulges in Figure 5.2 represent the orientations for idealized instantaneous responses to Jupiter's tide-raising potential. In fact, of course, the real material that composes Europa cannot respond instantaneously in this case, just as in the case of the circular orbit with non-synchronous rotation discussed earlier (Figure 5.1). In that case, we saw how the lag in the response caused tidal elongation to be offset slightly relative to the direction of Jupiter. In fact, for each of the traveling waves, the geometry is almost identical to that earlier example. Again, the orientation of the elongation is continually changed relative to the body of the satellite. Therefore, each of the two traveling waves must lag behind the orientation of the potential shown in Figure 5.2. In other words, the retrograde wave (clockwise-moving, left column) at any instant would be shifted slightly further counterclockwise from the black positions, and the prograde wave (counterclockwise-moving, second column) would be shifted slightly in the clockwise direction. With these lags, the symmetry of the elongation of Europa relative to Jupiter is broken.

Another way to represent the same lag geometrically is to add a very small wave traveling 45° behind each of the wave components, as shown by the gray bulges in Figure 5.2. (Simple trigonometry shows that adding such an extra wave is equivalent to a slight phase shift in the original.) Note that the lag components are oriented 45° from the direction of Jupiter at pericenter.

With these lags and the symmetry broken, Jupiter can exert a rotational torque on Europa, just as it would in the case of non-synchronous rotation discussed earlier. In this case, however, the torque varies over the course of each orbit; so, in order to calculate any expected long-term change in the rotation rate, we need to average the torque over each orbital period. To take the average, we need to add up the torque at each point in the orbit taking into account the changing position (direction and distance) of Jupiter, as well as the changing tidal deformation.

The calculation requires crucial assumptions about tidal response. It is not evident that the lag angle can be represented by a single parameter, or that the responses to the separate traveling waves can be treated independently, as in the above treatment based on the Darwin model of the tide. In an alternative treatment, developed by geophysicist G.J.F. MacDonald, the material response is incorporated in the theory not by lags in the separate wave components, but by a lag in the total tide. The total deformation of the body shown near the right in Figure 5.2 is considered to be offset by a constant angle, but in a direction (clockwise or counterclockwise) that corresponds to the instantaneous angular velocity relative to the direction of Jupiter (in Figure 5.2 shown by the epicyclic motion of Jupiter). Both models, the Darwin tide and the MacDonald tide, are probably oversimplified compared with the actual ways that the material of Europa (or any planetary body) responds to the changing tidal potential, but in our ignorance there is no basis for constructing any more detailed tidal model.

Despite our ignorance of the details of how the material of Europa responds to the changing tidal potential, it seems reasonable that if Europa were in synchronous rotation, the average tidal torque would tend to speed it up in the prograde sense (i.e., in the same direction as its orbital motion). Here is why: The strength of the tide-raising potential depends strongly on the distance from Jupiter (inversely as the cube of the distance, according to Eq. 4.1). Also, the torque exerted on any elongation of a planet varies inversely as the cube of the distance; so, the torque exerted by Jupiter on a tide that it raises on Europa depends extremely strongly on distance: It varies inversely as the 6th power of distance. Therefore, the dominant tidal torque is near pericenter. Consider the Darwin tidal model. If we look at the geometry at pericenter (top row in Figure 5.2), we see that the tidal lag in the two traveling wave components gives a lag that will result in a prograde torque, because Jupiter pulls most strongly on the nearest bulge, increasing the spin rate to faster than synchronous. We get the same result if we consider the MacDonald model. At pericenter, Jupiter is moving ahead (upper right in Figure 5.2) relative to Europa, so if the tidal response is lagging, the bulge would be behind where it is shown. Again, the pull of Jupiter on the nearest bulge at pericenter would act to spin up the rotation of Europa.

While it is conceivable that some unpredicted behavior of the material of Europa gives a response very different from such models, it seems most likely that the torque due to Jupiter acting on the lagging tides that it raises on Europa would tend to speed it up from synchronous rotation.

With these basic concepts about the behavior of tides, we can begin to understand the four types of possible effects that they might have on the satellite: rotation, stress, heating, and orbital evolution.

5.3 ROTATIONAL EFFECTS ON EUROPA

Europa, like most planetary bodies, probably formed with a spin that was not synchronous with its orbit, but which rather reflected the result of the impacts that occurred as smaller bodies in orbit around Jupiter accreted together to become the satellite. Then, as discussed in Section 5.1, whether the orbit had some eccentricity or not, with substantial non-synchronous rotation the lag in the large primary component of the tide (Figure 5.1) would have resulted in a large torque. Spin would have slowed to near synchronous in about 100,000 years, a blink of an eye compared with the 4.6-billion-yr age of the solar system. Rotation would have become precisely synchronous if Europa's e were zero, and this despinning would have occurred almost immediately.

However, as we saw in Section 5.2, the tides would establish and maintain non-synchronous rotation if there were significant orbital eccentricity and corresponding diurnal variation of the tide. Diurnal variation gives the needed tidal torque, which would come into play once the rotation rate becomes small compared with the epicyclical motion of Jupiter.

But even with eccentric orbits, the tidal torque might not be able to drive a satellite out of synchronous rotation if the satellite contains a frozen-in density distribution that is not spherically symmetric. Suppose, for example, that a density anomaly were buried below the surface of a solid Europa on its Jupiter-facing side. Independent of any torque on the tide, the pull of Jupiter would tend to keep that anomaly on the side toward the giant planet. (A similar effect helps lock the Moon into synchronous rotation relative to the Earth. The Earth's pull on density asymmetries in the Moon helps keep one face locked toward Earth.) For Europa, even the primary tidal elongation could provide the frozen-in distortion, if Europa were so cold and rigid that the elongation could not change fast enough to conform to the change in the tidal potential that would accompany non-synchronous rotation.

What makes non-synchronous rotation plausible for Europa, and equally so for Io, are two effects of the Laplace resonance. First, the resonance drives the necessary eccentricity needed for the torque on the diurnal tide. Second, the friction generated during the response to diurnal variation results in such substantial internal heating, especially of Io and Europa, that the satellites' structures might not be able to support any frozen-in asymmetry. A definitive determination of whether non-synchronous rotation should be expected is impossible from tidal theory alone, because the strength of diurnal tidal torque is so uncertain. Nevertheless, because of the synergistic effects of orbital resonance, tides, and internal heating (Figure 4.3), it is possible, indeed quite plausible, that Europa and Io rotate non-synchronously.

Unfortunately, a misconception about non-synchronous rotation has propagated within the scientific community. Non-synchronous rotation is frequently described as meaning that only the ice shell on top of the ocean rotates relative to Jupiter, while the rocky mantle and metal core remain in synchronous rotation, locked to Jupiter. In fact, tidal torques operate on the solid interior of Europa as well as on the crust; so, it is plausible that they both rotate non-synchronously. It is also conceivable that the silicate interior is locked to the direction of Jupiter by a mass asymmetry, while the ice crust, uncoupled from the silicate by an intervening liquid water layer, rotates non-synchronously due to tidal torque. How well the crust's rotation is coupled to the mantle's through the global ocean is completely unknown. Presumably, if they rotated at different rates due to different tidal torques, there would be some friction due to the viscosity of the ocean, tending to keep them together. In any case, in terms of observable effects, it is only important that the crust may rotate non-synchronously relative to the rotation of Jupiter, whether or not the rock below the ocean comes along with it.

Independent of the tidal effects on rotation, the whole global shell may slip and slide over the ocean in any direction relative to the interior (Figure 5.3). This effect is called "polar wander", because, if a pole were planted in the ice at the north pole of the spin axis, it would move away from the spin axis as the ice shell slips around. If the ice were fairly uniform in thickness and density, little force would be needed to reorient the shell in any direction. In principle, oceanic currents could reorient the

Sec. 5.3] **Rotational effects on Europa** 69

Figure 5.3. As the ice shell of Europa slips around the body, the poles may move, as they do on Earth when the Arctic ice cap shifts position.
Andrew C. Revkin/*The New York Times*. See Chapter 12 and Figure 12.15.

crust in that case. Centrifugal force could also conceivably play a role in the following way. If some extra mass were attached to a spot on the surface, it would spin out toward the equator, dragging along the entire shell. Such mass variation could develop if tidal heating is not perfectly uniform, allowing the ice to get thicker in some regions than others. As we delve into the geological record, Chapter 12 will discuss evidence for major reorientation of the entire icy shell. Such events may be frequent, sudden, and fast compared with non-synchronous rotation.

Whether the crust changes its orientation relative to Jupiter by systematic, tidally driven, non-synchronous rotation, or by other reorientations of the crust, or both, the effects on the observable geology would be profound. The shell would be reoriented relative to the fixed, primary component of the tide, with its kilometer high bulges in the sub- and anti-jovian hemispheres. Stretching the ice shell where it passes over the bulges, and compressing it where it passes between the bulges, would introduce stresses bound to dominate the tectonic record.

6

Tides and stress

Cracks and other associated features, which dominate more than half of the surface of Europa, record the history of stress on the icy crust. This record, as laid out in images, includes the entire geological history of Europa. Unfortunately, this is not saying much if geological history is defined as the period during which the currently-visible surface was created. In fact, the visible surface records only the most recent 1% of the 4.6-billion-yr age of Europa, a small part of the total history. Moreover, even during the short age of the surface, the continual slicing and dicing by the tectonics and disruption by chaotic terrain has left an interpretable record covering only the last few million years. Nevertheless, even this set of features, the most fresh and recognizable crack patterns, can be interpreted in terms of tidal stresses that have operated for much longer than the age of the current surface. Ockham's razor implies that the processes recorded in the most recent tectonics represent the same types of continual tectonic modification that has gone on for much of Europa's history; there is no compelling reason to believe otherwise.

The upper part of the ice crust is very cold because heat that reaches it from the interior is rapidly radiated into space. At the surface the temperature is about −170°C. So, this upper portion of the ice is brittle and elastic. The elasticity means that it acts in a springy way, while the brittleness refers to its breakability when a certain stress limit (its strength) is exceeded. Near the bottom of the crust, the ice is much warmer, approaching melting point where it meets the ocean. The warmer ice is a viscous solid, capable of flowing slowly (one might say *glacially*, both in the sense of flowing ice and of low speed). Somewhere in-between there is a transition from brittle–elastic to viscous ice, but how deep is uncertain, although probably a kilometer or two down according to the evidence I discuss later. The exact depth of transition is uncertain for many reasons. We do not know the temperature profile going down. We do not know much about the rheology (flow characteristics) of the type of ice there, which depends sensitively on unknowns, like the ice grain size and the amounts and types of contaminants. And the transition temperature (and thus the depth of the transition from elastic to viscous behavior) depends on the rate of strain of the ice. When ice is distorted quickly, it behaves elastically and may break in a brittle fashion; but, when it is strained slowly, it flows.[1]

On Earth the lithosphere is defined as the outer part of the crust that behaves elastically. On Europa too we have come to refer to the elastic portion of the crust as the lithosphere. This terminology is confusing because *litho-* literally means rock, but on Europa it means brittle–elastic. Such weird terminology is acceptable, as long as it is defined clearly and used consistently.

As the ice crust rides over the changing tidal shape of Europa, and specifically the elongated shape of the ocean, the lithosphere stretches and compresses and shears elastically, depending on where it is. Below it, a few more kilometers of viscous ice follow along, but do not build up elastic stress. The tidal effect of the ocean is only to push upward on the ice; it is too slippery to push laterally on the bottom of the ice. Of course, the ocean might push laterally if there were currents involved, but we have no information about currents. Tidal potential alone probably cannot drive currents in such a thick ocean.

In order to determine the stress field in the surface ice, we model the crust as an elastic sheet overlying the tidally-deforming body and the tidally-deforming ocean. We assume that this sheet is decoupled from the rest of the body in the sense that there is no shear stress between it and what lies below, a reasonable assumption given the likelihood of a global liquid water ocean. We also assume that most of the 150 km of H_2O on Europa is liquid so that the remaining ice layer is thin compared with the size of the body, allowing us to treat the ice as a two-dimensional sheet. Specifically, we ignore its thickness and we ignore any stress components normal to the surface.

We also assume that the ice sheet is continuous and uniform in its elastic properties. In reality, continuity is disrupted whenever a crack forms, and real material is unlikely to be uniform on a global scale. However, as a first step toward understanding global stress fields, these approximations are reasonable.[2] Knowing how tides affect the stress on such a crust (or precisely on the lithosphere), we can interpret the tectonic record of observed cracks and ridges.

For a mental picture of this physics problem, imagine a thin rubber balloon stretched over a slippery spheroid, which keeps changing shape underneath it. The stress in the rubber will continually change in response to the changing shape of the slippery body that it encases.

Tidal stress can be separated into two major parts. The first is the stress that would result as non-synchronous rotation reorients the icy shell relative to the direction of the primary tidal elongation. The second is the stress caused by the continual diurnal change in the tidal figure. Of course, the diurnal tide may also

---

[1] The dependence of rheology on strain rate can be appreciated by playing with a small blob of what is known as Silly Putty in North America, which displays the full range of viscous/elastic/brittle behavior at temperatures and strain rates of everyday experience, which makes it a perfect toy, as well as a powerful pedagogical tool.
[2] Like all physicists, we would rather work with a spherical cow than a real one.

be indirectly responsible for the first part of the stress because it is the most likely cause of non-synchronous rotation.

We first consider theoretically the effect of non-synchronous rotation on stress in the crust. Suppose the crust were in a relaxed state, with no stress, floating over the surface of the ocean on a tidally-elongated Europa. In the sub- and anti-jovian regions, the crust lies over the tops of the tidal bulges. Now, suppose the tidal bulges were reoriented relative to the crust by non-synchronous rotation. In other words, the crust rotates as a whole, while each tidal bulge beneath it keeps pointing in the same direction relative to Jupiter. This could happen if the whole body were rotating non-synchronously or if only the crust rotated non-synchronously. In the process, the crust must distort, stretching to accommodate the changing shape below it. In the elastic lithosphere, stress builds up.

We calculate the stress by exploiting the equilibrium solution for an elongated (or flattened) spherical shell derived by geophysicist F.A. Vening Meinesz in the 1940s. Envision a shell that is flattened at its poles by a fraction f. It does not matter whether the poles are the north–south poles of the spin axis, or the poles of an axis aligned with a tide raiser (as in Figure 4.1); but, if we are considering the elongation due to tides, the flattening value f would be negative. The stress in such a shell is given by the following equation:

$$\sigma_\theta = \tfrac{1}{3} f \mu [(1+\nu)/(5+\nu)](5 + 3\cos 2\theta) \qquad (6.1\text{a})$$

$$\sigma_\varphi = -\tfrac{1}{3} f \mu [(1+\nu)/(5+\nu)](1 - 9\cos 2\theta) \qquad (6.1\text{b})$$

Here, σ_θ is the stress along the direction of the meridian lines relative to the pole, and σ_φ is the "azimuthal" stress perpendicular to it. A positive stress is compressive and a negative stress is tensional.[3] The elastic response of the material is represented by the shear modulus μ and Poisson's ratio ν. The angle θ is the distance from the pole of the flattening.

We can check that this formula is reasonable by considering the stress 90° from the axis (i.e., $\theta = 90°$). For example, suppose a planet is stretched by tides, as in Figure 4.1. From the stretching, we would expect σ_θ to be in tension at a location 90° from the axis of the tidal pull (i.e., it should be negative). Referring to Eq. (6.1a), f is negative (because stretching of the body is negative flattening) and $\cos 2\theta = -1$, so we confirm that σ_θ is negative as expected. At the same place, σ_φ should be in compression (with a positive value), because the belt around the narrow part of the elongated planet is shortened by the tide. Using Eq. (6.1b), again with negative f and $\cos 2\theta = -1$, we find that σ_φ is indeed positive (compression) as expected.

A direct application of this formula in planetary science was the investigation by planetary geophysicist H.J. Melosh (of the University of Arizona) of the stress that may have built up in Mercury and the Moon. Melosh considered that, under the initial rotation, the poles of the planet would have been flattened as its equator bulged out by centrifugal force, just as the equators of all the fast-rotating planets, like the Earth, are now bulged out. Then, when the spin of Mercury or the

---

[3] By symmetry there cannot be any shear along these axes, so these are the "principal stresses".

Moon slowed down, the amount of flattening decreased. While the Vening-Meinesz formula gives the stress that would result from flattening, we can simply reverse the signs to find what stress develops if a planet goes from a flattened to a more spherical state.

With appropriate care, we can apply the same formula to calculate tidal stress. The elongation of the shape of a planet by tides is similar in shape to the flattening that occurs if a planet is spinning rapidly (like Earth) in the following sense. In both cases, the shape can be described by the formula for the elongation, Eq. (4.1), except that for the flattening an extra minus sign is put in front. Also, the axis of symmetry points in a different direction. The axis of symmetry of spin flattening is the same as the spin axis, with poles at the north and south. The axis of symmetry of a tidal elongation is perpendicular to the spin axis, along a line pointing toward the tide-raising body (really, not quite toward it, if we take into account the tiny response lag).

The similarity means that the Vening-Meinesz formula for flattening stress (Eq. 6.1) can also be used to calculate tidal stress. All we need to do is change the sign of the stress, so that tension becomes compression, etc., and rotate the whole stress field by 90° so that it is aligned with the tidal axis of symmetry. In effect, we redefine the angles φ and θ in Eq. (6.1) so that they are referenced to the appropriate axis of symmetry.

6.1 TIDAL STRESS DUE TO NON-SYNCHRONOUS ROTATION

We can also apply the same formula to the problem of non-synchronous rotation, in which the stress develops in a shell that is reoriented relative to an elongated figure. The analysis is fairly straightforward in principle if we think of the reorientation of the tidal bulges relative to the crust in the following way. It is equivalent to first removing the tidal bulge (stressing the crust) and then adding back the tidal bulge in the new orientation (stressing it further in a different way). This equivalence is possible because the lithosphere is assumed to behave as a linear elastic medium. In other words, like an ideal spring, the stress change in a period of time only depends on the initial and final geometry, not on how things changed in-between. The advantage of calculating the stress using the two steps, removing the tidal bulge and then adding it back in a different direction, is that we can simply apply the Vening-Meinesz/Melosh theory. The stress in each step is calculated directly from Eq. (6.1), and then we add them together.

In principle, this transformation is straightforward. In practice, it requires tedious application of spherical trigonometry. This work was first done in the mid-1980s by Paul Helfenstein, then a student at Brown University, working with Prof. Marc Parmentier there. They were interpreting the global lineament patterns on Europa that had been revealed at low resolution, similar to Figure 2.2, by the *Voyager* spacecraft. Then, after I described the possibility of non-synchronous rotation, they calculated the stress patterns for comparison with observations. (What they found, and subsequent results from the *Galileo* mission, is in Chapter 9.)

As *Galileo* images began to come to Earth in the late-1990s, it was obvious that the more detailed record of tectonic features needed to be compared with the theoretical models. I knew exactly how to reproduce the stress calculations that had been done by Helfenstein and Parmentier. At least, I knew the Vening-Meinesz formula, the relevant physics, and how to do the spherical trigonometry. Indeed, I had been selected to be on the *Galileo* Imaging Team precisely because I had made the case in my 1976 proposal that what would be seen at the Galilean satellites would largely be governed by tides.

However, the arrangement within the Imaging Team was that those who did the image sequence planning would have first dibs on reporting discoveries and publishing results. This system ensured that the most politically-powerful team members would control the science analysis, as well as enlarge their contracts to pay for graduate students to do the planning work. Accordingly, I and my research group at the University of Arizona were locked out of the planning process. The process was at its most blatant during a team meeting in Washington, D.C., where the team members who controlled the sequence planning were complaining about how much work it was (they wanted more money to pay more students to do the planning). When I volunteered the efforts of myself and my own research group, the offer was turned down. It would have meant sharing resources and sharing the rights to lead the initial image interpretation. I was not welcome. So, I and my students and associates were set up to be bench-warmers in the data analysis, at least during the crucial first scientific pronouncements and publications.

Responsibility for planning the spacecraft's several encounters with Europa went to two of the most powerful members of the Imaging Team, and arguably the most powerful players in planetary science: Ron Greeley at Arizona State University and Jim Head at Brown. The job was divided between alternate encounters: the Arizona State group planned half and the Brown group did the rest. This arrangement meant that Greeley and Head each would control the press reports, the scientific conference presentations, and the initial publications in the scientific literature. Their initial impressions of the images would become the authorized results from the *Galileo* mission as reported to the world beyond the Imaging Team. I found it curious that discoveries would be reported and documented in units that were based on separate encounters with the satellite, because it seemed to me that reconstructing the story of Europa would depend on an integration of all of the lines of evidence. However, both Greeley and Head were geologists and they followed a tradition of working first with particular regions or subsets of the information.

Both Greeley and Head also had large academic empires to govern, so they assigned much of the work on image sequence planning and scientific analysis to students and their postdoctoral associates. Eventually, one of those people emerged as spokesman for the Imaging Team for Europa. Robert Pappalardo had been a student of Greeley's at ASU and moved on to a postdoctoral job with Head at Brown. Although young and inexperienced, Pappalardo had both of the key players as mentors, so his role was secure.

These machinations were amusing rather than annoying, primarily because I was involved in plenty of other interesting projects, and also because the political forces at work were far beyond anything I could affect. My original proposal to NASA in 1977 to join the *Galileo* Imaging Team had been based on the idea that tides might be important and that interpretation of the images might require my expertise, in addition to that of the traditional photogeologists. That notion had proven to be correct (at least for Io) when *Voyager* flew past the Jupiter system in 1979. By the mid-1990s, it was evident that interpretation of what we saw at Europa really would require understanding of the effects of tides. I figured that once Greeley, Head, and Pappalardo made their initial qualitative pronouncements, there would still be an opportunity for me and my students and associates to figure out what processes were behind the appearance of this strange satellite. I and my group were waiting impatiently for the initial interpretation to be completed, so that we could go ahead and finally apply tidal theory.

I almost lost that opportunity, because by that time the importance of tides had become more widely recognized. I had been talking about tides and their potential effects over the course of almost 20 years of Imaging Team meetings. Helfenstein and Parmentier had shown that tidal stress could help explain global crack patterns on Europa based on *Voyager* data. So, Pappalardo and his mentors were well aware of the importance of this effect. Although I had been appointed to the team for my expertise on tides, they chose not to invite me to collaborate, keeping me locked out of the initial science interpretation. Instead, Pappalardo asked Paul Helfenstein for help. In addition to being smart, Paul is one of the nicest people in planetary science. He agreed to give Pappalardo the software that he had developed to calculate the tidal stress due to non-synchronous rotation.

Eventually, for a strange reason, my group did get the opportunity to do the stress analysis after all. The software that Helfenstein gave Pappalardo was in an obsolete medium, a tape format for which no reader still existed. My students Greg Hoppa and Randy Tufts were monitoring the situation with amusement, but also with considerable frustration, because we knew that we could do the calculations from scratch. Finally, once Pappalardo was convinced that no usable tape reader existed, he allowed my group to proceed with the analysis.

For me, getting the results was easy: I told Greg Hoppa to compute them. Of course, I also showed him Eq. (6.1) and how to use it. Then, I sat back and waited. In a few weeks Greg returned to me with a complete set of tidal stress plots, which eventually became the basis for explaining most of the major tectonic patterns on Europa.[4]

---

[4] Pappalardo, familiar with Greg Hoppa's thesis and publications, has continued trying to calculate tidal stress. At the 2003 Lunar and Planetary Science Conference, Pappalardo and his students presented a poster showing diagrams of tidal stress. It was surprising that they did the same cases that had been published by Greg Hoppa years earlier. More surprisingly, these new plots were incorrect, except for one which was an unattributed photocopy from Greg's thesis. At the 2004 LPSC they reported that "when a significant degree of non-synchronous rotation ($\sim 5°$) is included in the model, it quickly dominates the stress pattern", without citing Greg's 1998 thesis which stated, "The stresses associated with $> 5°$ of non-synchronous rotation are much greater than the stresses associated with the diurnal tides so they swamp out the effect of the diurnal tidal stress."

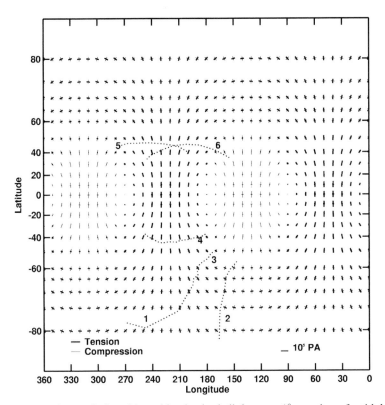

Figure 6.1. Map of stress induced in a thin elastic shell due to a 1° rotation of a tidal bulge of the magnitude of Europa's primary tidal component. Crossed lines indicate the orientation and magnitude of the principle components, with bold lines indicating compression and fine lines tension. Note the scale bar for the magnitude of the stress. The locations of several major Europan lineaments are indicated by dotted lines: (1) Astypalaea Linea; (2) Thynia Linea; (3) Libya Linea; (4) Agenor Linea; (5) Udaeus Linea; (6) Minos Linea. In several of my research group's papers we labeled Udaeus (which was not named until 2003) incorrectly as "Cadmus Linea" because we thought, on the basis of low-resolution images, that this lineament was an extension of Cadmus Linea; later images showed it is not.

Figure 6.1 shows the surface stress field that would accumulate as non-synchronous rotation reoriented the bulge by an angle of 1° toward the east, as first computed by Greg Hoppa in 1997. This display format is something that I invented as a way to show the results of his calculations. The plotting format used by Helfenstein and Parmentier made it cumbersome to extract the necessary information. Our format shows for each location the magnitude and direction of the principal stresses, which contains all the information about the stress state. In the theory of elasticity, the principal stresses represent the tension or compression (negative tension) on orthogonal planes, along which there is no shear stress. If you want to know the tension, compression or shear stress on any other plane through the material, you can calculate them from the principal stresses.

We plotted the directions and magnitudes of the principal stresses on a Mercator projection, which allowed us to show the principal axes of stress in their true azimuthal orientations. In other words, the directions relative to the vertical and horizontal axes are the same as the azimuthal directions relative to north, south, east, and west on the sphere of Europa. The coordinates are fixed to the lithosphere and zero longitude corresponds to the direction of the long axis of the elongated figure of Europa before the reorientation of the tidal bulge. (As Europa rotates eastward, the bulges move westward relative to the crust.)

The stress patterns make sense. Along the equator, in the quadrants moving eastward towards the peaks of the tidal bulge at longitude 0° or 180°, there is tension, while in the quadrants moving away from the bulges, there is compression. At the poles, the principal stresses are equal in magnitude but one is tension and the other is compression, which means that in the polar regions the stress is characterized as nearly pure shear along any direction 45° from the principle stresses. As the crust is stretched over the changing orientation of the tidal elongation, it seems reasonable to find such shearing near the poles.

The details of the stress pattern plotted in Figure 6.1 for non-synchronous rotation apply to a specific case. As shown by Eq. (6.1), the stress depends on both the rigidity μ and the Poisson ratio ν of the elastic lithosphere. Rigidity is a measure of elastic resistance to stress, while ν, combined with μ, describes the elastic resistance to compression or tension. Tides cause a deformation of the lithosphere, as it must accommodate to the changing shape, over which it rides. If the lithosphere were very flabby, it could be deformed without building up much stress. The elastic constants determine how much stress actually builds up during the distortion (or "strain") of the material. For the stress plots shown here, we used values of $\mu = 3.52 \times 10^9$ Pa for the shear modulus and $\nu = 0.33$ for the Poisson ratio, based on laboratory measurements of ice in the literature.

From Eq. (6.1), we know that the magnitude of the stress field is proportional to $\mu(1+\nu)/(5+\nu)$. Thus, for example, if μ were very small, the lithosphere could accommodate to the changing tidal shape, with its elastic material undergoing strain as necessary, but little stress would develop. The values we have adopted are good estimates for typical ice, but the actual properties of the material of Europa or how they act on a global scale, are unknown. We are also assuming that the properties are uniform over the globe, and we are ignoring any cracks that would break the continuity of the elastic envelope. Any of these effects could modify the global stress field significantly.

Figure 6.1 is specific to the case of stress change due to 1° of rotation (i.e., 1° of displacement of the lithosphere relative to the shape of tidal elongation). For any other angle of rotation, as long as it is small, the pattern would be similar, with amplitudes approximately proportional to the angle. In each case, symmetry would be offset from the cardinal longitudes (0°, 90°, 180°, 270°) by 45°, as it is in Figure 6.1. For example, the tension zones are centered at 45° west of the sub- and anti-Jupiter longitudes. More precisely, however, the pattern is actually shifted eastward from that orientation by a distance equal to exactly one-half the rotation angle. Thus, if strain could accumulate over rotation by 30° (an example that would be possible only if the material were sufficiently elastic and strong, because the material would be extremely strained), the stress patterns would be similar except oriented eastward by 15°, such that maximum tension would be 30° west of the sub- and anti-Jupiter longitudes. (The displacement of the pattern by half of the rotation angle is a consequence, familiar from trigonometry, of taking the difference between two identical sinusoidal functions that differ only in that they are offset only in phase.)

The stress patterns shown in Figure 6.1 could also be applied to a case of polar wander, but only if the shift is about an axis aligned with the direction of Jupiter. A fairly uniform shell uncoupled from the interior is susceptible to reorientation relative to the spin axis. Suppose the ice shell slips as a whole, so that a location on the ice formerly near the pole moves toward the equator along a meridian 90° from Jupiter. In Chapter 12, I describe evidence for one such event in the recent history of Europa, and likely there have been many more. In that event, because of the particular direction of polar wander, the position of tidal elongation relative to the shell did not change, so the tidal shape probably did not contribute much stress. Instead, it was the polar flattening that got reoriented relative to the ice shell. The stress pattern due to this reorientation would be the same as non-synchronous rotation relative to tidal elongation, except that now the bulge is replaced by flattening (changing the sign of the stress at each location) and the reorientation is about an axis lying on the equator. As long as appropriate care is taken for those geometric corrections, the pattern shown in Figure 6.1 could be applied to polar wander.

Getting back to non-synchronous rotation, the case that is actually shown in Figure 6.1, the magnitude of the tension for 1° of non-synchronous rotation is about 10^5 Pa, or about equal to the pressure of the Earth's atmosphere at sea level. This stress is comparable with the plausible tensile strength of the ice. If failure occurred, it would probably be tensile, because ice is relatively strong in its resistance to other modes of failure, such as shear. Even at high latitudes, where there is a substantial differential between principal stresses, meaning that there is considerable shear stress in some directions, the failure mode would probably not be shear, because the principal stresses are nearly equal and opposite. Tensile cracking is the most likely mode of failure. Figure 6.1 includes the locations of several major large-scale lineaments of interest, confirming that they are orthogonal to the direction of tension, consistent with the expectation that tension would be the dominant determinant of failure.

Because tension would exceed the strength of the material after more than about 1° of rotation, it is unlikely that more rotation can occur before stress is relieved by cracking. Of course, if the rate of rotation is slow enough, stress could also be prevented from building up by viscous relaxation of the ice through most of the crust. Either way, it is unlikely that the stress field due to non-synchronous rotation would exceed the magnitude shown in Figure 6.1. In the literature on Europa, occasional interpretations of particular tectonic features have invoked far greater amounts of stress, assuming that such stress could have built up during many tens of degrees of rotation. However, such build-up is implausible, because the stress must be relieved much sooner, when it exceeds the strength of the ice. Even though the ice could resist failure in regions where compression dominates, if it cracks elsewhere under tension, the assumption of global continuity is violated and the stress patterns would change considerably from what is shown in Figure 6.1. Europa's surface is dominated by the record of the cracking that relieves the stress.

6.2 TIDAL STRESS DUE TO DIURNAL VARIATION

Next consider the stress due to the diurnal tide. If Europa were rotating synchronously, the diurnal variation would be the only source of tidal stress. Greg Hoppa computed the diurnal tidal stress at various points in the orbit as shown in Figures 6.2(a–d). These diagrams show the stress relative to the "average figure" of Europa. By average figure, we mean the shape of the primary tide, which would represent the response to Jupiter if there were no orbital eccentricity (i.e., if Jupiter were at the center of its orbital epicycle relative to Europa—Figure 4.2).

At pericenter (Figure 6.2a), when Europa is closest to Jupiter, the stress patterns are exactly the same as those generated by distorting a spherical shell into a tidally-elongated shape, because only the amplitude differs from the average configuration. Here, with the distance from Jupiter at a minimum, the amplitude of the tidal bulge is a maximum. At the sub- and anti-Jupiter points the lithosphere is stretched tightly over the tops of the tidal bulges, giving tension that is isotropic (equal in all directions) as shown at longitudes 0° and 180° along the equator. Along the belt 90° from these points, the body gets narrow at pericenter; so, we see a corresponding compression along that belt.

Figure 6.2 also shows the stress 1/8 orbit after pericenter (Figure 6.2b), 1/4 orbit after pericenter (Figure 6.2c), and 3/8 orbit after pericenter (Figure 6.2d). The stress map for any other orbital position 1/2 orbit from the positions shown is identical to the corresponding case shown in Figure 6.2, except that the signs of the principal components are reversed (tension becomes compression and vice versa).[5]

Because ice is most prone to failure in tension, we are especially interested in locations and conditions of maximum tension. The stress pattern at apocenter (which would be given by Figure 6.2a with signs reversed), has equatorial regions of north–south tension, which extends over the poles. In other words, this band of tension runs around the belt 90° from the Europa–Jupiter axis. It makes good physical sense that there would be tension around this belt at apocenter (Europa at its farthest point from Jupiter), the point in the orbit where the tidal amplitude would be at a minimum. At that point, this belt around Europa would be stretched, because the tidal elongation is less than average.

It is important to remember that the diurnal tide is not simply due to the changing distance from Jupiter, but also to the changing direction of Jupiter. At 1/4 orbit before or after pericenter, the tidal amplitude is identical to the average, but

---

[5] The stress pattern in Figure 6.2c is approximately equal and opposite to Figure 6.1. This similarity is because Europa's eccentricity is 0.01 (as discussed in Chapter 4) which causes the tidal bulge to move by ~1° as it tries to trace the path of Jupiter relative to Europa.

Sec. 6.2] Tidal stress due to diurnal variation 81

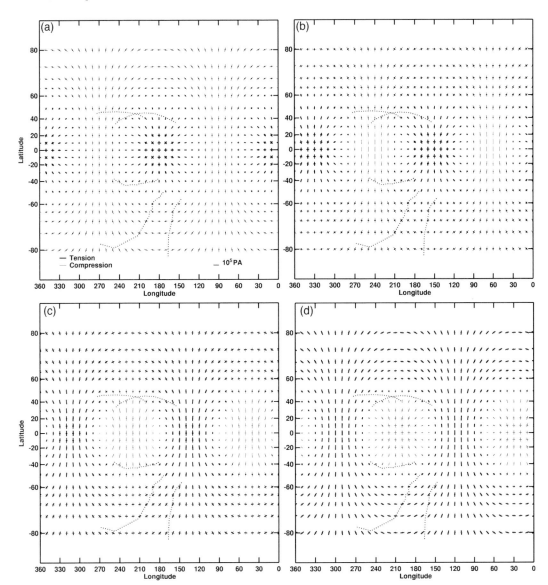

Figure 6.2. Maps of stress induced due to diurnal variation of the tide. The stress field is shown at intervals of 1/8 of an orbit, with bold lines indicating tension, and fine lines indicating compression as in Figure 6.1. Maps are shown covering only 1/2 of an orbit, from pericenter (a) in steps of 1/8 orbit. The stress field is anti-symmetrical over time (identical except for a reversal of sign at times 1/2 a period apart). Thus, for example, the stress at apocenter would be the same as in part (a), except that the bold lines would now represent compression, and the fine lines tension. Note that 1/4 of an orbit before or after pericenter (e.g., in (c)), the stress is identical to an equivalent amount of non-synchronous rotation (cf. Figure 6.1).

the orientation is shifted by about 1.5°, equivalent to non-synchronous rotation (compare Figure 6.2c with Figure 6.1, noting that the stress pattern is practically identical, except for the difference in sign and that the magnitude is 50% larger, in proportion to the angular offset of the tidal bulge).

6.3 TIDAL STRESS: NON-SYNCHRONOUS AND DIURNAL STRESS COMBINED

On Europa, tidal stress results from a combination of diurnal and non-synchronous effects. Figure 6.3 shows the combined stress of 1° of non-synchronous and diurnal stress over an entire orbit, starting at pericenter (Figure 6.3a), in steps of 1/8 orbit. Shortly after my own research group was given permission to go ahead with a study of tidal stress, we realized that, given the likely strength of the ice, neither diurnal stress nor non-synchronous stress alone could be sufficient to explain the crack patterns we observed on Europa. Instead, it seems likely that a background stress builds up as the primary tidal bulge gradually migrates due to non-synchronous rotation. The diurnally-varying component of the tidal stress is superimposed on that slowly-increasing background stress field, until one day the maximum tensile stress exceeds the strength of the ice. According to Figure 6.3, that event probably occurs about 1/8 of an orbit after apocenter.

These stress patterns (Figures 6.1–6.3) lay the basis for interpretation of many of the major lineament patterns seen on Europa as cracks resulting from tidal stress. As we will see, these theoretical results have allowed us to understand large-scale global patterns (Chapter 9), the formation of double ridges (Chapter 10), shear displacement of the crust along cracks (Chapters 12 and 13), and the distinctive cycloid-shaped linear patterns that are ubiquitous but unique to Europa (Chapter 14). We will see that the observed tectonics provide a record indicative of non-synchronous rotation. Most importantly, the remarkable agreement of this tidal stress theory requires that there be a global ocean under the ice to provide a tidal amplitude large enough to overcome the strength of the ice crust, and the processes involved require that cracks penetrate from the surface down to the liquid water ocean below.

Sec. 6.3] **Tidal stress: Non-synchronous and diurnal stress combined** 83

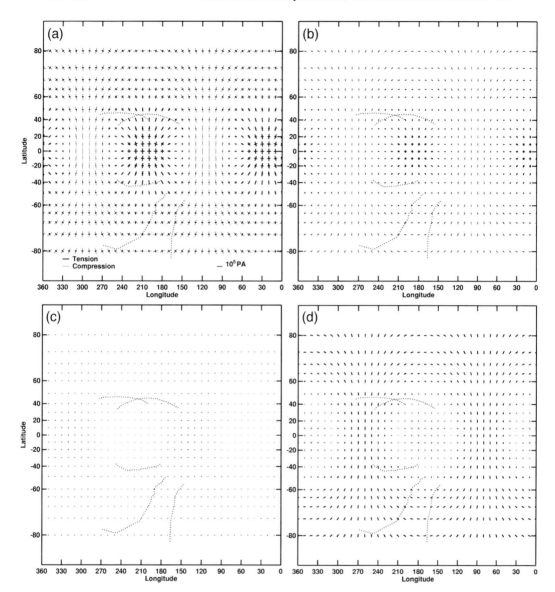

Figure 6.3. Maps of stress due to "diurnal" variation of the tide, added to the stress that had accumulated during 1° of non-synchronous rotation. The parts (a)–(h) show how the stress pattern changes over an entire orbit, starting at pericenter (a), in steps of 1/8 orbit. Maximum tension occurs approximately 1/8 of an orbit after apocenter.

84 Tides and stress [Ch. 6

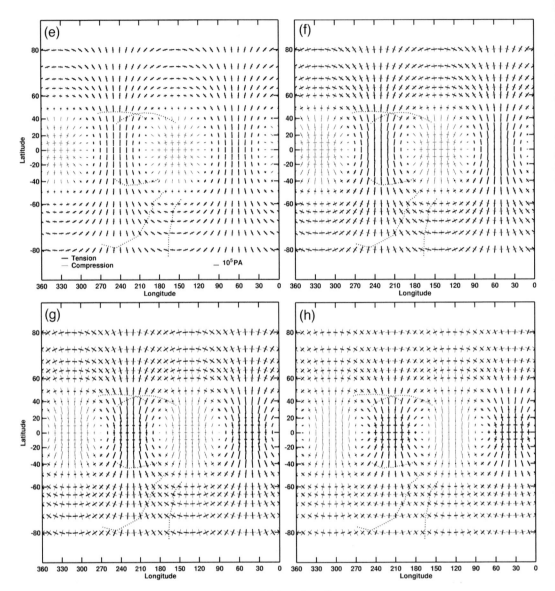

Figure 6.3 (*continued*)

7

Tidal heating

As the figure of a satellite is distorted periodically due to diurnal tides, friction may generate substantial heat. When Stan Peale and his colleagues realized in 1979 that Io had a substantial orbital eccentricity, they were able to estimate the heating on Io, which led to their legendary prediction that *Voyager* images would show major thermal effects (Chapter 4). Only a few days later, huge amounts of ongoing volcanism were discovered there. The *Voyager* images of Europa were not so dramatic, but it was clear that this satellite had been an active body recently, and there was no reason to believe that the activity had stopped. The contrast between Europa's appearance and that of inactive, heavily-cratered objects was obvious; so was the clear trend among the Galilean satellites correlating the internally-generated geological activity with decreasing distance from Jupiter, and thus with increasing tidal strength. Farthest out and not part of the eccentricity-driving Laplace resonance, Callisto was the archetype of a dead bombarded planet. Closer to Jupiter, and with a resonance-driven eccentricity, Ganymede displayed large Callisto-like areas, but other broad regions showed tectonics and resurfacing. Closer still was the completely resurfaced Europa, and closest to Jupiter was the spewing Io.

Given the massive current activity on Io, it was natural to calculate the likely effects of tidal heating on the other satellites in the resonance. The relevant parameters could only be guessed at, especially the dissipation parameter Q, which is supposed to represent and characterize all of the diverse frictional effects within the satellite. Conservative assumptions about unknown parameter values, which tended to minimize tidal heating, were generally adopted, because doing so made a stronger, more convincing case for the predicted thermal effects. A substantial fraction of the heating was assumed to come from the distortion of the ice layer. Even with such minimal-heating models, tidal friction could maintain a liquid water ocean, but only if one already exists: Without a global ocean, the tidal stretching of the satellite would not be very large (only about 3% of what it is with an ocean, as discussed in Chapter 4), so the crust would not be stretched much, and there would not be enough heating to create an ocean. In that case, the presence of liquid water appeared to be an all-or-nothing proposition, in the sense that it could not exist without a global ocean. This result meant that if there is significant liquid anywhere, there must be a global ocean.[1]

Those early estimates may have given heating rates much lower than what actually occurs within Europa. The classical theoretical calculations gave internal heat production of about 1.6×10^{12} W, with steady-state flow of heat out of the body, at a corresponding rate of about $0.05\,\text{W}/\text{m}^2$. If we could measure the heat flux by astronomical infrared remote sensing, we could check the accuracy of that estimate. However, radiation from the Sun hits the surface at a rate of $50\,\text{W}/\text{m}^2$, so reradiated solar energy from Europa's surface swamps out the much smaller flux of tidal energy, making direct detection of the tidal heat very uncertain.

We do have measurements of the heat escaping from Io, most of which radiates from a few very hot spots associated with the largest volcanoes. The total internal heat radiating out from Io is at least 10^{14} W. Less than 1% of that amount could be due to internal heating from radioactive constituents, and the rest must be generated by tidal friction. This measurement of heat flux is a lower limit. There could be much more heat radiating over broad areas of the surface between volcanoes, but camouflaged in the reradiated solar energy.

The tidal heating rate of $>10^{14}$ W in Io provides a basis for estimating the plausible tidal heating rate within Europa, if we scale this rate to Europa's orbital and physical parameters. For a uniform spherical satellite in an eccentric orbit around Jupiter, the tidal energy dissipation rate is given by:

$$dE/dt = \tfrac{21}{2} G^{-1} R_s^5 e^2 n^5 (k_2/Q) \qquad (7.1)$$

where G is the gravitational constant, R_s is the radius of the satellite, e is the orbital eccentricity, and n is the orbital mean motion (the average angular velocity). Here Q is the globally effective dissipation parameter and k_2 is the tidal Love number that describes the amplitude of the portion of the gravitational field of the satellite due to tidal elongation. If Europa and Io were compositionally and structurally identical, we could assume that Q and k_2 were the same for both bodies. However, Europa's water layer is very different from anything on or in Io, where there is no evidence for water. We can avoid that problem by supposing that Europa only consists of its rocky interior and core, ignoring the ocean and ice that lie above it. In that case, we can assume that the Q and k_2 are reasonably similar to Io's.

Equation 7.1 can then be used to scale Io's dissipation to the heating in the interior of Europa, by taking into account the ratios of the values of the satellites' radii and orbital parameters. The heating rate in Europa, relative to that in Io,

---

[1] It also suggests that, if an ocean existed recently, it had existed for a long time. And if it had existed for a long time until recently, there must be one now: it would be surprising if it froze solid just before our spacecraft arrived there.

would be:

$$dE_2/dt = (R_2/R_1)^5 (e_2/e_1)^2 (n_2/n_1)^5 \, dE_1/dt \tag{7.2}$$

where the subscripts 1 and 2 refer to values for Io and Europa, respectively. For Io, R_1 and e_1 are 1,821 km and 0.0041; for Europa, R_2 and e_2 are 1,450 km and 0.010, where the radius is that of the rocky interior. According to the Laplace resonance, the ratio of the mean motions n_2/e_1 is very near 2. Taking the lower limit for heating of Io of 10^{14} W, Eq. (7.2) gives a heating rate in Europa of 6×10^{12} W.

With this amount of internal heating, $0.2 \, \text{W/m}^2$ must be continually removed through the surface of the satellite. This value is nearly four times greater than the classical conservative estimates from the years immediately after the *Voyager* encounter.

In fact, the internal heating could be even greater for numerous reasons:

(1) Remember, we based this estimate on the measured heat from Io. The actual total heat flux from Io may be somewhat greater.
(2) In addition, radiogenic heating within Europa probably contributes about $0.01 \, \text{W/m}^2$.
(3) Also, we ignored the water layers (ocean and ice) on Europa, which would increase the tidal heating in a couple of ways:
   (a) The great heights of the tidal bulges in the water would produce a gravitational field that would actually enhance the tidal elongation of the rocky interior. In effect, the mass of the tidal bulges of the ocean would gravitationally pull up on the rock beneath. This effect would increase the internal tidal dissipation relative to our estimate.
   (b) Dissipation in the ice itself due to tidal distortion could in principle be significant, although, given the large amount of more deeply generated heat and the consequent thinness of the ice, this effect may not be crucial on a global scale. Dissipation in the ice could be due to the stretching described in Figure 6.2, as well as by enhanced friction due to shear along cracks in the ice.
(4) Depending on the viscoelastic state of Europa's rocky interior, the heating relative to Io could be much greater than our estimate above.

Paul Geissler, who originally joined my research group at the University of Arizona to help me with remote sensing of the Earth during *Galileo*'s fly-bys on its way to Jupiter, is an extremely capable and creative geophysicist. Working with one of my graduate students, Dave O'Brien, Paul considered a fairly standard viscoelastic model for the behavior of material in the rocky interiors of Europa and of Io, which includes the temperature dependence of the viscosity and rigidity. Paul found that the actual heating rate is probably a factor of 1.5 greater than what we obtained above (where we scaled relative to Io, assuming the same Q and k_2), resulting in the greater heating rate of 9×10^{12} W or $0.3 \, \text{W/m}^2$. Total heating could be even greater if the other factors enumerated above were taken into account.

We can estimate the thickness of the ice by first assuming that heat is carried outward by thermal conduction. For heat flowing through material, the heat flux F (heat per time per area of the material) is proportional to the temperature difference (ΔT) across the material and inversely proportional to its thickness h:

$$F = k\,\Delta T/h \qquad (7.3)$$

The constant of proportionality k is called the thermal conductivity, and it is a physical property of the material involved. For ice the value of k is about 3.5 W/m/°C. The temperature difference is that between the surface temperature at $-170°$C and the temperature at the base of the ice, where it meets the liquid ocean, which is $0°$C according to the definition of zero on the centigrade scale. With a flux $F = 0.3\,\text{W/m}^2$ and the values of k and ΔT, we can solve Eq. (7.3) to obtain $h = 2$ km.

In this conductive model the ice thickness varies inversely with the amount of heat being transported. If the internal heating rate decreased, the ice would thicken until the rate of heat transport decreased into a new equilibrium. If the tidal dissipation were as small as had been estimated in the post-*Voyager* years using conservative guesses for the value of Q, at about $0.05\,\text{W/m}^2$, the ice would be about 12 km thick according to Eq. (7.3).

The ice could be much thicker than that and still transport the same amount of heat if the ice were convecting, because convection carries heat more quickly and efficiently than conduction. Convection would require fairly thick ice to function, but the exact requirements depend on the very poorly-known properties of the ice, especially its viscosity which describes its flow characteristics. Results vary, depending on assumptions. Bill McKinnon, of Washington University in St. Louis (Missouri, U.S.A.), showed that the ice would need to be thicker than 10 km for fine-grained ice, and >20 km if the grain size were greater than 1 mm. Other estimates require ice thicker than 30 km. Recently, the planetary geophysicists Tilman Spohn (University of Muenster, Germany) and Jerry Schubert (UCLA) found from their quantitative modeling that any convecting layer would probably be inconsistent with the existence of a liquid ocean, a result that strongly favors the thin conducting ice layer.

For a given rate of internal heat production, if the ice layer were transporting the heat out by efficient convection, it would be thicker than if the transport were by relatively inefficient conduction. Thus, either a thin-ice model (thinner than 10 km) with conductive heat transport or a thick-ice model (thicker than $\sim$20 km) with convective heat transport could be consistent with a heat flux value in the middle of the range of plausible values.

If heat production were near the lower values, generating $0.05\,\text{W/m}^2$, the ice would be thick enough that convection would have a good chance of being activated, which would let the ice be substantially thicker than 20 km. Remember, the lower estimates of heating rates were deliberately conservative to strengthen the argument that tides could maintain an ocean. This thickness was the canonical value at the time of *Galileo*'s arrival at Jupiter. Consequently, the initial authorized interpretations of Europa's apparently active geology were based on the notion that the processes were dominated by solid-state convection and other solid-state flow in the ice. In that view, widely publicized and promoted under the authority of the *Galileo* project as fact, the ocean would have been isolated from the observable surface and would not be linked to the geological processes that governed the observed surface.

Comparison with Io gave a basis for estimating heat production on Europa, which at 0.3 W/m^2 (or even a bit higher) proved to be much greater than most of the earlier assumptions. For so much internal heat, a steady-state flux would require the ice to be so thin that convection seems unlikely. In that case the ice would most plausibly be conducting and only a couple of kilometers thick. Even for a heating rate closer to the old value of about 0.05 W/m^2, conduction (with ice thinner than about 10 km), and not necessarily convection (with ice thicker than about 20 km), is quite plausible.

Thermal transport models may still be too dependent on uncertain parameters for them to definitively discriminate between thin conductive or thick convective ice. Even if the relevant parameters were known, both possibilities might provide a stable steady state. In that case, Europa's actual current physical condition may depend on its thermal history. If the heating rate has increased toward its current value, the ice would likely be thick and convecting. If the heating rate has decreased from significantly-higher rates, then the current ice would more likely be thin and conducting. Because tidal heating is driven by orbital eccentricity, the history of Europa's orbit may be crucial.

8

Tides and orbital evolution

8.1 ORBITAL THEORY

Over the long term, tens of millions of years or more, the tidal deformation of Europa contributes to changes in the resonantly-coupled orbits of the Galilean satellites, which in turn modifies orbital eccentricities, which changes the amplitudes of diurnal tides.

In the current Laplace resonance (introduced in Chapter 4), pairs of orbital periods are near the ratio of 2/1:

$$P_2/P_1 \approx 2 \quad \text{and} \quad P_3/P_2 \approx 2 \qquad (8.1)$$

where the subscripts 1, 2, and 3 refer to Io, Europa, and Ganymede, respectively. The average angular velocities of the orbits (the "mean motions", n) are inversely proportional to the periods (the faster the motion, the shorter the period), so we also have:

$$n_2/n_1 \approx \tfrac{1}{2} \quad \text{and} \quad n_3/n_2 \approx \tfrac{1}{2} \qquad (8.2)$$

These equalities are approximate, not exact. The quantity:

$$\nu \equiv n_1 - 2n_2 \qquad (8.3)$$

is a measure of how close the system is to the exact 1/2 commensurability. It would be zero if the ratio of mean motions were exactly 1/2, but ν is not quite zero. In fact, ν is very small compared with the mean motions: $\nu = 0.74°/\text{day}$, which is much less than the mean motions, which have values of about 200, 100, and 50°/day, for Io, Europa, and Ganymede, respectively.

Not only is the other combination of mean motions $n_2 - 2n_3$ also small, but it has exactly the same small value ν. So, the Laplace relation among these orbits can be described by the equation:

$$n_1 - 2n_2 = n_2 - 2n_3 = \nu \qquad (8.4)$$

Figure 8.1. Conjunction of Io with Europa always occurs on exactly the opposite side of Jupiter from conjunction of Europa with Ganymede. The three satellites can never all be lined up together on the same side of Jupiter, or even come close.

The behavior described by this equation is remarkable. Consider what it means to have a pair of mean motions in an exact ratio of $1/2$ (e.g., if $n_1 - 2n_2 = 0$). At the moment that the faster one, on the inner orbit, passes the other, we say they are in "conjunction", because viewed from Jupiter they would be at the same point in the sky (i.e., at the same celestial longitude). After that point, the slower one falls behind. Conjunction occurs again only after the faster satellite gains a whole lap, which happens after the inner one has made two complete orbits and the outer one has made one. This second conjunction occurs at exactly the same longitude, relative to Jupiter, as the first conjunction did. And every subsequent conjunction occurs at the same longitude.

Now, if the ratio were not quite $1/2$, the direction of conjunction, as viewed from Jupiter, would not be fixed, but instead would slowly migrate. For example, if ν (defined in Eq. 8.4) were slightly greater than zero, n_1 would be slightly too large for the exact $1/2$ ratio. The inner satellite would be going a bit too fast, so it would catch up with the slower one slightly further back on the track at each conjunction. Conjunction would migrate backwards at rate ν.

According to Eq. (8.4), the conjunction of Io with Europa and the conjunction of Europa with Ganymede migrate at exactly the same rate, so the longitudes of the conjunctions are always separated by the same angle.

This angle is exactly 180°. Conjunction of Io with Europa always occurs on exactly the opposite side of Jupiter from conjunction of Europa with Ganymede. The three satellites can never all be lined up together on the same side of Jupiter, or even come close.[1]

---

[1] This geometry is important not only from a scientific standpoint, but also from the mission-planning point of view. The *Voyager* fly-bys could only do a close fly-by of three out of the four Galilean satellites, because the Laplace resonance keeps them apart. Similar considerations also constrained *Galileo* planning, especially in light of the high-gain antenna failure, which restricted the return of data.

Europa plays a key role in this ballet. Remember Europa's eccentric orbit, with perijove (the closest point to Jupiter), and apojove (the farthest point from Jupiter). It happens that the longitude of conjunction of Io with Europa always occurs in line with the orientation of Europa's apojove, and the conjunction of Europa with Ganymede always occurs in line with Europa's perijove, 180° away. The conjunctions occur at opposite ends of the major axis of Europa's elongated orbit.

The gradual migration of the conjunctions at the slow rate $\nu = 0.74°$/day means that the orientation of the major axis of Europa's orbit must precess at the same rate, in order to keep the conjunctions occurring at perijove and apojove. The precession of Europa's orbit is driven by the combined effects of the oblateness of Jupiter's shape and the gravitational effects of the other satellites, but the rate of precession depends strongly on the eccentricity e_2 of Europa's orbit. The more circular the orbit, the more readily Europa's major axis can be reoriented. Therefore, the precession rate depends inversely on e_2. Maintaining the behavior of the Laplace relation depends on the precise tuning of e_2 to maintain the alignment of Europa's major axis with the slowly-migrating orbital conjunctions.

As Laplace showed, this configuration is stable, maintained by the perturbations that the gravitational pull of each satellite apply to each other satellite's orbit. Ordinarily, a satellite's effects on the others would be minimal. However, because the geometry is so repetitious and periodic, the effects are cumulative and strong, hence the term "resonance". These enhanced perturbations act so as to maintain the geometry. If one satellite were to be slowed down, the motions of the others would adjust somewhat, as would the orbital eccentricity of Europa in such a way as to keep the conjunctions occurring at the opposite ends of the major axis.

The important implication for understanding the satellites is that the closer the system is to exact resonance ($\nu = 0$), the greater the forced eccentricities. Not only does the resonant behavior affect the eccentricity of Europa, but it pumps the eccentric trajectories of all three of the resonant satellites, Io, Europa, and Ganymede. The forced eccentricities are approximately inversely proportional to ν, although that simple rule breaks down as ν gets very close to zero.

In general, resonances play a role in many of the most important interactions among planetary bodies, because of the way they can enhance the mutual gravitational effects. The Laplace resonance is one of the more complicated resonances because three large satellites are involved (e.g., the eccentricities of the other satellites are involved, as well as Europa's). Here, I consider only the effects most important to Europa's evolution.

Suppose the mean motions of the inner two satellites were decreased slightly, so that ν decreased from 0.74°/day down to a value closer to zero. In other words, the mean motion ratio gets closer to exactly 1/2. In that case, the eccentricity of Europa is pumped up, slowing the precession to match the slower migration of the conjunctions. I usually describe this type of evolution as going deeper into the resonance. Any process that can change the semi-major axes of satellites could drive the system either deeper into resonance or further away from exact resonance, depending on which way it pushes. Possible sources of such a change might have been the gas drag and collisions experienced by the satellites during their initial formation in a nebula around Jupiter. Another process, still operating today, must be the effects of tides.

Tides raised by Jupiter on any of the satellites eventually result in loss of energy from the satellite's orbit. The frictional heating by diurnal tides has to come from somewhere and, ultimately, the source can be traced to the orbital energy. Thus, the satellite's orbit gets smaller and, in accordance with Kepler's third law, its mean motion gets faster. The effect is greatest on Io, because it is closest to Jupiter. Thus, the friction of diurnal tides tends to drive Io in toward Jupiter and the satellite system out from deeper resonance.

The satellites also raise tides on Jupiter, which rotates much faster than any of the orbital motions. Just as tides raised by Jupiter on a rapidly-spinning satellite tend to slow the spin of the satellite (Figure 5.1), a satellite that raises a tide on Jupiter tends to exert a torque slowing Jupiter's spin. The equal and opposite reaction is an increase in the orbital energy of the satellite. Again, the effect must be strongest for Io, but this effect tends to drive Io outward from the planet and the satellite system into deeper resonance.

A variety of scenarios for the history of the orbital resonance have been proposed. So far, most models have ignored tides raised on or by any of the satellites except Io, because Io is closest to Jupiter and the tidal strength is therefore so much stronger. Considering only the tides raised on Io or by Io on Jupiter, the main issue involves the competing effects of these two tides, the first tending to drive the system out of resonance and the second tending to drive it into deeper resonance.

Chuck Yoder (of JPL), who had been a student of Stan Peale's at the University of California in Santa Barbara, suggested that the orbits were originally driven into resonance by the torque due to the tidal distortion of Jupiter by Io. In his scenario, Io was first moved outward until it became locked into resonance with Europa, and then the two moved out until they reached resonance with Ganymede. Then, the system evolved more deeply into the Laplace resonance with the ratios of orbital periods becoming closer to exactly 2/1, and the forced orbital eccentricities increased. As a result, eventually, diurnal tides on the satellites became significant. Yoder pointed out that tides on Io would exert torques that tend to resist the orbital evolution into deeper resonance, suggesting that the system might have reached an equilibrium with the effects of tides on Jupiter balanced by those on Io.

Yoder's scenario was widely accepted. First, from a mathematical point of view, Yoder's model was incredibly elegant, and thus appealing to the handful of celestial mechanicians who understood it. I always thought it was one of the most beautiful pieces of theory in my field during my career. I described it in what I hope were more accessible forms in various reviews. Then, a couple of studies of possible scenarios of the evolution toward resonance provided further arguments supporting Yoder's resonance capture model. Two young researchers at Caltech, Renu Malhotra and Adam Showman, both of whom are now on the faculty in my department at the University of Arizona, showed how a particular resonance capture scenario might explain Ganymede's free eccentricity as a by-product. And Mandy Proctor, an undergraduate math student at the University of Maryland, working with Professor Doug Hamilton (and now doing graduate work with me at the University of Arizona), showed that as the system evolved during capture into resonance, gravitational interactions would pump up, and explain, the observed orbital inclinations of some of the tiny inner satellites of Jupiter.

The resonance capture scenario was so appealing that most people overlooked the problem that, as far we know, it could not work in the real world. Yoder's analysis depended on a balance between, on one hand, the tidal dissipation in Jupiter, which is necessary to have a torque that adds energy to Io's orbit, and, on the other hand, tidal dissipation in Io which takes energy out of Io's orbit. Given how deep the system is in resonance, the model required a particular ratio between the tidal energy dissipated in Io and in Jupiter. Now, we know how rapidly heat is escaping from Io, so the model specifies the tidal heating rate within Jupiter. The problem is that no one has been able to figure out how there could be that much friction in Jupiter.

For that reason, in 1982, I argued that the history of the system may not have involved the Yoder equilibrium at all. There probably is much less tidal energy dissipation in Jupiter than is required for Yoder's resonance capture model to work. More likely, the tides on Io dominate over the tides on Jupiter. In that case, the system must be evolving out from deeper resonance, while Io moves inward toward Jupiter.

Given that likelihood, I suggested several scenarios which would have led to the system's currently moving out of resonance. One idea was simply a primordial origin for the resonance, in which the satellites entered resonance during the satellite formation process. Stan Peale and others did not like the idea; it was too hard to abandon Yoder's theoretical model, especially because no compelling reason to believe that the process of satellite formation would establish a resonance. A convincing mechanism had not yet been demonstrated.

Recently, however, new ideas about the formation of the Galilean satellites have included just such a process. The revised models of formation were motivated by a problem with earlier models of satellites' formation. The key issue involves the structure of Callisto, whose density profile (inferred from its effects on the *Galileo* spacecraft's trajectory) shows that it is only partially differentiated. It was not heated sufficiently to allow the iron, rock, and water to separate by density into distinct layers like those within Europa. Even with the minimum mass for the circum-jovian nebula, with just enough material to build the Galilean satellites, accretion would have been so fast and the temperatures would have been so high that Callisto should have become fully differentiated. Robin Canup and Bill Ward, of the Southwest Research Institute in Boulder, Colorado, have addressed this problem by proposing that the satellites formed later, near the end of Jupiter's formation period, while dust and gas were flowing from the solar nebula into Jupiter through an "accretion disk" around the giant planet. They calculated that the satellites could form slowly within the environment of the accretion disk, and that the conditions would be consistent with the amount of internal differentiation now found in each of the satellites.

Most important for the history of the orbital resonance, the gravitational interaction of the growing satellites with the accretion disk would cause their orbits to migrate. Their semi-major axes would gradually change, because they would induce spiral density waves in the accretion disk (similar to the spiral waves in galaxies or in Saturn's rings) and their gravitational interaction with the waves would change the satellites' orbits.

This migration of orbits provided a viable mechanism for a primordial origin of the Laplace resonance, as demonstrated recently by Stan Peale and his postdoctoral associate Man Hoi Lee. Subsequent to the formation in resonance, tides on the satellites would modify the orbits, moving away from deep resonance, increasing ν to its current value, and decreasing the orbital eccentricities from higher early values, as I had described in my 1982 paper. It was personally gratifying that my conjecture was proving plausible. Even better was that Peale and Lee (2002) acknowledged that I had it right all along, which partially compensated for the 20 years of disrespect that I had experienced on that issue.

In another scenario that I proposed in the early 1980s, the system could episodically move into and out of deep resonance. In that scenario, a cycle starts with Io in deep resonance with a large forced eccentricity, and energy dissipation due to the diurnal tide results in partial melting, which raises its tidal amplitude and promotes even more dissipation (as in the current state). With the tides on Io dominant, the system is driven out of resonance, the eccentricity damps down, diurnal heating decreases, and the body refreezes. At that stage, tides on Jupiter dominate, driving the system back into deep resonance, where the next cycle repeats the process. Given Io's current rate of tidal dissipation, the duration of the current phase, in which the system is moving out from deep resonance and the orbital eccentricities are decreasing, would be $\sim 10^8$ years. In 1986 Greg Ojakangas (a graduate student at Caltech, who later worked with me as a postdoc) and his advisor David Stevenson considered this scenario from a geophysical perspective and concluded that the scenario was indeed viable.

In considering a variety of possible histories, I noted as another alternative that the system might simply have formed on the other side of deep resonance (i.e., with ν less than zero). Tidal evolution could then have driven it through the resonance to the other side ($\nu > 0$) where it is today. I was curious about how the satellites would behave as they passed through the deep resonance with ν passing through the value zero. Extrapolating from our understanding of the behavior of the resonance with the current value of ν, with the eccentricities inversely proportional to ν, the eccentricities would have become infinite, implying elongated orbits that would have crashed into Jupiter. That simple result did not seem plausible, so I investigated the dynamics of deep resonance. I discovered very interesting behaviors, in which as ν approached zero, the geometries of the conjunctions adopted various asymmetric configurations. For $\nu < 0.2°/\text{day}$, Europa's e would be >0.04 (much greater than its current value 0.01) and conjunction could be locked to longitudes oblique to the major axes of the orbits.

The history of the orbits of the Galilean satellites remains a matter of speculation. The fact that tides on the satellites are strong and must have affected the evolution of the orbits has opened up a range of possible scenarios. The possibility that the evolution of the resonance was affected by tides raised on other satellites, besides Io, remains to be explored in depth.

Assuming that Io tides are dominant, unless some mechanism can be proposed to provide substantial tidal dissipation within Jupiter, or otherwise push Io outward from the planet, it seems the Galilean satellites are currently evolving from deeper resonance. The system could have formed in deep resonance, it could be involved in episodic evolution in and out of deep resonance, or it may have passed through deep resonance on its way to the current condition. Europa's eccentricity, along with that of the other satellites, would have been gradually decreasing toward its current value. Given the tidal dissipation apparent in Io from both thermal measurements and imaging of volcanism, significant change in orbits through modulation of the resonance appear to have taken place fairly quickly, in ~100 million years.

That timescale is strikingly similar to the probable 50-million-yr age of Europa's surface. At the time of formation of the oldest features currently visible on Europa, the satellite's orbital eccentricity would have been substantially larger. All the tidal effects that are important on Europa may have been even stronger, perhaps repeatedly so, in the past. The internal tidal heating would have been much greater, because energy dissipation goes as the square of the eccentricity according to Eq. (7.1). A natural question is whether the age of the surface has been determined by erasure that occurred during the deep resonance, or at least whether there is evidence for change in the geological record over the past 50 million years.

8.2 POLITICS TAKES CONTROL

In fact, the canonical *Galileo* reports claimed to have found such evidence. There were purported to be changes over time in the style of tectonic features, which were interpreted as implying a thickening of the icy crust, consistent with decreasing tidal heating. Moreover, the party line had chaotic terrain as the youngest type of surface on Europa, implying that formation of chaotic terrain was only a recent process (i.e., late in the 50-million-yr age of the surface). The interpretation of that purported observation supported my idea of evolution from deep resonance by invoking a notion that chaotic terrain is formed by solid-state convection. The story was that the ice was too thin in the past to allow convection, only relatively recently becoming thick enough for convection and subsequent chaos formation. Again, geological evidence seemed to support a decrease in tidal heating, consistent with evolution of the Galilean satellites out from deep resonance during the short geological history of Europa.

As the leading advocate for orbital evolution from deep resonance, and having taken considerable flak for that position until Peale and Lee came around, I would have found it very gratifying if the geological record had provided supporting evidence for such a change. Unfortunately, as the party line was being developed, none of these geological arguments made any sense to me. First, the geological time sequences seemed to be based on *ad hoc* interpretations of the images. Mapping exercises that purported to show changes in tectonic style or thermal processing were ambiguous and subjective.

The argument based on chaotic terrain made the least sense to me. As I and my students studied the images, we saw no evidence that chaotic terrain was especially recent. To be sure, Conamara was a very fresh feature, but it was selected as a target for detailed imaging precisely because it was so fresh and obvious. Yet, even at such a fresh feature, we could see the cracks and ridges that were already beginning to cross it (Figure 2.10). As we surveyed the surface, we could see that over much of the surface, buried under densely-packed strands of newer ridges, lay very old areas of chaos. The putative freshness of most chaotic terrain was an observational selection effect: It is simply easier to spot the more recent chaos.

Moreover, even if chaotic terrain formation were only a recent process, the argument that it implied thickening of the ice over time seemed completely backwards. The appearance of chaotic terrain seemed to argue for melt-through exposure of the ocean. This interpretation would suggest that the ice had gotten thinner with time, not thicker.

In fact, even before I had gotten into the fray, there was a battle among the Imaging Team geologists over this point. The imaging of Conamara Chaos, with its remarkable appearance, was a major early accomplishment of the Imaging Team. The veteran planetary geologist Mike Carr of the U.S. Geological Survey was the team member assigned to lead the preparation of a paper for the influential and prestigious journal *Nature* reporting on this discovery. Along with most people who looked at those images of Conamara, Carr's interpretation, as described in early drafts of the *Nature* paper, was that we were seeing a site of a relatively recent melt-through event, with oceanic exposure. At the same time, however, Bob Pappalardo and his mentor Jim Head had begun to promote their view that everything on Europa involved thick ice and solid-state processes. The keystone of their case was the putative existence of a class of features they called "pits, spots, and domes", although that taxonomy has proven to be premature and its generalizations incorrect (see Chapter 16, and the detailed discussion in Chapter 19). With the backing of his mentors, Pappalardo was assigned to lead the preparation of an Imaging Team paper for the same issue of *Nature* as Carr's, reporting on this supposed evidence for thick ice, in direct contradiction to the interpretation of Conamara. Having become the spokesperson for the Imaging Team regarding Europa, Pappalardo had a platform of authority from which to promote his own views. Considerable pressure was brought to bear during the preparation of the Carr et al. paper, so that, while portions of it reflect the interpretation of oceanic exposure, the paper finally favors a solid-state formation process for chaotic terrain. From that time forward, thick ice became the obligatory canonical model.[2]

Pappalardo expressed surprise that I did not jump on the thick-ice bandwagon. The story that chaotic terrain was a sign of thick ice and that it only formed recently seemed to support my old idea, based on orbital evolution, that there would have been more tidal heating in the past. If it were true, my predictions based on celestial mechanics would have been proven correct. Unfortunately, I saw no evidence that

---

[2] Cynthia Phillips, then a graduate student at the University of Arizona, and now at the SETI Institute, had prepared a beautiful mosaic image of Europa, which was expected to be on the cover of that issue of *Nature*. She was devastated when the issue came out with a comic cartoon of a giant pig on the cover. Cynthia's image processing at the University of Arizona's Planetary Imaging Research Laboratory (PIRL) contributed to many of the most revealing versions of *Galileo* images, including images used in this book.

chaos was an especially recent phenomenon. That impression was an observational selection effect: Old chaos is simply harder to identify.

The fact that there is no clear evidence for change in geological processes over the recorded geological history does not mean the orbits did not evolve. It may be that evidence for such change has been destroyed by more recent resurfacing, or perhaps much of what we see on the surface represents processing that slowed tens of millions of years ago. Equally plausible is that the processes recorded on the surface have been continuous over the surface history, and continue currently.

Studies of orbital dynamics are continuing. We need to understand how tides raised on Europa and Ganymede may have affected the evolution, especially if we take into account the accompanying changes in the geophysical states of the satellites, and how they feed back into orbital evolution. Such effects could be similar to the feedback from melting on Io that I hypothesized in my episodic heating model. We also need to consider evidence from direct measurements of the rates of change of orbital periods. We have a useful record of observations that goes back to Galileo's original 17th-century data on timing the orbits, and important continuing work by Kaare Aksnes (University of Oslo) and Fred Franklin (Harvard-Smithsonian Center for Astrophysics).

These orbital studies, along with detailed investigation of what is really seen on Europa, may eventually play key roles in elucidating the geological and geophysical history. We may also learn which way the system is evolving onward from the present. Europa has had an ocean and an active surface in the recent past, probably continuing up to the present. However, it may well freeze solid in the future, depending on how its orbit continues to evolve.

Part Three

Understanding Europa

9

Global cracking and non-synchronous rotation

9.1 LINEAMENTS FORMED BY CRACKING

The first direct connection between geological features on Europa and tides was made by Paul Helfenstein and Marc Parmentier, who identified sets of large-scale linear features on *Voyager* images that seemed to be consistent with orientations roughly perpendicular to the local direction of maximum tension that would have been generated by a small amount of non-synchronous rotation. These large-scale patterns can also be seen in the global Mercator projection shown in Figure 9.1. A number of those lineaments, prominent enough to warrant official IAU names, are marked on the tidal stress plots shown in Figure 6.1, where they can be seen to be indeed orthogonal to the tension.

The orthogonal correlation of the major lineaments with tension suggests that these features initiated as tensile cracks in response to non-synchronous rotation. In fact, for ice, mechanical failure would be expected to be tensile according to laboratory investigations of its strength. In the far north the shear stress is substantial. Shear stress is indicated by the fact that the principal stresses are nearly equal and opposite, and the shear is at a maximum along directions oriented 45° from the principal axes. Even there, ice is more likely to fail in tension.

The correlation of major global-scale lineaments with the stress field expected from a small amount of non-synchronous rotation implies that since the cracking occurred there has been very little non-synchronous rotation. This result means either that the non-synchronous rotation is very slow compared with the age of Europa's surface features, or that the lineaments are fairly recent compared with the rest of the surface. As we will see, the latter explanation is more likely correct, but we cannot be sure.

The most important implication of the general correlation of lineaments with tidal stress is that these lineaments represent cracks. This result is not surprising. Even a naive look at the globe of Europa (e.g., as shown in Figures 2.2–2.4) gives the

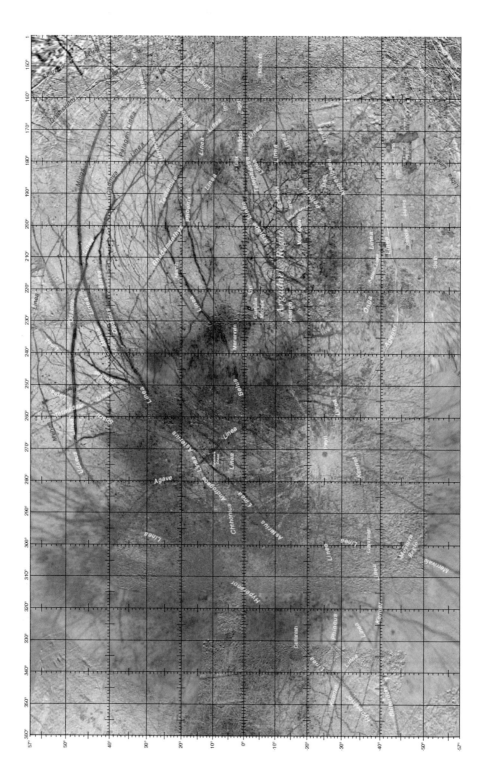

Figure 9.1. A map in a Mercator projection of Europa, constructed by the U.S. Geological Survey from a mosaic of mostly *Galileo* images (and some *Voyager*). For orientation relative to likely tidal stress fields, compare the major lineament systems with those marked on Figure 6.1. Two dark arcs near about 45°N are Udaeus and Minos Linea, which are (5) and (6) in Figure 6.1.[1]

---

impression that the lines crossing the surface are cracks. Yet most of the major linear features that define these large-scale patterns do not appear as cracks when we examine their morphology. In our tour of the surface (Chapter 2), we saw that they are usually "triple-bands", which on close inspection prove to be complexes of ridges with dark margins. The correlation with tensile stress confirms the naive impression that, at their heart, these features originated as cracks, and it places a key constraint on any models of how ridges and surface darkening develop.

9.2 THE TECTONIC RECORD OF NON-SYNCHRONOUS ROTATION

9.2.1 *Voyager* analysis

While the general trend of the major lineaments is consistent with the stress patterns predicted by a small amount of recent non-synchronous rotation (Figure 6.1), the agreement is not perfect. Consider the long dark lineaments in the northern hemisphere in the longitude range 160° to 260° in Figure 9.1. In this view, many of these lines (e.g., like Minos) follow curved paths that reach their northernmost extreme near about 200°W. This longitude is about 25° east of where the non-synchronous tensile stresses should have created such crack trajectories, according to Figure 6.1. The general impression that cracks are about 25° too far east relative to the expected non-synchronous stress was first noted by Alfred McEwen[2] in 1986, who quantified it by mapping numerous major global and regional lineaments seen on *Voyager* images and comparing their directions with predictions from tidal stress.

McEwen reported that most of the global-scale lineaments did fit the theoretical stress pattern that Helfenstein and Parmentier had computed for a small amount of non-synchronous rotation (the pattern in Figure 6.1), but only if the lineaments were shifted westward by 25°. The implication is that most of these cracks formed due to the stress accumulated during a modest amount of non-synchronous rotation, but that, after the cracking, the surface of Europa rotated another 25°. (Actually, McEwen suggested that the cracking may have been due to stress that accumulated during the 25° of non-synchronous rotation, but, as I discussed in Chapter 3, it is

---

[1] The names of lineaments are drawn from Homer's *Odyssey*, related to Europa in classical mythology, and other features are named after Celtic myths, heros, and place names, all as assigned by the IAU Nomenclature Committee.

[2] McEwen was then a graduate student at Arizona State University and also working at the U.S. Geological Survey's (USGS) Astrogeology Center in Flagstaff, Arizona. He became a member of the *Galileo* Imaging Team, and is currently my colleague on the Planetary Sciences faculty at the University of Arizona, where he directs the Planetary Image Research Laboratory, whose students and staff have produced some of the most spectacular and revealing versions of *Galileo* images.

unlikely that the ice could have resisted cracking before such great stress built up.) It is not clear why so much cracking would have happened in one short interval, with no cracking having occurred during the subsequent 25° of rotation when stress would have built up again. McEwen addressed that point by suggesting that the visible cracks may have formed over the course of several tens of degrees of non-synchronous rotation, continually relieving stress accumulated over short intervals during that time, and that only on average do the orientations fit the tidal stress pattern offset backwards by 25°.

In that case, the crack record may potentially be deciphered to determine the relative age of individual features, based on what longitude shift would put them in best agreement with the theoretical tidal stress. Indeed, individual features do not all fit the 25° offset pattern. For example, the orientation of Libya Linea is nearly perpendicular to maximum tension in the current non-synchronous stress pattern as shown in Figure 3.7. Thus, it fits recent non-synchronous stress better than it would have fit in the past (i.e., better than if its position were further to the west when the crack first formed). Similarly, the orientation of Thynia Linea (Figure 6.1) along most of its length fits more recent non-synchronous stress better than it would have fit the non-synchronous pattern in the past. These observations proved to be just the first insights into how the crack patterns could be exploited to determine the dynamical history of the satellite.

9.2.2 Orbit G1 and the Udaeus–Minos region

The first systematic evidence for a time sequence of lineament formation that could be correlated with non-synchronous rotation resulted from Paul Geissler's expertise at multi-spectral remote sensing, combined with his ability to link interesting observations with fundamental theory. During each orbit around Jupiter, the *Galileo* spacecraft generally spent a good fraction of its time farther out than the orbits of the Galilean satellites, and then swooped into the neighborhood of the satellites' orbits. Generally, the geometry of the satellites' positions precluded flying close to more than one satellite during each orbit. The swoops came to be labeled not only with the number of the orbit, starting with the first one after the spacecraft was inserted into orbit around Jupiter, but also with a letter identifying which satellite was encountered on that orbit. The first orbit was called G1 because it involved a close targeted encounter with Ganymede. In a sense the letter is redundant information. Only one satellite was targeted on that orbit. Yet, the letter serves as a reminder of which satellite was favored at that time. With orbit G1 targeted on Ganymede, *Galileo* did not get close enough to Europa to get high-resolution images, but it did get a set of images at a resolution of about 1.6 km per pixel covering a wide region around the Udaeus–Minos intersection, spanning longitudes from 180° westward to 250°, and latitudes from 20° to 75°N. Figure 2.5 shows one of them. The same region was imaged several times using the same black-and-white camera, but each time through a different filter.

These were just the kinds of data that Paul Geissler loves. He obsessively combined the images from different filters to produce various color products.

Usually a night worker, Paul was so excited that he was up all day perfecting these beautiful images. While the products were colorful, they did not show what the human eye would have seen. Europa in reality would appear white and bland. To show the information in the image, Paul had to exaggerate the color and the contrast. Even if the subject had been colorful, the camera was not sensitive to the same range of wavelengths of light as the human eye, and the filters did not correspond to the usual sets of colors (such as red, green, and blue on a television screen) that can be combined to approximate the way things look to the eye. For example, what is displayed on a monitor as a combination of red, green, and blue, may actually represent the appearance in green, red, and the near-infrared, respectively.

So, Paul's' images, in various versions, combining filters in various ways, showed Europa's ice in various shades of pink and blue, with the dark lines bordering the triple bands in shades of orangey-brown. When I first hired Paul to help with one of *Galileo*'s quick encounters with the Earth on its way to Jupiter, he had done the same thing with images of Antarctica that were taken from the spacecraft with the same camera used later to image Europa. For Antarctica, he had discovered that in his false color images the pinks and blues could distinguish between the sizes of the grains of ice in snow and clouds. For Europa, Paul could tell us how the ice grain sizes varied from place to place.

Those same color images showed something that got my whole research group excited. Paul noticed that, in color, it was relatively easy to distinguish which lineaments appeared to cross over on top of other ones. These cross-cutting relations reveal the order in which the linear features were created. It is reasonable to assume that the older line represents older cracking. In principle, that assumption could be wrong. A younger crack might have sprouted ridges and dark lines before an older crack did. But we invoked Ockham's razor and made the simple assumption: Older lines mean older cracks.

Then, Paul noticed some patterns. The sequence of lineament ages correlated to some extent with evolution in color. The dark margins of the "triple bands" in Figure 2.5 seemed to undergo aging that gradually returns them toward the coloration of the surrounding terrain. Such observations allowed age sequencing based on color even where detailed cross-cutting information was lacking, and there was a systematic variation in their character. The oldest lineaments appear fairly bright and white in Figure 2.5, as end members of a fading sequence that Paul had detected. The intermediate-aged ones were the triple bands with their dark margins. And the youngest were faint thin lines, which we interpreted as cracks that had not yet had time to develop ridges or dark margins.

The most striking pattern was an orderly sequence of azimuth directions that changed systematically with age. To demonstrate this sequence, Paul produced a set of maps that showed separately the oldest, the intermediate-aged, and the youngest lineaments (Figure 9.2). It was evident that the most recent lineaments trend roughly northwest to southeast, older ones run more nearly east–west, and the oldest run southwest to northeast. Going forward in time, the orientations had systematically rotated clockwise. Moreover, the trend was followed even among the intermediate-aged "triple-bands" (second panel in Figure 9.2). Examining the intersection of the

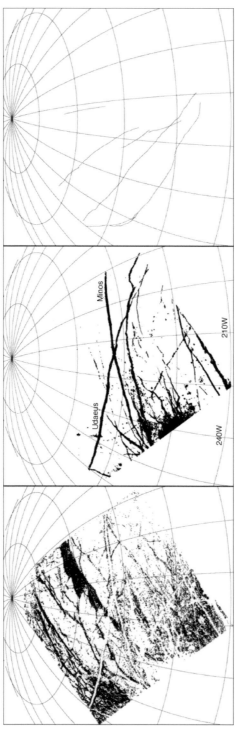

Figure 9.2. Three sets of lineaments identified by Paul Geissler as having formed in a time sequence of a single region, from left to right, based on cross-cutting relationships. More recent lines cross over older ones. The oldest are on the left, intermediate aged ones in the center, and youngest at the right. As time advanced, the orientation (in azimuth) seems to advance systematically in a clockwise sense. Even among the medium-aged ones, Udaeus is younger than Minos, confirming the clockwise trend (cf. Figure 2.5). Lines of latitude and longitude are spaced 15° apart.

triple-bands of Udaeus and Minos near the center of the region, we can see that Udaeus crosses over Minos (see Figure 2.5), exemplifying the general trend: The younger feature is oriented further clockwise than the older one.

Because the broad global patterns of lineaments had suggested that they represent cracking due to tidal stress, the changing orientation of lineaments in the Udaeus–Minos region suggested a systematic clockwise rotation of the direction of tension over time.

At one of our weekly research group meetings, where my students and post-doctoral associates discuss our progress and plans, Paul insisted that we needed to compare this clockwise trend with the tidal stress plots that Greg Hoppa had been computing. It was a very good idea. We sketched the positions of Udaeus and Minos on an early version of the non-synchronous stress plot, just as they are shown in Figure 6.1. A remarkable pattern jumped out from this comparison: If the ground in the G1 images had moved from the west to the east over the last few tens of degrees in longitude to its current location, it crossed a part of the stress diagram where the direction of the local tidal tension would have rotated in the clockwise sense, just as the observed record of lineament orientations had changed. The tension would have changed over a range of directions orthogonal to the directions of the lineaments Paul had mapped.

This discovery provided compelling evidence for non-synchronous rotation. Apparently, in the past, the Udaeus–Minos region had been further west, when the non-synchronous tidal stress caused the older cracks to form. Cracking would have relieved the stress. Later, this region, along with the entire surface, moved eastward with Europa's non-synchronous rotation, relative to the direction of Jupiter. On Earth, even as the planet rotates each day, the longitude of any given piece of ground remains at a fixed longitude. New York tends to remain at longitude 74°W all day long. But on Europa, longitude is defined relative to the direction of Jupiter, so non-synchronous rotation changes the longitude. As Europa rotated non-synchronously, the real estate around Udaeus and Minos moved eastward relative to the Jupiter-driven stress field, experiencing ever-changing tidal distortion.

Immediately after the earlier cracking, continuing non-synchronous rotation would probably not have built up stress. Even as the geometry of the shell was distorted by the change in the orientation of the primary tide, as long as the cracks remained open and active, they could minimize accumulation of elastic stress. But apparently after 10–20° or so of non-synchronous rotation, the earlier cracks were ineffective. Evidently, the ground had moved far enough to the east that the accumulating strain in the region was no longer oriented so as to work the existing cracks. These cracks may even have begun to anneal by refreezing once the geometry of the tidal working had changed. At that point, continuing non-synchronous rotation could once again build up the stress field depicted in Figure 6.1, leading to a new set of cracks in the Udaeus–Minos region. But, by then Udaeus–Minos was further to the east than it had been. According to the stress field (Figure 6.1), the new cracks would be oriented further clockwise than the earlier ones. As this process repeated itself, the observed record of cracks developed, with its sequence of clockwise change in orientations.

This discovery made for a very exciting debate in my lab. Paul and my students Greg Hoppa and Randy Tufts overcame every objection I could raise and convinced me that we had good systematic evidence for non-synchronous rotation. The result was quickly accepted within the *Galileo* Imaging Team, and then generally across the planetary science community when we described it in a couple of papers for which Paul was the lead author. It was so compelling because the data showed several sets of lineaments that were progressively and nearly continuously more clockwise with decreasing age, over a few tens of degrees of rotation. The appeal of the interpretation lay in its elegant linkage of several lines of quantitative research. It confirmed my own much earlier predictions, based on tidal theory, that Europa might be rotating non-synchronously; the observational evidence involved beautiful images that Paul produced using state-of-the-art image processing and quantitative multispectral analysis; and it involved discovery of unexpected systematic trends that fit Greg's computed tidal stress field.

I had reservations about this model though. They came from inspection of the positions of Udaeus and Minos on the stress field (Figure 6.1). Remember, these lineaments were both near the middle of the age sequence discovered by Paul Geissler, with Udaeus crossing over, and thus younger than Minos. It is clear that Minos can be shifted back to the west by about 20–25°, and it will fit the tension field perfectly, with its curvature on the Mercator projection keeping it perpendicular to the tension. (Actually, Udaeus and Minos track nearly along great circles on the sphere, and the apparent curvature is an artifact of the Mercator projection, but comparison with the stress field is independent of the projection.) The required westward shift for Minos, in order to fit the stress, is an example of what McEwen had noticed in the mid-1980s.

My concern started with consideration of Udaeus. Because it formed more recently than Minos, our scenario would require that it formed east of the position at which Minos formed. That requirement is met, but too much so. To fit the stress field, it really needs to be at least 10° to the east of its current position. If we accept the scenario developed in our Geissler et al. papers, Udaeus would appear to have formed in the future, after Europa rotated ahead of its current position. Something was wrong.

The same problem comes up if we consider the azimuths of the lineaments plotted by Paul Geissler in the Udaeus–Minos region. The orientation of the oldest lineaments (azimuth 30° N of E) is consistent with the direction of tension in the non-synchronous stress field (Figure 6.1) at a location backwards in rotation by about 40° (i.e., at a longitude 40° to the west of the current position relative to Jupiter). Slightly to the west of the current position, the tension is maximum in the north–south direction, consistent with the generally east–west orientation of the middle-aged lineaments. However, this scenario fails to explain the orientation of the most recent cracks, whose orientation would not be produced unless Europa were to continue to rotate somewhat further ahead from its current position.

One solution would be to abandon the idea that the crack record showed rotation during only the past few tens of degrees of rotation. We could invoke the periodicity of the stress diagram, noting that it repeats every 180° of longitude. For example, we could attribute the orientation of the oldest cracks in Paul's sequence to formation $180° + 40°$ to the west, rather than $40°$ west. Then, the most recent ones could have formed $180° - 20°$ to the west, rather than invoking the absurd $20°$ of rotation into the future. However, that explanation would raise the question of why no more recent cracks formed during the past $160°$ or so of rotation.

It was preferable to find an explanation that would preserve the elegance of the correlation of the continuous crack azimuth changes with the changes in stress over the past few tens of degrees of rotation. The dilemma could be resolved if we could find a good reason that the stress pattern would have been shifted further west by $10°$ or more. We considered several possibilities. Building up more than a few degrees of non-synchronous stress would have exactly the opposite effect from what we needed (shifting the pattern eastward), so it would not provide a solution. Moreover, it was not very plausible because the stresses involved would have been so great that the crust would have cracked too soon, relieving the stress.

We turned to consideration of diurnal stress. The stress pattern near apocenter (Figure 6.2 with the signs of stress reversed) would resolve the problem, because the pattern is similar to non-synchronous rotation, but shifted about $30°$ west. But why would the surface crack at that particular point in an orbit? The magnitude of stress is fairly low then, and tension in the region would have been much greater earlier in the orbit.

Instead, consider a combination of the two sources of tidal stress: non-synchronous and diurnal. As non-synchronous stress builds up during the many years that it takes to rotate by about a degree, the surface also undergoes much more rapid changes in periodic diurnal stress. There is a gradual, monotonic increase in the strength of the non-synchronous stress field (Figure 6.1), as the daily oscillations are superimposed on it. Eventually, one day, the maximum diurnal stress added on top of the building non-synchronous stress would exceed the strength of the surface material and cracking occurs.

To explore this scenario, we examined the combined diurnal and non-synchronous stress fields. We considered the non-synchronous stress corresponding to no more than about $1°$ of rotation; otherwise, the non-synchronous stress would dominate and not give the needed shift of the pattern to the west; also, we know that the ice probably is not strong enough to support the stresses that would build up during more rotation. One degree of non-synchronous rotation gives stress comparable in magnitude to the diurnal (Figure 6.3). In fact, at 1/4 orbit after pericenter, the diurnal stress nearly cancels out the non-synchronous (Figure 6.3c). That result should have been predictable because the diurnal tidal distortion at that point in the orbit (Figure 4.2) is the same as for rotational displacement. At apocenter, Figure 6.3a shows that the combined stress field gives substantial north–south tension ($>10^5$ Pa) and with a pattern such that the maximum tension is at $65°$ and $245°$ (instead of at $45°$ and $225°$ as in Figure 6.1), giving just the necessary westward shift that our scenario required. However, the maximum tension is still not reached at this point in the orbit.

The maximum tension in the region of interest is reached 1/8 orbit after apocenter (shown in Figure 6.3f) and remains at about that level for the next 1/8 of the orbit. Here we find tension $>1.5 \times 10^5$ Pa and a westward shift of the pattern relative to pure non-synchronous rotation (Figure 6.1) of about $10°$. This stress magnitude is close to a plausible value of the tensile strength of the ice crust, based on estimates of the strength of sea ice (as low as 3×10^5 Pa according to Mellor, 1986) and on scaling to the thicker crust of Europa which would reduce strength further. The shift of the stress pattern to the west is more consistent with the scenario in which the Udaeus–Minos intersection region has moved across the stress field from west to east. It avoids the problem that, with non-synchronous stress alone, the most recent cracks would have to have formed either nearly 1/2 of a rotation period ago, or some time in the future.

To summarize, our model has cracks forming in response to the maximum tension during diurnal tidal variations, during the orbit when the combined diurnal and accumulated non-synchronous stresses exceed the strength of the crust. In the region under consideration, as well as over much of Europa, the cracks associated with large regional- to global-scale lineaments formed orthogonal to the tension. When non-synchronous rotation carried the region to a place where the stress field was sufficiently differently-oriented, new sets of cracks formed. Accordingly, the orientations and relative ages of cracks in the Udaeus–Minos intersection region fit the following sequence of initial cracking relative to the rotational orientation: The oldest cracks mapped by Paul Geissler (Figure 9.2) fit the theoretical tidal stress pattern (Figure 6.3) if they were formed at a time $60°$ backwards in rotation; Minos formed about $40°$ ago; Udaeus appears to have formed relatively recently ($<10°$ ago) because in its current position it fits the stress field in Figure 6.3 reasonably well; the freshest cracks in this region, which run more nearly north–south (farther clockwise) than Udaeus would not have formed until Europa reached its current rotational position. At an earlier time, the tension would have been less nearly-perpendicular to the direction of these lineaments.

9.3 HOW FAST DOES EUROPA ROTATE?

This rotational displacement occurred recently in the age of the current surface, because the features that were prominent enough to be mapped are among the youngest tectonic lineaments visible on the surface. Paul's sequence of changes in color as lineaments age showed that as they get older they blend back into the background. The background, as it appeared on the G1 images at 1.6 pixels/km, evidently consists of older features that make up most of the current surface. The features involved in Paul's analysis must have formed during a tiny percentage of the age (<50 million years according to the paucity of craters) of the surface, probably well within the past million years. This result suggests that Europa's non-synchronous rotation period (the period relative to the direction of Jupiter) must be less than a few million years.

Non-synchronous rotation certainly could not be very fast. Ridge building and border coloration must have occurred after cracking and within $10°$ of rotation; otherwise, the next-formed lineaments would not cross over these structures and markings. Such geological processes must take time, so the rotation occurs on a geological timescale, at least a Europan one. Slow rotation is also consistent with our predictions of non-synchronous rotation. Tidal torque would act on the rotation if it were synchronous, but, once the rotation was only slightly faster, the torque would go to zero. We have no way to know the values of the relevant geophysical parameters, so we cannot make a precise theoretical prediction of the steady-state rotation rate; but, if it is not synchronous, we would expect it to be only slightly so.

So far we have constrained the rotation period to less than a few million years and long enough for ridges to grow. Can we be more precise than that? Something that I had in mind from the beginning of the *Galileo* project was to try to measure the satellites' rotation from direct observation. In principle, if we take a picture of a satellite at two different times, we should be able to see if it rotates. In practice, there are problems. Depending on illumination, viewing angle conditions, and resolution the exact position of a reference feature on the surface may be difficult to define. The exact position is well-defined in a high-resolution image, but it is useless, because in such a close-up view there is nothing to use as a reference for the exact location.

Shortly after Greg Hoppa asked to work with me as a PhD student, I suggested that we try to solve this problem. We decided that the best strategy was to compare *Voyager* and *Galileo* images, because they would give us the longest time between observations, and thus the best chance to see motion. We also decided to use the terminator (the day/night boundary) as a position reference, because its direction in space was well-defined at any time by the precisely known position of Europa relative to the Sun. With Paul Geissler's help, Greg found moderate-resolution images of a particular region taken by both spacecraft 18 years apart, each of which showed the position of the terminator at the time the picture was taken (Figure 9.3).

Since we knew the terminator position at the time of the *Voyager* image, and given the exact time interval between the two images, we could predict where the terminator should have been on the *Galileo* image if it had been rotating synchronously. To our surprise, the actual *Galileo* terminator was about 1° east of the prediction, as marked on Figure 9.3. This result implied that Europa had rotated slightly slower than synchronous during those years, contrary to the tidal theory. (Actually, to fit this observation, Europa could have been rotating faster than synchronous at a rate that would have given one or more extra rotations, but that would imply a great deviation from synchroneity that, if correct, would have been obvious from other images, or even from Earth-based observations.)

As a theorist, I would have been happy to start dreaming up strange and imaginative reasons that it made perfect sense for Europa to rotate slower than synchronous. But, before that, we needed to check whether the observations were really telling us what we thought. I pointed out to Greg that the exact position of a terminator could be shifted by topography. Ridges and rafts could cast long shadows near the terminator that would extend the dark, night region slightly into the day hemisphere. Greg created some simulations that quantified the effect and shifted the terminators westward (see Figure 9.3).

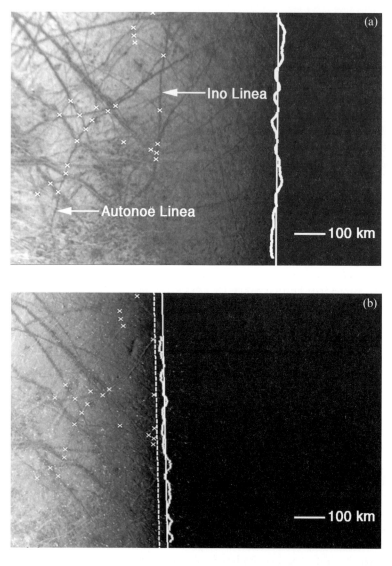

Figure 9.3. The images from *Voyager* in 1979 (a) and *Galileo* in 1996 (b) selected by Hoppa et al. because they showed the same region with the terminator (day/night line) nearby. The images are reprojected to show the same viewing geometry. The wiggly lines show the edge of the dark side, with the corresponding straight line showing a straight line fit to the curve. Matching reference points were marked with an ×. With synchronous rotation, based on the position of the terminator in the *Voyager* image, the terminator should have been at the dotted line in the *Galileo* image, but it is not. The discrepancy can be explained by the fact that only the northern part of the *Voyager* terminator lies in "bright plains" (where the wiggly terminator line shifts eastward), while nearly all of the *Galileo* terminator does. This region lies between 0° and 30°N with the west edge near 172°W longitude (cf. features recognizable in Figure 9.1).

Then he made one of those breakthrough observations that show when a student has become a real scientist. He considered a geological map, which had been prepared at the USGS by Larry Soderblom and Baerbel Lucchitta, based on *Voyager* images. The map showed that the location of the *Galileo* terminator crossed "bright plains" terrain, using their *Voyager* issue terminology, also evident in Figure 9.1. The *Voyager* terminator crossed the surface farther east, where the USGS map (and Figure 9.1) indicated predominantly "brown mottled terrain". With much more data in hand, we now recognize the so-called "bright plains" as predominately densely-ridged tectonic terrain and the "mottled terrain" as chaotic terrain. Greg Hoppa reasoned that the topography of these terrains was so distinct that they might produce differing amounts of shifting of the terminator. Comparing the terminator positions that ran through different kinds of terrain was like comparing apples and oranges, or, more precisely, like comparing chaos and tectonics.

However, the northernmost part of the terminator in the *Voyager* image did pass through the "bright plains' area on the USGS map. There, the *Voyager* terminator was a bit farther east than where it passed across chaotic terrain. As Greg Hoppa noted, that shift was enough to put it in agreement with the *Galileo* image's terminator, consistent with synchronous rotation. We had been hoping to measure the rate of non-synchronous rotation, and, while it was disappointing not to have it, at least we were somewhat relieved that we were no longer finding non-synchronous rotation in an implausible direction.

Paul Geissler pointed out that all was not lost if we took into account the uncertainty in Greg's measurements. No measurement is exact, and Greg had calculated the uncertainty in the measured rotation, mostly due to the wiggliness in the terminator line, to be about $0.5°$. In other words, Europa might have rotated as much as an extra half-degree in 17 years, relative to synchronous rotation. At that rate, the period for one full rotation, relative to the direction of Jupiter, would be about 15,000 years. Our result did not measure or even detect the non-synchronous rotation, but it told us that it had to be slower than one rotation per 12,000 years.

Earlier results from photometric observations of the variation in light from the full disk of Europa, as viewed from Earth, had shown that one hemisphere was slightly darker on average than the other, and that the dark hemisphere had remained on the trailing side for over 50 years, implying slow rotation, with a period $>1,000$ years. The Hoppa et al. result showed that the rotation must be even slower, by a factor of 12, but it still admitted the possibility that Europa rotated synchronously.

Another line of evidence independent of tidal tectonic theory gave an upper limit to the period. Because Europa moves in its orbit quickly around Jupiter, we would expect more and larger impact craters on its leading hemisphere as it sweeps through space. This effect had been quantified and predicted by Gene Shoemaker and Ruth Wolfe of the USGS in 1982. Even with the small number of craters on Europa, the asymmetry should have been marked. Yet no such asymmetry has been observed. It appears that during the history of the surface, Europa rotated significantly, so that no one side was always leading. The uniformity of craters tells us that the rotation period must be less than about 10 million years.

However, as we have seen, the evidence for non-synchronous rotation from the tectonic record of crack orientations, as developed by Paul Geissler, also gave an upper limit to the period. Because that record accumulated during the most recent few percent of the age of the surface, we infer that the rotation must have a period shorter than a few million years, at least as strong a constraint as the limit from the crater distribution.

9.4 LARGE-SCALE TECTONIC PATTERNS—SUMMARY

Up to this point in our research we had concentrated on the large-scale global and regional tectonic patterns marked by the dark lines (mostly the double dark lines known as "triple bands") that could be seen on full-disk images. Several very important conclusions had been reached and were generally accepted not only by the people thinking about Europa, but even by those who were pontificating about it.

First, lineaments on Europa seemed to be associated with tensile stress, so they were reasonably presumed to be based on cracks. Their locations, shapes, and orientations were all initially determined by the formation of cracks which eventually developed the observable markings that make them evident even in the lower resolution, full-disk images. The observation that the most recent lineaments in the sequence mapped by Geissler et al. were less developed and more thin and crack-like supports this picture.

Second, most lineaments have ridges, from which we can infer that ridges form from cracks. The fact that ridges generally come in pairs seems consistent with such an origin. In some way, the ridges must form along both sides of a crack.

Third, both the orientation sequence of the large-scale lineaments in the Udaeus–Minos region[3] and the crater distribution suggested that the rotation period (relative to the direction of Jupiter) is less than about 10^7 years, while the *Voyager/Galileo* comparison showed that the period must be at least 10^4 years. Ridges must have been able to form before they rotated to a significantly different location (or else the local diurnal stress regime would change), which implies that they grew in less than 300 to 300,000 years, depending on the rotation period. Ridge formation has not only been one of the dominant resurfacing processes, but it has the potential for providing a constraint on the rate of non-synchronous rotation.

---

[3] In Chapter 15, I describe why we modified this interpretation of the Udaeus–Minos sequence as we developed more evidence regarding the tectonic record.

10

Building ridges

Ridges appear in nearly every image of Europa's surface. Even within chaotic terrain, rafts often display portions of earlier ridged terrain that predated the chaos (e.g., Figure 2.10). The densely-ridged terrain in Figure 2.9 shows excellent examples of simple double-ridges. But, of the thousands of Europan ridges that I have surveyed, my favorite is shown in Figure 10.1. Just as road cuts give terrestrial geologists windows on what lies below the surface, this slice shows us what lies within and below a double-ridge, imaged at very high resolution.

How could such structures have formed? *Voyager* had shown mobile plates of surface ice, and post-*Voyager* theorizing had shown that these plates might have been moving over a liquid ocean. As *Galileo* approached Jupiter, we began thinking that the surface of Europa might more resemble the Arctic ice cap than anything else on Earth. Ron Greeley (Arizona State University) had engaged an expert on the physical structure of Arctic ice to advise on what we might find. The consultant, Max Coon of Northwest Research Associates, briefed the Imaging Team on Arctic ice processes and structures in anticipation of what we might see on Europa. Coon pointed out that ridges in the Arctic can form as large plates of ice press together, suggesting that similar processes might be at work on Europa.

Diurnal tides, combined with that Arctic analog, suggest a way to systematically build ridges. Consider a new crack, formed, for example, when the non-synchronous stress had built up enough that, together with the maximum diurnal stress at its daily peak, the total had exceeded the strength of the ice. Even after the stress had been relieved by cracking, the body of Europa would continue to undergo its daily distortion. As long as the crack did not anneal by refreezing, it would open and close on a daily basis. Then, the diurnal tensile strain would be accommodated by periodic opening of the cracks, rather than stretching of the elastic layer of ice.

Figure 10.1a. A double-ridge, curving down from top center, rides over other ridges coming from the upper left, and is sliced off revealing its interior, much like the way a road cut on Earth reveals the strata and materials below the surface. This is part of the highest resolution image of Europa ever taken (resolution is 6 m/pixel). The area shown here is about 2.4 km across in an oblique view, much like looking out the window of an airplane. The terrain in the foreground is probably chaos, and its formation was responsible for cutting off the ridges and revealing their interiors. The whole image is shown in Figure 10.1b.

Ridge building, as I envisioned it, occurs over many small daily cycles (Figure 10.2):

(a) When the daily crack opening occurs, liquid water from the ocean below rushes up to fill the gap. The ice crust is like a giant raft floating on the ocean, so the rising water can go as far up as the float line. This level is defined by the buoyancy of the ice, and it is about 90% of the way up from the ocean, because ice is about 10% less dense than the liquid water in which it floats. Ice-skaters on northern lakes are familiar with the effect: Water flows up to the float line, but not to the surface, so that when it freezes it leaves a crack deep enough to catch a skate blade. Similarly, in crack openings in the Arctic (called "leads"), new ice forms at the surface of the exposed water. On Europa, the

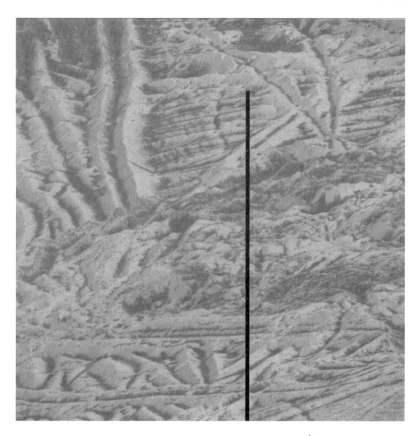

Figure 10.1b. The highest resolution image ever taken by *Galileo*,[1] at 6 m/pixel, shown in its entirety. Taken during orbit E12, it was part of a sequence designed to scan a portion of "mottled terrain" (i.e., chaos), southeast of Manannán crater near 13°S, 235°W. In fact, the sequence mostly got tectonic terrain, specifically lots of double ridges. The view is oblique (not straight down) as if looking sideways at the ground from an airplane window. In this frame, densely-ridged terrain is seen in the foreground (bottom 1/4 of picture) and background (top 1/3 of the picture), with chaos in-between. The chaos has cut off some ridges, giving the road cut-style cross-section enlarged in Figure 10.1a.

---

[1] A tabloid newspaper might call this spectacular image "The Picture That The U.S. Government Did Not Want You To See." There is plenty of fuel for paranoia: The press-release version of this image shows only the part to the right of the black strip of missing data in Figure 10.1b, well away from the road cut area; the catalog of images released on behalf of the Imaging Team does not include this image sequence in its diagrams of the "footprints" on Europa's surface of *Galileo* images. All *Galileo* images are available to the public in principle, but such access is not useful if you do not know that an image exists. People rely on public releases and image catalogs. It does seem strange that information about this image, and especially its left-hand side, has been especially hard to find. The power-brokers of the *Galileo* Imaging Team, who did control public information releases and catalog preparation, might prefer that this image remain obscure: Its detail supports the interpretation of ridge formation described in this chapter. Is the U.S. Government really hiding this picture? In fact, the government could hardly care less, and is probably too uncoordinated to hide it in any case. The real reason Figure 10.1a was cut off from the rest of the highest resolution image (and generally forgotten) was that it was downloaded separately from the spacecraft's tape recorder and assigned a separate image catalog number.

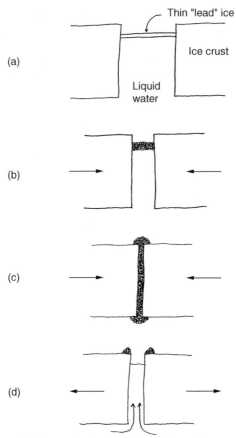

Figure 10.2. A model for ridge formation by diurnal working of a pre-existing crack. (a) After the crack opens and fills with liquid water, ice quickly forms at the fresh surface. (b) The crack then closes, forming a narrowing, but vertically thickening, raft of crushed ice. (c) Some of the slurry is squeezed onto the surface as the crack closes. (d) The crack reopens, leaving a double-ridge of surface deposit, and liquid water enters renewing the cycle.

freezing process must be very dynamic. The freezing would be accompanied by boiling in the near vacuum, so the lead ice might well be very weak, porous, and irregular. Ray Reynolds, a NASA geophysicist who, along with Stan Peale, had predicted tidal heating in the Galilean satellites, estimated during the early post-*Voyager* years that such newly-exposed water would freeze at the surface within a few hours, to a thickness of about one-half meter. After that, the freezing is somewhat slower, limited by the rate of conduction of heat from the bottom of the new lead ice out to the surface.

(b) After a few hours, as tidal stress reverses, the crack begins to close. Lead ice is fragmented in the press. As the ice is crushed, some new liquid is exposed, so additional freezing and boiling goes on, increasing the amount of ice near the top of the crack. The thickness of the raft is enhanced by the continuing freezing of newly-exposed water. Even more significantly, the raft of buoyant crushed ice becomes thicker inversely with the closing width of the crack. It must become very thick as the crack closes.

(c) As the crack approaches closure, the top of the now-thick raft of crushed ice (perhaps as a slurry with boiling interstitial liquid) is squeezed onto the surface, building a ridge. Simultaneously, even more material would be squeezed down below the ice, but for the moment we assume that it remelts quickly in the relatively warm water.

(d) In the next diurnal cycle, the crack reopens, with some debris falling back into the crack, leaving ice debris in ridges at the surface along both sides of the crack, as the process begins to repeat itself.

We can roughly quantify this model using the known scale of diurnal tides. The diurnal part of the tidal elongation of Europa is about 60 m (remember the height of the tidal bulge at each end is about 30 m) and this stretching is accommodated over the 3,100-km diameter of Europa, so the strain of the surface (its fractional distortion) must be typically 2 parts in 100,000, or ~2 m per 100 km.

A key point is that tides define the daily strain, so the value 2 m/100 km does not depend on the elastic parameters or the structure of the ice. In fact, the dependence goes just the other way: The amount of elastic stress that this tidal distortion exerts on the lithosphere depends on the elastic parameters of the ice. We have already seen that, for typical elastic parameters for ice, the diurnal tidal stress is at most a bit over 10^5 Pa, which follows from the strain estimated above, given the Young's modulus E of ice, which is about 10^{10} Pa (stress = $E \times$ strain). If the ice is cracked, the tidal stress must be less. But, even if the ice is cracked, the diurnal tide forces the surface to strain, to accommodate tidal distortion. This strain will open and close the cracks over the course of a day.

How much each crack opens and closes would depend on crack spacing. Linear ridge structures are densely packed over much of Europa, their cross-cutting and evolution of colors (cf. Chapter 9) shows that they represent a wide range of ages spanning the age of the surface, with the freshest ridges tens to hundreds of kilometers apart. If we assume the spacing between active cracks at any one time is typically about 100 km, the open cracks would be a couple of meters wide. Given the thickness of expected lead ice, it follows that the cross-section of crushed ice produced in each crack in each diurnal cycle would be at least $1\,\text{m}^2$ (i.e., about $1\,\text{m}^3$ per linear meter along the crack) per 85-h Europan day, which is about 0.01 Earth years. If only 10% of this material is squeezed out onto the surface (most is squeezed back down into the sub-crustal ocean), enough volume would be cumulatively deposited on both sides of the crack to make ridges 1 km wide and 100 m high in only about 10,000 yr. Even including an efficiency factor for an imperfect process, it seems plausible that ridges would grow in less than 30,000 yr.

This calculation shows that, even though each cycle of the diurnal tide has only a small effect, the variations are frequent enough to have played a significant role in Europan geology. The calculation of formation time applies to the larger ridges on Europa. Much more common are small ones, which are not as high or wide. They required far less material per kilometer of length, so they would have formed proportionately more quickly.

The efficiency of extrusion calculated above could be affected by a number of factors. The irregular walls of a real crack might prevent the idealized quantity of crushed ice from reaching the surface. On the other hand, bumpy walls (as well as any crushed ice that failed to be extruded either up or down below the crust in previous cycles) might contribute additional material to the raft of crushed ice, especially if the diurnal variation of stress includes a periodic shearing phase as well as simple opening and closing. Moreover, the continual freezing of additional exposed water, during the 40 hr that the ice is crushed by the closing crack, could enhance the production of material for extrusion.

After my initial sketches of the process (Figure 10.2), various artists created their own versions. *American Scientist* magazine created Figure 10.3 (see color section) to accompany an article, and NHK television in Japan created a very slick animation (Figure 10.4, see color section). The artwork does a great job of describing the ridge formation model.

However, a problem with all of these illustrations is that, in order to show the flow of water and the rafting of ice, they needed to make the cracks much wider than they really are. Remember the ice must be at least a few kilometers thick, but the cracks are probably no more than a few meters wide. If these sketches were done to scale, the ice would be 1,000 times thicker than the crack.

Whether the postulated processes could actually work in such a thin crack is open to question. Such a thin sheet of liquid would be susceptible to freezing as it flowed past the cold ice around it. However, each day's new charge of liquid would warm the adjacent ice somewhat. Questions of the dynamics of the flow of fluid remain open to further consideration. Other outstanding issues involve details of how each day's newly-extruded crushed ice gets added to the pile on each side. It probably would not rise high enough to flow up and over the tops of the growing ridges. More likely, it pushes out on the bases of the ridges, making room for itself as it forces the previously-extruded ice further from the crack. Some of it would probably fall back into the crack, to be re-extruded again the next day.

Whatever we may learn as we consider these details more carefully, at least at the present this model for ridge formation seems to be the best explanation for the character of the ridges. It is consistent with their appearance as piles of loose, crushed material (e.g., as seen in Figures 2.9 and 10.1). Most compelling was how naturally it followed from the periodic opening and closing that we expected from the diurnal tide.

10.1 OTHER RIDGE FORMATION MODELS

Another model of ridge formation was developed shortly after ours by two graduate students who were working with another professor in my department, Jay Melosh. Elizabeth (Zibi) Turtle and Cynthia Phillips in 1998 worked with computer software that Jay had developed to simulate tectonic processes. They considered what would happen if two crustal plates squeezed together as a crack closed, with a bit of extra material wedged into the crack. They noted that such a condition might occur in the likely event that the raft of frozen, crushed ice from a frozen oceanic intrusion were not completely squeezed out as the crack closed. The simulations showed that a modest lip of upwardly-pushed ice would form on either side of the crack. This effect almost certainly contributes to the development of ridges. My guess would be that each ridge consists of a pile of extruded crushed ice on top of a tectonic lip generated by the process envisioned by Zibi and Cynthia.

When Zibi and Cynthia prepared a presentation summarizing their findings at a meeting of the American Geophysical Union, they created a small storm within the Imaging Team. Mike Belton decreed, under pressure from one of the powerful team members, that they would not be allowed to publish because they were not members of the Imaging Team. The rationale was that Imaging Team members had first rights to the data because of our decades-long work in support of the mission. Although NASA policy has now changed toward more open access to mission data, the *Galileo* policy had been standard for most large space missions up to that time. Numerous pictures and descriptions of double-ridges had already been published, going back to *Voyager* images. Yet the Imaging Team was claiming control of all publication rights on the subject. Whether it made sense or not, such control was enforceable. Most planetary scientists depend on NASA for funding or information and, if the powers in control are displeased, a career could be in jeopardy. Consequently, most mainstream scientists try to abide by the policies of space missions, even if they are not parties to them.

Eventually, we were able to arrange for Zibi and Cynthia to present their clever idea at a couple of AGU meetings, but after their difficult experience they tabled that line of research and moved on to other things. Nevertheless, the process they envisioned probably plays a role in ridge formation and I would expect it to accompany the process of extrusion by diurnal tides. Their model deserves further consideration.

By 1998, Jim Head and Bob Pappalardo were firmly committed to the canonical thick-ice model, with viscous solid-state convection and diapirism in a tens-of-kilometer-thick layer of ice. To explain ridge formation they proposed that long linear diapirs (low-density blobs within the ice crust) have risen up though the hypothesized thick, warm, viscous layer of ice that lay below the brittle–elastic lithosphere, tilting it upward along the sides of cracks. In this model, tidal working of the original crack in the lithosphere periodically worked the underlying ductile material, heating it enough to create the low-density diapir. Then, as the diapir rose up below the crack, it pushed up the lips, creating the double-ridges.

One of the most impressive pieces of evidence that they presented was the identification of triple-ridges. Their example was a 6-km-long segment shown in Figure 10.5. In this ridge system, the central ridge for such features was interpreted as the top of the diapir, rising up between the two upturned lips of the crack.

That interpretation would have been very reasonable if such triple-ridges actually existed. But, Head and Pappalardo must have forgotten to look at the extensions of this ridge system in the area beyond the postage stamp-sized cut-out

124 **Building ridges** [Ch. 10

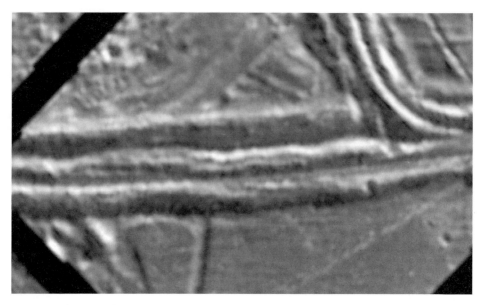

Figure 10.5. The archetypical example of a triple-ridge, a class of feature that does not exist, as cropped and displayed by Head and Pappalardo (the full context and correct north–south orientation are shown in Figure 11.1). The portion of the ridge system shown here is about 6 km long from left to right.

in Figure 10.5. Just looking a few kilometers beyond tells a very different story (Figure 10.6). The supposed triple-ridge is really part of a complex of multiple ridges, probably sets of double-ridges, that have been considerably modified by overlapping one another and by an adjacent patch of chaotic terrain, just above the "triple-ridge" in the figure. The chaotic terrain has clearly disrupted ridges in this area. In one place, the formation of chaotic terrain disrupted one ridge of a double

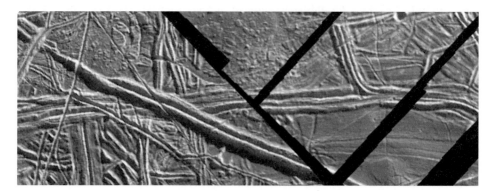

Figure 10.6. The terrain a few kilometers beyond the postage stamp cut-out in Figure 10.5. The straight black slashes are gaps in data acquisition.

Sec. 10.1]	**Other ridge formation models**	125

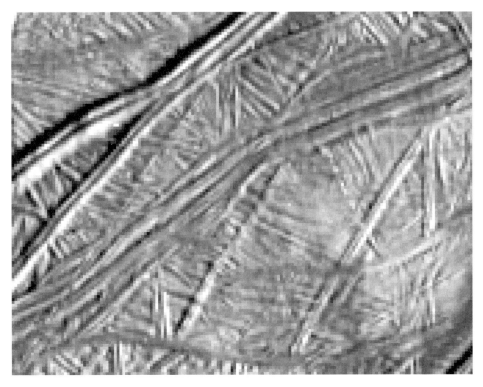

Figure 10.7. The complex of double-ridges running from the lower left to the upper right in this cut-out was mapped as a "triple-ridge" by Jim Head's Research Group at Brown University. It is no such thing. Note that this area appears in Figure 2.12, and lies just northeast of Conamara.

ridge, leaving only a single ridge for a couple of kilometers of its length. The triple-ridge cited by Head and Pappalardo appears to have been created when a multi-ridge complex was reduced to three ridges along a short part of its length due to disruption by chaos. This feature is not a characteristic product of ridge formation.

The only other example of a triple-ridge was mapped by Head, Pappalardo, and their student Nicole Spaun near Conamara Chaos. The feature identified in their map as a triple-ridge (Figure 10.7) is clearly just a set of roughly parallel and very typical double-ridges. The triple-ridges that were such an appealing part of the hypothesis of linear diapirs simply do not exist.

The diapir model had the ridges forming as the lithosphere on either side of the crack tilted back from the upward force of the diapir rising below. One would expect to find the pre-existing terrain on the outward-facing slopes of the ridges. Head and Pappalardo showed a couple of examples of what they interpreted as such terrain. The best one is Figure 10.8. However, the fact remains that of the tens of thousands of kilometers of double-ridges that have been imaged, places where features on the flanks are lined up with nearby ridges are very rare. While one can advocate a theory

126 **Building ridges**

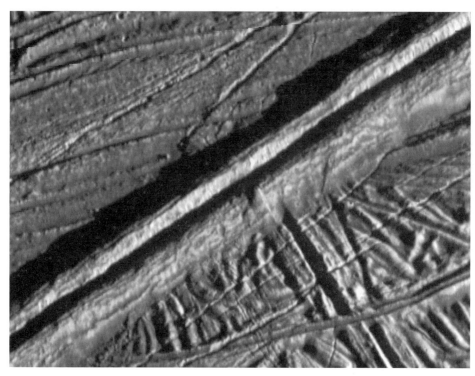

Figure 10.8. A segment of a double-ridge selected by Head and Pappalardo as an example of surrounding terrain extending up the flank of the ridge. In the center of the ridge, a block-like shape on the slope seems to be aligned with an older ridge that is perpendicular to the younger, bigger ridge. Otherwise, the large ridges appear to be piles of loose material dumped on top of the older terrain. This is a portion of the same ridge shown in Figure 2.9. The large ridges are about a kilometer wide.

by displaying selected sites and representing them as typical examples, that approach takes advantage of the reader's assumption that the presentation is as objective as possible.

I have been intrigued by numerous places where ridges do seem to run up the flanks of later ridges. Some examples are found in a set of high-resolution images taken near the south pole (Figure 10.9). I have been surprised that Head and Pappalardo have not cited them as examples in support of their theory, because, more than in most places, it could be argued that the adjacent terrain seems to continue up the ridge flank. However, in the few places where there is such an appearance, it looks more consistent with my model. As crushed ice was piled onto the growing ridges, it slid down the outer slopes. Where it was supported by the pre-existing underlying topography, raised ridges of loose material on the slope were formed. This picture seems consistent with the observation (e.g., in Figure 10.9) that the ridges on the flanks all run perpendicular to the main ridge, rather than following the direction of the underlying older ridges. Thus, the terrain on the flanks,

Sec. 10.2] Downwarping, cracking, multi-ridge complexes, dark margins 127

Figure 10.9. Near the south pole, this bright flank of a ridge (running from the upper right corner down to the lower left corner) has extensions of older ridges running up its flank.

even in these rare cases, is completely consistent with our model of ridge formation by tidal extrusion of crushed ice.

In addition to the lack of expected central ridges of upwelled material, and the absence of old terrain on the flanks of the ridges, it has been hard to understand how such a model could explain the uniformity along great lengths of ridges. It is difficult to envision a diapir that could be so uniform along many hundreds of kilometers. Consequently, the linear diapir model for ridge formation has not been widely embraced.

10.2 DOWNWARPING, MARGINAL CRACKING, MULTI-RIDGE COMPLEXES, AND DARK MARGINS

A great strength of the tidal–tectonic extrusion model is that it provides a natural explanation for some of the characteristics of larger, more "mature" ridge systems. Such ridges often have fine cracks running along parallel to them on either side. Examples are obvious in the case of the simple, archetypical, but large, double-ridge shown in Figure 2.9. Because our ridge formation model piles material on top of the surface, the brittle, elastic lithosphere must be bent down under the weight. These cracks evidently formed in response to tensile stress along the top of the downward bend. Similar cracking, paralleling the margins of ridges that load the ice, is also known in the Arctic ice cap.

In some places, the downwarping of the burdened lithosphere is obvious. A double-ridge near Conamara was one of the first where this effect was recognized (Figure 10.10). The geometry of the bending of a weighted elastic sheet is well understood theoretically. The lithosphere is pushed down right below the weight, rises up slightly a bit away farther out, and then flattens out to the normal height at greater distance. The distance at which the rise occurs depends on the thickness of the elastic layer. As soon as he noticed this ridge, my student Randy Tufts realized that it could be used as a probe of lithospheric thickness. He found that this elastic layer had to be about 300 m thick. What is elastic depends of course on how slowly the stress is applied. In this case the stress is of very long duration, so the thickness of the ice that responds elastically (e.g., to tidal stress on a daily basis) would be significantly greater.

Another thing to bear in mind is that this ridge is in a special location, running right up to a very large fresh patch of chaotic terrain, Conamara Chaos. In fact, the portion imaged at high resolution (in Figure 10.10) lies on rafts that have broken off from the shore, although they have not been displaced much. Because Conamara, like other chaotic terrain, appears to have formed by melting from below, the elastic layer might have become temporarily thinner under this ridge at that time.

In any case, an elastic thickness of 300 m is reasonably consistent with the expected average temperature profile for heat-conducting ice of a few kilometers thickness. But it is also consistent with the temperature profile for convecting ice, which would increase in temperature quickly with depth in the top couple of kilometers, and then stay fairly uniformly warm down through the thick viscous convection layer.

Elsewhere downwarping is less obvious, but its effects can be pronounced. We have already noted the cracks that form on top of the downwarping. Moreover, the common multi-ridged complexes may be a related phenomenon. We saw that the large-scale ridge systems that show up in global and regional imagery as the "triple-bands" actually are composed of complexes of roughly parallel double-ridges. The large complexes that cross just north of Conamara are classic examples (Figure 2.8). Aside from such well-known and prominent examples, other similar complexes are very common as well. The group of ridges in Figure 10.7 (incorrectly mapped by Jim Head and his colleagues as a so-called "triple-ridge") is another example, in that case a more subtle complex of double ridges.

Such complexes may well be a result of the marginal cracking that accompanies downwarping under ridges. Once marginal cracks are formed, they may open farther by subsequent tidal stress, and activated by diurnal tides, creating additional nearly-parallel sets of lateral ridges (Figure 10.11).

The downwarping of the lithosphere around large ridges may also provide a basis for explaining the dark margins that are so common. Once the surface of the original ice crust is pushed down by the weight of the ridge to the level of the water line (independent of the class of the ridge structure), liquid water might spread over the surface. The ridges, being built of crushed ice, may be porous enough that the

Sec. 10.2]   Downwarping, cracking, multi-ridge complexes, dark margins   129

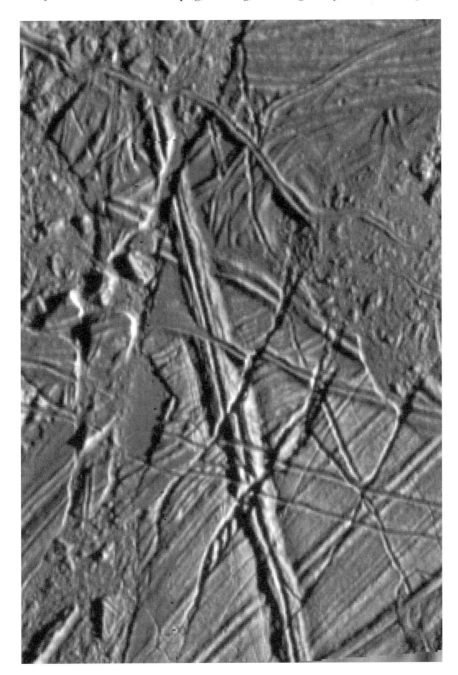

Figure 10.10. This double-ridge runs into Conamara, where it was cut off by formation of chaotic terrain. Shown here in a high-resolution image (~30 m/pixel), it also appears at the bottom center of Figure 2.8. This ridge pair has weighted down the lithosphere below it, causing downwarping on either side.

130 Building ridges

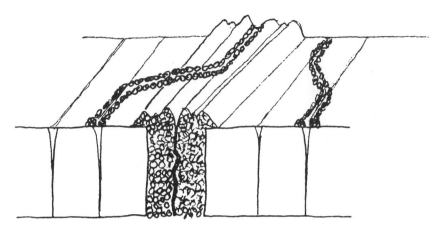

Figure 10.11. A mature ridge (in this case, one in which there has been some dilational spreading of the main crack, but that is not essential) weighs down the lithosphere, forming parallel cracks along the flanks. These cracks are later exploited by tidal stress, producing additional ridges nearby, nearly parallel or intertwined with the others.

liquid water could spread laterally through the base of the ridge, and then flow across the surface beyond the ridges. This process might explain the emplacement of dark material as impurities carried in oceanic liquid water, perhaps including organic material, along with salts detected in the dark areas in the infrared. The relative lack of such darkening and coloration in the ridge material may be due to purification of the water as it froze in the cracks during our diurnal scenario. Association of the dark margins with oceanic substances is consistent with the impression that chaotic terrain, which is similarly darkened and colored, represents sites of melt-through from below.

Other mechanisms have been proposed for creating the dark margins, and they are consistent with the linkage of the surface to the ocean through the cracks. Early in *Galileo*'s imaging of Europa, Ron Greeley suggested that the dark material might be sprayed out from the cracks, perhaps as a result of oceanic exposure. This idea is appealing, because it would also explain the darkening, with apparently similar substances, of the surface at chaotic terrain. Also, if the characteristic spray distance is about 10 km, as indicated by the width of the dark margins bordering "triple-bands", it would also explain the 10-km-wide margins that extend beyond the edges of patches of chaotic terrain, which define the ~10 km lower limit to the size of the dark spots (the "lenticulae").

The main problem with this explanation for the dark margins is that the spray would be expected to darken the ridges as well as the neighboring area, yet the ridges themselves are bright, forming the bright central portion of the triple-bands. Presumably, the ridges are bright because the darkening agents were distilled out during the freezing of the lead ice that eventually was squeezed onto the ridges. Perhaps the final part of ridge building occurred after most of the spraying had been completed. The spray hypothesis is very interesting and definitely needs to be considered more carefully. It could be an important part of what gives Europa its distinct appearance, but we just do not understand well enough how, or even whether, it could have worked.

Another idea for creating dark areas was suggested by Sarah Fagents, a volcanologist at Arizona State University. Her idea was that warming from below may have concentrated impurities as sublimation removed water ice from the surface, rendering certain areas darker. This model would help explain why some areas around chaotic terrain are not only darkened, but in some cases have their topography softened, as discussed in Chapter 16. Similarly, because ridges formed where cracks have allowed oceanic water to regularly flow to the surface, heat is brought up to the ice bordering the cracks. It seems possible that the darkening along the margins of highly-developed ridge systems is a consequence of such thermal effects.

10.3 CRACKING THROUGH TO THE OCEAN

Major ridge systems correlate with tidal stress, so they must be related to cracks. Moreover, the currently most plausible model of ridge formation requires that those cracks penetrate to liquid water. However, tidal stresses probably cannot drive cracks very deep.

Consider the overburden pressure under the surface. At any depth the weight of the overlying ice is equal to $\rho g h$, the product of its density $\rho = 1 \, \text{gm/cm}^3$, the gravitational acceleration $g = 1.3 \, \text{m/sec}^2$, and the depth. So, 100 m down, the pressure would be about $10^5 \, \text{kg/m/sec}^2$, which is 10^5 Pa, about the same as atmospheric pressure at sea level on Earth. What is more relevant, it is about the same as the tidal stress. Because the overburden pressure acts over a very long time, the material responds as if it were a fluid, so the pressure is isotropic, in other words the same in all directions. (On Earth you can test for isotropic pressure using your ears whenever you change altitude or dive under water: The pressure is independent of how you tip your head.)

Therefore, at any depth below 100 m, the hydrostatic pressure prestresses the material so that tidal tensile stress of 10^5 Pa can at most relieve compressive stress. It cannot induce cracking, because the stress is overwhelmed by the hydrostatic burden from above. On the other hand, as a crack that initiates at the surface propagates downward, additional stress is concentrated at the base of the crack, but even that concentration is unlikely to drive a crack from the surface much deeper than a few kilometers. Several other factors could affect the depth of crack penetration. If liquid water got into the crack (e.g., from the walls of the crack as they are warmed by friction), it would provide internal pressure that would push outward against the walls, balancing the hydrostatic pressure in the ice. If solid material fell into a crack, it would help wedge it open and drive the crack deeper.

Another factor that could come into play would be shear along a crack, which must occur during the diurnal tidal cycle. Shear displacement has a very important role in the tectonics of Europa, as we will discuss in more detail in Chapter 12. At this point, we simply note that shear stress would be concentrated at the bottom of a crack that had not yet penetrated to the ocean. Strain rates at the bottom tip of the crack could be high enough that even warm ice at depth would act elastically at that point, allowing brittle crack. The entire problem is complicated by the fact that the ice is warm down near the ocean, and crack propagation in such material is not well understood.

Similarly, if cracks initiated at the base of the ice, they would be filled with ocean water, which would mitigate the resistance of hydrostatic pressure in the ice. Initiating a crack at the bottom would be difficult because the ice at the bottom would be warmer and less brittle than the ice at the surface. Even if a crack could get started, water could only help propagate the crack about 90% of the way to the surface (i.e., to the float line). But, with dissolved gases that might be released from the water, the crack could conceivably be driven through to the surface.

If these various poorly-understood processes, or something we have not yet thought of, play roles, it is conceivable that cracks penetrate far below the 100 m limit seemingly imposed by hydrostatic pressure in the ice. On the other hand, it is very difficult to imagine how any crack at the surface could extend down more than a few kilometers. Therefore, if our proposed mechanism for ridge formation is correct, it implies that the ice must be fairly thin, certainly less than 10 km. The canonical model of thick, convecting ice is not consistent with this highly-visible linkage of the ocean to the surface.

11

Dilation of cracks

The simple double-ridges that dominate much of the tectonic terrain on Europa show no signs that the cracks have dilated significantly. By dilation, I mean an opening process, where the crust on either side is moved apart and new surface is created between them. For most double-ridges, except for the diurnal opening and closing by a few meters, the plates of crust on either side of double-ridges have generally not moved apart. The extruded crushed ice has been emplaced on top of the adjacent surface rather than filling a dilating gap.

However, a second class of ridge system does seem to be associated with the dilation of cracks. These features have much wider ridges on both sides of the central groove. These ridges are grooved along their lengths, such that they usually appear to comprise multiple sub-ridges, and these structures are generally symmetrical on cross section about the central valley (Figure 11.1a). In some cases, where the ridges are only modestly wider than simple double-ridges, they might simply be more mature versions of simple double-ridges, in which a longer duration of diurnal pumping has extruded more material, which has spread laterally and symmetrically due to the symmetrical nature of the process.

However, in cases like that in Figure 11.1a dilation is usually involved. Reconstruction of the regions around such features (as in Figure 11.1b) show that these are indeed sites of dilation of the crack.

The morphology of these features seems consistent with what one might expect if slow dilation accompanied the ridge-building process. In fact, dilation may be driven by the same process of diurnal pumping if the extrusion of crushed ice is not completely efficient. In that case, some fraction of the crushed ice may remain jammed between the walls. With each diurnal closure, the original walls (and the original extruded ridges) of the crack may be ratcheted gradually apart. Whether that mechanism or something else causes the slow dilation, additional extrusion of crushed ice slurry will be between the original ridges, gradually producing the symmetrical ridge sets with the crack at their center (Figure 11.2). The geometry is similar to the

134   **Dilation of cracks**   [Ch. 11

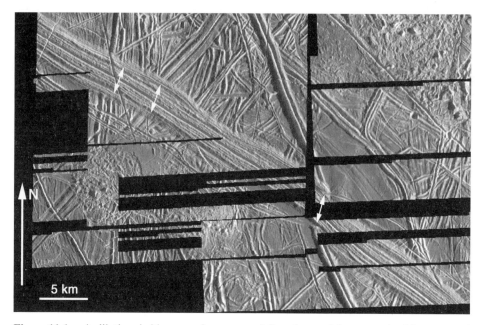

Figure 11.1a. A dilational ridge runs from upper left to lower right, recognized by a central groove with multiple, symmetrical ridges on both sides. Double-headed arrows show matching points of older lineaments that were cut off and dislocated by the dilational ridge system. The black gores in this image represent areas where we have no data. (Note the putative "triple-ridge" to the upper right of center is the same one in Figures 10.5 and 10.6, a product of modification by the chaos to its upper right.)

symmetrical emplacement of volcanic material around terrestrial seafloor spreading centers, such as that under the currently dilating Atlantic Ocean Basin.

Dilational "bands", where cracks have opened even wider, along with corresponding displacement of adjacent terrain, are very common on Europa. Their dilational character is readily demonstrated by reconstruction of the surrounding terrain (e.g., as shown in Figure 11.3). A great concentration of these features is located in the "Wedges" region (south of the equator running west from longitude 180° for about 60°; see Figures 2.4 and 9.1). Several of these cracks dilated at an angle so that the bands are wedge-shaped, which led to the informal name of the region.

This region was clearly imaged by *Voyager*, and the character of the displacement of the crust was recognized early on by Paul Schenk, then a graduate student at Washington University in St. Louis, and his thesis advisor Bill McKinnon. After Paul demonstrated at a conference how these features could be reconstructed, it seemed obvious to me that the crust was highly mobile. However, that notion was too radical for the referees judging Schenk and McKinnon's paper for the planetary science journal *Icarus*, and publication was delayed for several years, a classic example of being too correct too soon. *Galileo* images later confirmed their discovery many times over.

Dilation of cracks 135

Figure 11.1b. The region in part (a) has been reconstructed by cutting out the dilational ridge, and moving the terrain from either side together. Older linear features are now restored to their original straight, continuous configurations.

Markings and reconstruction by Randy Tufts.

The sickle-shaped band in Figure 11.3 lies at the northwest edge of the Wedges. Reconstruction of this and many other bands is practically seamless. Clearly, this band formed recently relative to the features around it. The crust of Europa has been highly mobile up until the most recent part of its short surface history, and there is no reason to believe that such active displacement is not continuing today.

Figure 11.2. Ridge formation during dilation builds multiple, symmetrical ridges. Here, too, cracks have formed on the adjacent terrain due to the weight of the ridge.

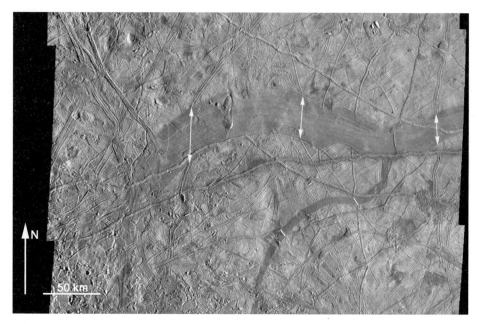

Figure 11.3a. A typical dilational band, the "Sickle" lies at the northwest corner of the Wedges region, and is recognizable on a global-scale image (Figures 2.4 and 9.1) at longitude 240°W, just south of the equator. A narrower curved band lies below it. The Wedges region, filling the lower center of Figure 2.4, contains a cluster of dilated cracks, many highly curved like the Sickle, many forming boxy patterns, and many dilated into wedge-shaped bands. Double-ended arrows show the "piercing points" of features identified by Randy Tufts to help with reconstruction.

The morphology of the Sickle is typical of dilational bands. The boundaries are parallel and matched, allowing for reconstruction along the original crack line (Figure 11.3b). The interior of the band is relatively smooth and flat, with only fine, low-amplitude furrowing parallel to its length. Often a central groove shows where the most recent spreading has been occurring. The appearance is similar to dilational ridges (e.g., Figure 11.1), except that this band is much wider, and the ridges are much flatter, appearing only as subtle furrows.

Why do some dilating cracks develop ridges, as in Figure 11.1, while others take on the smoother character, like the Sickle? The most probable explanation is that whatever process drives dilation, it goes faster in the case of a smoother band. If dilation is slow, the diurnal opening and closing can pump ridge-building ice to the surface at the same time as the gradually widening crack fills with new ice. If dilation is fast, the crack widens enough each day that the diurnal closing cycle cannot slam it shut enough to pump up much material.

Consequently, there is a continuum of morphologies of dilational bands, ranging from smooth ones to dilational ridges, most likely reflecting a wide range of dilation rates relative to the rate of the small periodic diurnal displacement. Many ridges are

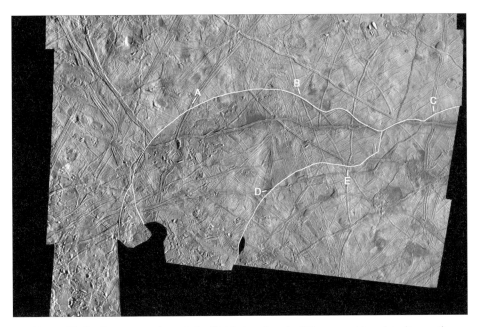

Figure 11.3b. Reconstruction of the Sickle band and of the curved band to its south.
By Randy Tufts.

of a hybrid character, where it seems during part of the dilation the rate was fast enough to emplace flat new ice, while at other times dilation was low enough to allow ridge growth. Such ridges move apart symmetrically as new material fills in during later dilation. The history is recorded by symmetrical ridges in otherwise flatter bands (e.g., Figure 11.4).

Dilational bands, like other types of surface features on Europa, appear to have formed throughout the recorded history of the surface. Although the most obvious examples are generally the ones that formed most recently, careful surveying can reveal older examples that have been partially resurfaced or modified by subsequent processes, such as later ridge or chaos formation. Randy Tufts found and reconstructed many examples in his thesis work on crust displacement (Figures 11.5 and 11.6). Reconstruction of such features emphasizes how constant, over the course of the past few tens of millions of years, the surface has been continually renewed by tectonic processes like ridge building and dilation, as well as by chaos formation.

These reconstructions do not tell us much about the process that drove the plates apart. In the case of dilational ridges, where dilation has been slow compared with the diurnal pumping cycle, it is plausible that the plates were wedged apart by ice that was stuck in the crack. That mechanism would explain why the closing phase of the diurnal cycle would apply the pressure needed to pump up new frozen ice to the ridges. If the plates are pulled apart by some force external to the crack, there would be minimal compression during the diurnal closure phase, explaining why most wide

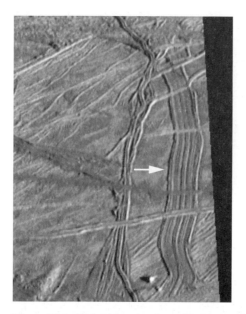 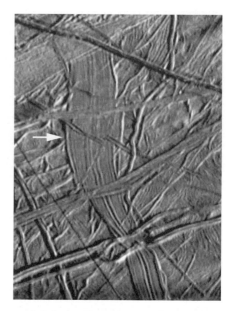

Figure 11.4. Hybrid bands where dilation rate has varied during formation, producing ridges symmetrically arrayed about the centerline.

bands are fairly flat with only shallow furrows rather than high ridges. Therefore, we would expect that whatever is driving the dilation is pulling the plates apart, rather than pushing them apart from within the crack.

Evidence that bands are created by broad regional forces, rather than locally to the crack, comes from a band complex with an unusual geometry, shown in Figure 11.7a. These bands are a couple of hundred kilometers east of the Sickle. In fact, they are both extensions of branches of the same band. The two main bands shown here are slightly darker than the older surrounding terrain, including the terrain that lies between them. Inspection of that intermediate terrain shows that it is divided by a shear displacement (known as a strike–slip displacement in geological jargon). The geometry of the displacement is sketched in Figure 11.7b and the previous configuration, obtained by re-closing the bands, is shown in Figure 11.7c. This geometry could not have developed if each band was pulled open separately by forces acting on either side of the bands. Evidently the two bands share a common displacement driver that must have been pulling the terrain to the north of the northern band away from the terrain to the south of the southern band.

Whatever drives the opening of bands must operate over a very broad regional scale. We can only speculate regarding the source of the force. One possibility may be oceanic currents. In that case, the ice must be fairly easy to displace. The implication would be that the ice must be fairly thin, or else the dilation would be resisted by the strength of the ice in the direction that the plates were displaced.

What happens in the direction that the plates are moved is an interesting question. Another way to think about that problem is in terms of surface area

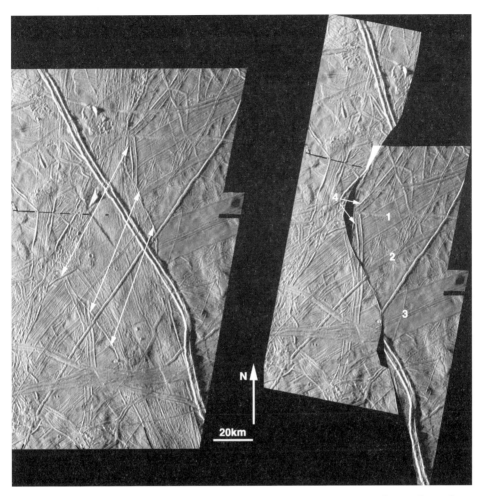

Figure 11.5. At left an old band near the large impact feature Tyre runs diagonally up from the lower right. The relative age of this band is shown by the several more recent ridges that run across it, and by places where chaos has formed within it. The double-ended arrows cross the old band and identify the piercing points of older lineaments that were separated when this band dilated. At the right, the band has been closed back up, aligning older lineaments. Three of those lineaments (labeled 1, 2, and 3) are bands that had dilated even earlier.

budget. If cracks have dilated as plates have pulled apart, then new surface area has been created. In order to balance the surface area, someplace area must be removed.

For several years after dilation had become well known, the mystery of surface area budget nagged us. Most of the other known resurfacing processes, such as ridge building, do not affect the overall area budget. They replace just as much surface as they destroy, because the new material is placed on top of the old.

The mystery of surface area budget lingered for several years as people speculated about processes that might yield the net destruction of surface area, to

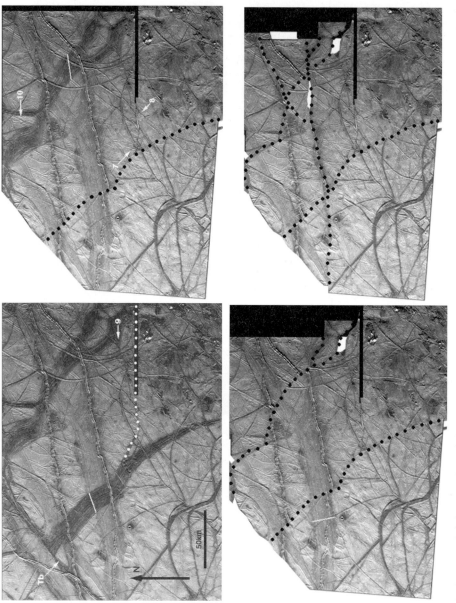

Figure 11.6. At upper left is a portion of the Wedges region crossed by several dilational bands. One shows an example of the wedge form that gave this region its name. A sequence of re-closing wedges can reconstruct the displacements of the crustal plates. At top right, the wedge-shaped band has been closed (to the dotted line). At lower left, a second band is closed. At lower right, going farthest back in time, an older east–west trending band is closed.

Randy Tufts' reconstruction.

Dilation of cracks 141

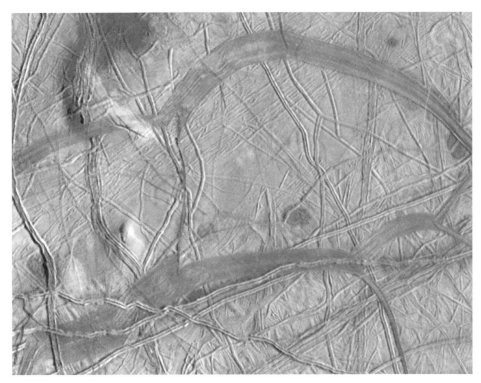

Figure 11.7a. A pair of parallel dilation bands, which are located east of the Sickle (Figure 11.3) at 225°W, 3°S and which can be seen near the center of Figure 2.4. A north–south running strike–slip (i.e., shear) fault connects the dilation bands. The width of this image is about 180 km.

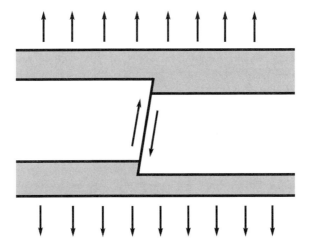

Figure 11.7b. A schematic of the dilational displacement that created the arrangement of crustal plates in Figure 11.7a.

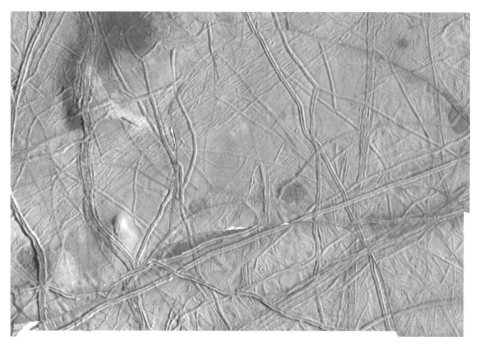

Figure 11.7c. Reconstruction of the band complex shown in Figure 11.7a.
By R. Greenberg.

counter-balance the creation of surface at dilational bands. On Earth, the creation of new area by dilation is balanced by processes that leave very distinct structures. One is subduction, where one tectonic plate (usually relatively thin sub-oceanic crust) slides down under another. Another is Himalaya-style mountain building, where thick continental plates have crashed together, squeezing up mountains between them. Nothing quite like those familiar signs of plate convergence revealed how dilation is balanced on Europa. In Chapter 12, I will describe how we stumbled on what I think is the answer.

Dilation is a major process for rearranging the surface and creating new surface area. Understanding what forces and processes drive it will be an important key to understanding the character of the satellites as a whole. Currently, what drives dilation is a major mystery.[1] We have learned several things from dilation, however: (1) the driving forces operate over large regions; (2) the crust is very mobile, able to slide readily over a slippery (i.e., low-viscosity) layer, known as a

---

[1] The Wedges, a region of considerable dilation, span a range of longitudes (west of the anti-jovian point) where tides may create tension (e.g., Figure 6.1). However, while tides can stress and crack the surface, they represent horizontal displacement of only a few meters, and cannot explain the tens of kilometers of dilation typical in the Wedges and of dilation bands elsewhere. The role of tides in creating the crack patterns (but not the dilation) in the Wedges is discussed further near the end of Chapter 14.

"decoupling" layer in geological terminology, and which is presumably the ocean; (3) the cracks that dilate into bands must penetrate to the ocean; (4) given that the cracks cannot have penetrated down more than a few kilometers, the ice must be fairly thin.

12

Strike–slip

While dilation of cracks moves the adjacent crustal plates apart, shear displacement along cracks is also common on Europa. We already saw an example of such shear near the center of Figures 11.1a and 11.1b, although those are somewhat unusual in terms of their relationship with the dilational bands at each end. In geological parlance, plates shearing past one another is called "strike–slip" displacement. "Strike" refers to the direction of a fault across the landscape. A pilot or navigator following along a fault line would call this direction the "azimuth" or the "heading". Strike means the same thing. Hence, "strike–slip" means that the displacement is in the direction of the strike, which is perpendicular to the direction that plates are displaced when a crack dilates.

On Earth, plate tectonics includes dilation, such as the parting of the plates on either side of the Atlantic Ocean or of the Red Sea, as well as strike–slip, such as the southward movement along the San Andreas Fault of the coastal area of southern California relative to the rest of North America. But strike–slip is rare elsewhere in the solar system, except on Europa, where it is very common.

Strike–slip was first noted on Europa when Paul Schenk and Bill McKinnon identified the mobility of crustal plates in the Wedges region. The displacement there, while primarily dilational, also included some moderate shear as well. The first identification of a predominantly strike–slip displacement, along a fault called Astypalaea Linea, was made by my student Randy Tufts, using 17-year-old *Voyager* images in 1996.[1]

---

[1] The main contribution of the *Galileo* mission to this discovery was that it paid Randy's salary and it stimulated him to look carefully at the older images of Europa from the earlier mission. In a similar way, the realization that tides could heat the Galilean satellites, which might have come at any time in the 20th century, was stimulated by the imminent arrival of *Voyager* at the Jupiter system, not by *Voyager* data itself. The obvious implication is that hyping non-existent missions might be a cheap way to get a lot of good research done, based on what we already have in hand. How many times could NASA get away with that?

146 Strike–slip   [Ch. 12

Figure 12.1a. Astypalaea Linea is the dark linear feature running from the lower left to the upper right in this image from the *Voyager* spacecraft, which has been reprojected to show the geometry as if viewed from straight overhead. The south pole is just below the area shown here. Note the dark parallelogram at the upper right (northern) end of Astypalaea, and the various widenings and narrowings along its length.

Our subsequent studies of strike–slip displacement have shown that it is common and widely distributed on Europa, and have demonstrated that it is probably driven by diurnal tides by a process we call "tidal walking". Moreover, our studies of strike–slip have identified sites of surface convergence, and have developed evidence that Europa's shell may have slipped around as a whole, moving former sites of the poles far from the spin axis. All of these disparate implications of strike–slip displacement are discussed in this chapter.

12.1 DISPLACEMENT AT ASTYPALAEA

Astypalaea Linea lies in the far south of Europa, branching southward from Libya Linea (shown in Figure 9.1) at nearly 60°S, passing within about 10° of the south

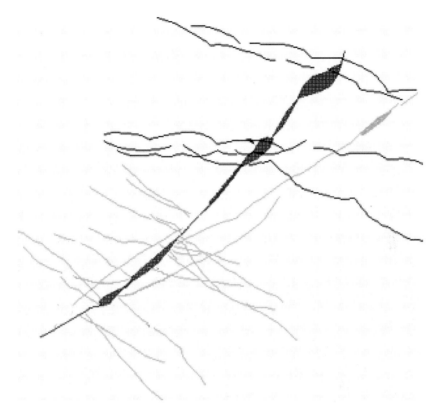

Figure 12.1b. A sketch of the key tectonic features in Figure 12.1a: The varying width of Astypalaea Linea itself, the more recent cycloid-shaped (chains of arcs) ridges, and the wispy, old, white lineaments that end at Astypalaea and were evidently cut off by it.

Based on a diagram by R. Tufts.

pole, and running for at least 800 km in all (Figure 12.1a, b). At its northern end it widens into a distinct parallelogram shape, and in the *Voyager* image it had various widenings and narrowings along its length. It is crossed by a family of more recent ridges (in fact, standard double-ridges) that follow "cycloidal" trajectories (i.e., chains of connected arcs, cf. Chapter 14) across Astypalaea. This cycloidal configuration is extremely important to understanding Europa, as we will see, but here it is irrelevant to defining the displacement along Astypalaea, because these ridges came later on. Most significant, as Randy Tufts noticed, are the wispy, old, white lineaments that end at Astypalaea and were evidently cut off by it.

Randy realized that by cutting this picture along Astypalaea, and shifting the east side northward (and where it ran past the south pole, the south side eastward), he could reconstruct the continuity of the wispy lines, and with the same displacement close up the parallelogram (Figure 12.1c). Clearly, the plates along Astypalaea Linea had slid (or scraped) along this boundary for a distance of 40 km. Furthermore, Randy noted that the parallelogram is a classic example of the geometry that

148 Strike–slip [Ch. 12

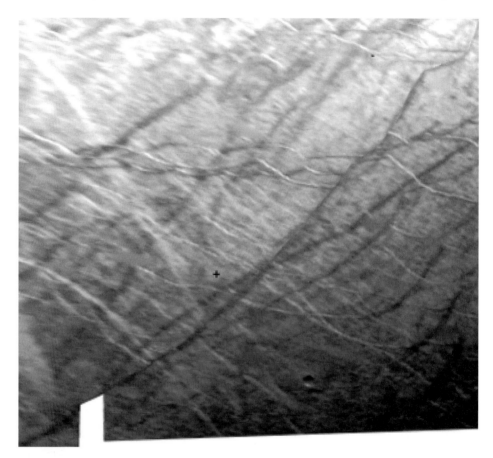

Figure 12.1c. Randy Tufts' reconstruction of Astypalaea Linea prior to the strike–slip offset. In the past, as shown, the wispy, old, white lineaments were continuous across Astypalaea and the angled portions of the fault had not yet been pulled apart by the shear.

must form if the original fault is not straight: The parts of the fault that do not run parallel to the shear direction must pull apart. The technical geological term for these features is "pull-aparts". In effect, the crack dilates along those portions. Randy suggested that the other widenings in Astypalaea, in addition to the most obvious northern parallelogram, might also be pull-aparts.

Pull-aparts associated with strike–slip demonstrate one way that dilational bands can be created. In fact, branching off of the northern end of Astypalaea (to the right, just off the upper right corner of Figure 12.1) is the dilational band Libya Linea. Evidently, as the huge region to the east of Astypalaea (lower right portion of Figure 12.1) moved southward (toward the lower left), it pulled Libya open. Strike–slip provides one demonstrated mechanism for pulling open at least some dilational bands.

12.2 TIDAL WALKING

But, then, what drives strike–slip on Europa? Randy Tufts began to address that problem as soon as he made his discovery of the displacement. He suspected that it might be related to the tidal stresses that my research group had been investigating, but his scientific training in geology was only tangential to those exotic processes and settings.

Fortunately, his background, experience, and talent gave him a much broader and useful range of expertise than his formal education alone. Beginning in high school, he had developed an interest in caves, and a resolution to discover one. This obsession led to undergraduate studies in geology, and a systematic program of exploration. His undergraduate years in the late-1960s and early-1970s had other distractions. Tufts became a prominent and effective student political activist. Not long after graduation, Tufts fulfilled his high-school goal, discovering with a friend the large and pristine Arizona cave now known as Kartchner Caverns, but always called by its discoverers simply "The Cave". For the next two decades, Tufts worked the difficult political problem of obtaining protection for The Cave, while keeping it a secret to ensure its preservation. At the same time, he pursued a career in social services and public policy. By the late-1980s, the careful political effort paid off and The Cave was on its way to protection as a state park.

Tufts became interested in Europa at about that time, having read about Europa and the possibility that there was an ocean there. He decided that he wanted to find out whether Europa was indeed habitable. Just as he had committed himself to finding a cave, he made a similar commitment to exploration of Europa. And just as his search for the cave included majoring in geology as an undergraduate, Tufts enrolled in the Geosciences graduate program at the University of Arizona to begin this new exploration. So, one day in the early-1990s this strange middle-aged man arrived at my office door, and told me he loved Europa. Eventually, he became my student and, along with Paul Geissler and Greg Hoppa, part of my core team working on Europa during the *Galileo* years.

Randy's political sophistication, honed over all aspects of his multifaceted career, came in very handy in our dealings with the *Galileo* Imaging Team. And his understanding of geology complemented the backgrounds in planetary physics of the rest of my research group. Randy's familiarity with the tectonics of deserts in the southwestern U.S. led to his discovery of the Astypalaea strike–slip fault. Just as Kartchner Caverns was always known around Randy as "The Cave", Astypalaea was known as "The Fault".

After discovering the strike–slip displacement along The Fault, Randy sat down with Greg Hoppa's charts of tidal stress and meticulously sketched how the stress at Astypalaea would vary over the course of an orbit of Europa. He and Greg shared an office, and they discussed the problem continually for a few weeks. Figure 12.2 (see color section) shows the diurnal variation of the stresses at the position of Astypalaea over the course of each orbit (the same information can be gleaned from Figure 6.1). At apocenter, tension runs across the fault line. Tidal distortion tends to pull the crack open. One-quarter orbit later, the principle stresses are equal

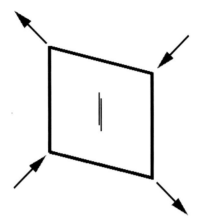

Figure 12.3. Equal and opposite principal stresses on an element of the surface result in shear stress at an oblique angle. This explains why the tension and compression at Astypalaea 1/4 orbit after apocenter or pericenter drives shear, as shown in Figure 12.2.

and opposite, with tension in one direction and compression perpendicular to it. The effect of such stress is to drive shear along a line 45° from the principal stress (as in Figure 12.3). In fact, that is exactly the orientation of Astypalaea. This shear would tend to drive the region to the lower right of Astypalaea down toward the lower left. (Shear in this direction is called "right-lateral"; if you were standing looking across the fault, the terrain on the other side would move to the right.) By 1/4 orbit later, at pericenter, the stress field has reversed itself relative to apocenter. Now the sides of Astypalaea are squeezed together by compression. Then, after another 1/4 orbit, the system is in shear, but now in exactly the opposite sense, left-lateral.

The shear effect is reversed on each cycle; so, at first look it appears that the strike–slip goes nowhere. But consider that the fault has just been closed before the left-lateral shear phase. Now friction in the tightly-closed fault is likely to prevent or at least minimize reversal of the right-lateral shear displacement that may have occurred half an orbit later. Thus, this sequence of stress can "walk" the fault in a manner closely analogous to actual walking, where we separate a foot from the ground, move it forward (in shear) just above the ground, compress it to the ground, and then try to shear it back. Friction prevents the foot from moving back. In exactly the same way, Astypalaea moved ahead in right lateral shear while the crack was held open by tides, and tried to move back in left lateral shear, while the crack was held too tightly compressed to move back. Each day on Europa, Astypalaea took a small step forward.

Remember each step was small. Tidal strain at a crack is a few meters at most, and the shear could slip back partially during each reverse shear phase. But with a small step each day, Astypalaea could shear very quickly. If the fault took a full 1-m step over each diurnal cycle ($3\frac{1}{2}$ Earth days), it would move 100 m in a year or the full 40-km displacement in only 400 years. The walking would be expected to have continued as long as the crack remained open and active.

The process of ridge building also seems to have depended on diurnal tidal working of an active crack. We saw that, if all the crushed ice in a crack each day were squeezed up onto the ridge, a fairly large ridge system could form in ~1,000 yr, comparable with the amount of time that it would have taken Astypalaea to walk its full displacement, if walking were perfectly efficient. Just as we assumed that ridge building is not so efficient and likely takes tens of thousands of years to build the larger examples of ridges, the walking process is probably similarly inefficient, due to back-sliding during the return phase, resistance from the surrounding crust, and other effects. Whatever the appropriate correction factors for the two processes, it seems likely that a major strike–slip displacement takes a comparable amount of time to building of a major ridge.

Such agreement is consistent with both processes depending on how long a crack stays open and active. As long as the diurnal tide continues to work the crack in a way that is appropriate for building ridges or for strike–slip displacement, the crack is likely to remain active and open. But with sufficient non-synchronous rotation, any crack is likely to move to a location where the stress regime is considerably different. Once that happens, the crack may become inactive and freeze solid. Thus, it seems reasonable that both strike–slip displacement and ridge formation are limited in duration by non-synchronous rotation. In that model, the size of the largest simple ridges and the distance of the greatest strike–slip displacement is controlled by the same effect; so, it is not surprising that both require similar amounts of time.

12.3 PREDICTING STRIKE–SLIP

What has proven to be most powerful about the walking model for strike–slip displacement is that it offers a way to make specific prediction for which way any crack should walk, either right-lateral like Astypalaea or left-lateral. The answer depends on the location on Europa, relative to the direction of Jupiter, and on its azimuthal orientation (its "strike").

Greg Hoppa and I did the required calculations for any location and azimuth over the surface, and made a map of the predicted sense of the displacement, which is shown at the end of this section (Figure 12.8). That map then formed the basis for interpretation of a survey that we made of strike–slip, wherever it could be found (Hoppa et al., 1999, 2000).

The interpretation revealed some important and surprising things about Europa, which follow from comparing the theoretically-predicted sense of the strike–slip displacement as mapped in Figure 12.8 with the observed displacement of faults on Europa, as shown in Section 12.4. The remainder of this section is a closer look at the theory of tidal walking, to give a more complete understanding of how the theoretical predictions shown in Figure 12.8 were computed.

Figure 12.2 shows the periodic diurnal stress sequence that drives strike–slip. At the heart of the model is the recognition that the maxima in the tension and compression across the fault are 1/4 cycle out of phase from the maxima in the

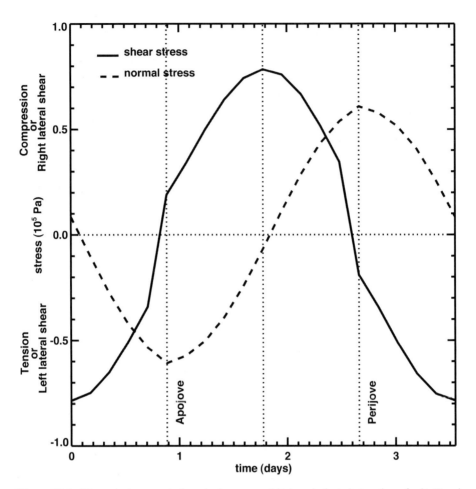

Figure 12.4. Diurnal stress variation during one orbital period at Astypalaea fault. Tension across the fault is defined as negative, compression positive (dashed line), while shear stress (solid line) in the right- and left-lateral directions are defined as positive and negative.

left-lateral and right-lateral shear along the fault. This phase relationship is more evident in Figure 12.4, which is an alternative representation of the same stress shown in Figure 12.2. It shows diurnal variation at this single location, with shear relative to azimuthal orientation of Astypalaea. Remember that the diurnal variations of tidal stress were calculated by assuming that the elastic shell is effectively intact, so a global stress equilibrium is always maintained. This condition requires that the shell is free to slide on the slippery decoupling layer, in this case the ocean, which is considered effectively frictionless. In applying this model to the case where there already is a local crack, we are assuming that the global stress model can be used to represent the regional strain around the crack.

Tidal walking can be envisioned by considering an elastic rubber sheet with a reference orientation line drawn on it. This sheet can be manipulated at its edges to produce sinusoidal variation in stress components similar to what takes place at Astypalaea, as shown in Figures 12.3 and 12.4. At 1/4 of an orbit before apocenter of Europa's orbit, there is shear stress along the reference line but no tension or compression across it. Then, as the shear stress diminishes, tension increases over the first quarter-cycle. The process continues until the system returns to its initial condition after one cycle ($3\frac{1}{2}$ days). An idealized graph of this sinusoidal change in stress is shown in Figure 12.5a.

Now suppose there is a crack in the elastic sheet along the reference line. Assume that while the crack is closed by compression, friction prevents any shear displacement of one side of the crack relative to the other side. While the crack is open under tension each side is free to move with respect to the other with no shear stress. Therefore, Figure 12.5a no longer describes stress at the crack, but it still does represent the strain imposed by the boundary conditions around the edge of the sheet. Therefore, starting just after $t = 0$ when the regional condition imposes tension across the crack (Figure 12.5a, dotted line), the crack undergoes right-lateral shear displacement (Figure 12.5b shows the shear displacement along the crack), but no shear stress accumulates along the crack (Figure 12.5c shows shear stress along the crack). Starting at $t = \frac{1}{2}$ (i.e., half an orbit after the crack opened), the crack is closed by regional compression (Figure 12.5a). After that closure, motion along the crack ceases (the line in Figure 12.5b goes flat), but left-lateral shear stress can now begin to accumulate (Figure 12.5c). At $t = 1$, exactly one full orbit after the crack first opened under tension, it now reopens again. As soon as it does, the shear stress along the crack drops to zero (Figure 12.5c), and the left-lateral regional shear causes the fault to spring with left-lateral displacement back to its original configuration at $t = 0$ (solid line in Figure 12.5b). In such a case there is no net shear displacement.

This discussion shows that, if the sheet were perfectly elastic, walking could not occur. As soon as the crack opened to take its second step, it would snap back to its original alignment.

For a more realistic case, the sheet will not respond as a perfectly-elastic material. Most real materials do not spring back completely after being stretched. We would expect some hysteresis to inhibit the sheet from fully returning to its original shape. It would only spring back part-way (as shown by the dashed line in Figure 12.5b). This effect results from partial relaxation of the accumulated stress during the compressive phase ($t = \frac{1}{2}$ to 1). The result is that a small right-lateral displacement step would occur just after $t = 1$, relative to the configuration one cycle earlier (just after $t = 0$).

Another way to look at the behavior diagrammed in Figure 12.5 is to envision the geometry of the rubber sheet as it goes through the cycle (Figure 12.6). In Figure 12.6 the condition labeled $t = 0$ corresponds to just after the crack opens at the left in Figure 12.5. Because the crack is open, it has snapped to the left-lateral displacement, corresponding to the shear imposed on the region. Between $t = 0$ and $t = \frac{1}{4}$ right-lateral displacement occurs along the open crack, as the shear stress on the

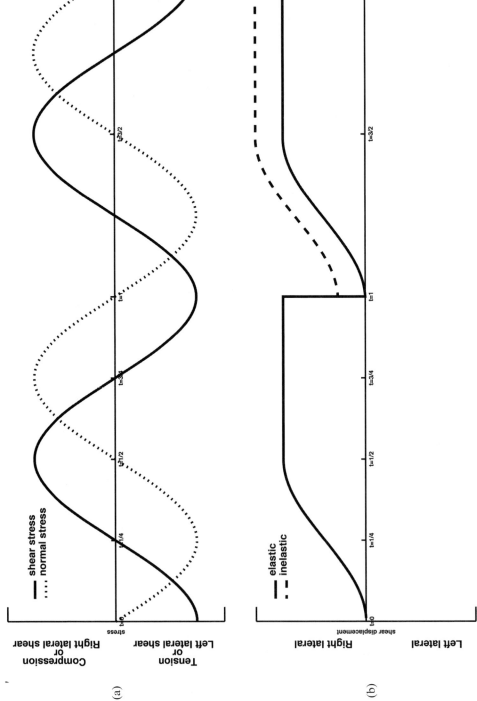

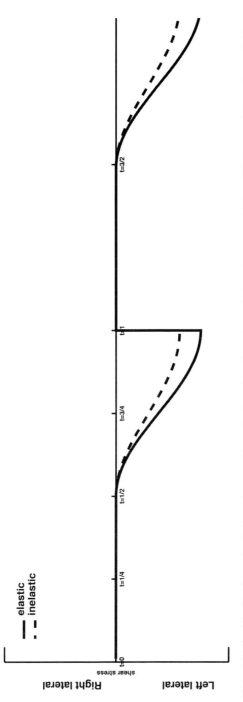

Figure 12.5. (a) The idealized resolved stress for a rubber sheet undergoing stress that varies sinusoidally with time, with a phase relation similar to that of Astypalaea (Figure 12.2). (b) The corresponding displacement along a crack in the rubber sheet. The solid line represents the displacement for a perfectly elastic material. For an elastic sheet, right-lateral motion occurs along the crack while the normal stress across the sheet is under tension and the shear stress is increasing. Lateral displacement stops when the normal stress across the sheet is compressive. No net displacement occurs for a perfectly-elastic sheet (solid line), but net right-lateral displacement is possible if there is some hysteresis in the sheet (dashed line) preventing the opposite sides of the crack from completely springing back to their original position. (c) Accumulated shear stress along the crack in the sheet: no stress is accumulated while the crack is held open under tension; left-lateral stress is accumulated on the crack only when the normal stress across the crack is compressive (holding it closed). In an elastic sheet (solid line) the accumulated shear stress along the crack allows both sides of the crack to spring back to their original position (no net displacement) when the normal stress becomes tensile (opening the crack). However, in a somewhat inelastic sheet (dashed line), some of the accumulated shear stress is dissipated or relaxed while the normal stress across the crack is compressive; thus, when the normal stress becomes tensile (opening the crack) the sides of the crack are unable to spring back (in shear) to their original position.

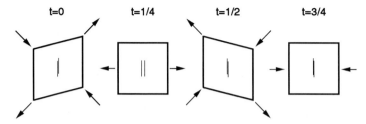

Figure 12.6. The crack in the middle of a rubber sheet responds to the distortion imposed by the stresses in Figure 12.5a, which is similar to the diurnal sequence at Astypalaea.

region returns to neutral. Right lateral displacement continues until $t = \frac{1}{2}$ after which the crack is closed by regional compression. Between $t = \frac{1}{2}$ and $t = 1$ shear stress accumulates, but the crack is closed during this time, so there is no displacement. When the crack reopens at $t = 1$ the two sides spring back to their original position (same as shown for $t = 0$) for an elastic sheet or part of the way back for a partially inelastic sheet.

The analogy of this process to human walking is very close. When a woman lifts up her foot, she may take a step forward; similarly, a fault may take a step when tidal stress opens up a crack. Motion of the foot temporarily stops when it is placed back on the ground; lateral motion along a crack ends when the fault closes due to compression. When the foot is lifted again it does not spring back to its original position because the main body has moved forward; stress at the fault is relieved by displacement of adjacent plates. The cycle then repeats in the next step.

The application of the "walking" model to Europa depends on how closely the icy lithosphere corresponds to our rubber sheet analog. Our model requires that Europa's crust behaves elastically, so that stress is proportional to strain. Although the lower portion of the ice crust likely behaves viscously (so that stress is proportional to strain rate, like a Newtonian fluid), the upper surface is rigid over timescales much larger than an orbital period. Thus, any elastic distortion of the lithosphere (the upper brittle–elastic part of the ice) would be damped by the viscous portion. The damping would contribute to the hysteresis needed to prevent the lithosphere from springing back fully on each new tensile (opening) phase.

As long as the crack penetrates all the way to liquid water the viscous ice would not interfere with motion opening or shearing the crack itself. Thus, for surface cracks to develop strike–slip offsets by "walking", they must penetrate to an extremely mobile layer that allows for significant displacement on an hourly timescale. Thus, in applying the "walking" model to Europa we are assuming that the lithosphere is elastic and not overdamped by a viscous layer below it and that cracks penetrate to a highly mobile layer like an ocean.

Based on this model the direction of "walking" (left-lateral or right-lateral) can be predicted for any location or orientation on the surface. In the rubber sheet analogy, right-lateral displacement occurred because the direction of the change in regional shear stress during the tension phase was right-lateral. Similarly, right-lateral motion was driven at Astypalaea because the shear stress moved right-laterally during the tension phase (Figure 12.4). If the curve in Figure 12.4 had had a net downward motion during the tension phase, the fault would have walked in the left-lateral sense.

Depending on the location and crack orientation, the resolved stress field may be more complicated than at Astypalaea. Consider, for example, the tidal stress associated with the Wedges region (15°S, 195°W) for a fault oriented in the east–west direction there (Figure 12.7). This fault would be predicted to move in the right-lateral direction because the shear stress is more right-lateral when the fault closes than when it opens. Aside from that similarity to Astypalaea, the shear stress behavior here is much more complicated. During the time that the fault is open (tension across the fault) the displacement moves back and forth. In the human walking analogy, this behavior can be represented by a woman who lifts her foot and shuffles it forward and back before deciding where to put it down. The critical issue though is whether she puts her foot down in front of or behind where she lifted it. Likewise, the critical issue in Figure 12.7 is that the shear stress curve ends higher (more right-lateral) at the end of the tension phase than at the beginning.

Based on this criterion, Greg Hoppa plotted the predicted distribution of left-lateral and right-lateral faults on Europa as a function of position and azimuth (Figure 12.8). At latitudes poleward of 30°, this tidal stress model predicts that there should be a preponderance of left-lateral faults in the northern hemisphere and a preponderance of right-lateral faults in the southern hemisphere, regardless of their orientation. Closer to the equator, this model predicts equal numbers of left-lateral and right-lateral faults near the sub-Jupiter and anti-Jupiter points. Near the apex of the leading and trailing hemispheres (i.e., longitudes 90° and 270°W) an abundance of left-lateral faults is predicted in the northern hemisphere at all latitudes (right-lateral for the southern hemisphere). Overall, this model predicts that there should be more left-lateral faults in the northern hemisphere and more right-lateral faults in the southern hemisphere.

12.4 SURVEYING STRIKE–SLIP ON EUROPA

Armed with this theory, Greg Hoppa surveyed several locales where images were available that were suitable for identifying strike–slip displacement. In the Astypalaea region (marked "a" in Figure 12.8), all strike–slip was found to be right-lateral, consistent with predictions. Even taking into account non-synchronous rotation, at whatever longitude the displacement occurred as the terrain rotated eastward, it should have been right-lateral, and it was. This agreement was good first evidence for the predictive ability of the theory of tidal walking.

At the other sites that were considered, the sense of the observed offsets (right- vs. left-lateral) would have varied as the terrain rotated eastward with non-synchronous rotation. In fact, the distributions of the strike–slip offsets at each of these sites do not fit the predictions for the current location of the terrain. Hoppa et al. interpreted that result in terms of non-synchronous rotation, showing how what was observed at each location could be consistent with displacement having

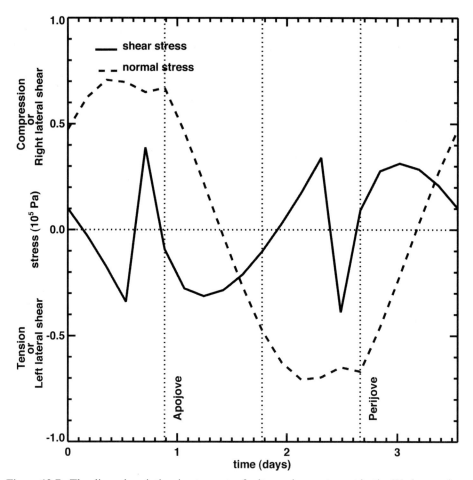

Figure 12.7. The diurnal variation in stress at a fault running east–west in the Wedges region. Although it is irregular, the fact that the shear curve makes a net step upward during the time that the crack is open (in tension) means that such a fault would be predicted to displace in the right-lateral sense.

occurred at various times during the past few tens of degrees of rotation. For example, a preponderance of right-lateral offsets in the Wedges region ("b" in Figure 12.8) was taken to mean that the displacements occurred when that terrain was at 270° or 90° longitude.

However, we were soon to discover that some more complicated things must have been going on. In the autumn of 2000, the NASA Space Grant program at the University of Arizona assigned an undergraduate student to do a research project with me. By that time we had an extensive set of *Galileo* images of Europa, and a complete and systematic survey of strike–slip displacement was long overdue. I assigned the student, Alyssa Sarid, the task of identifying and mapping strike–slip faults in a systematic way.

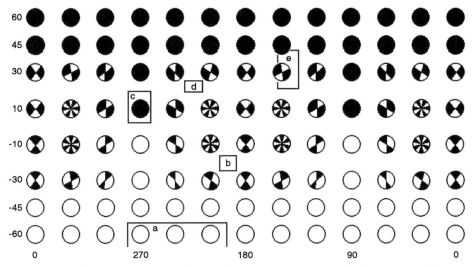

Figure 12.8. The theoretical predictions of the sense of strike–slip displacement (right-lateral in white and left-lateral in black) as a function of location (latitude and longitude) and azimuth, based on the model of tidal walking. Locations labeled "a–e" were surveyed by Hoppa et al.

Ideally, we would have wanted to survey the entire surface of Europa, but we were faced with a shortage of images. As the *Galileo* spacecraft had approached the Jupiter system, it was sent a command to open its larger radio antenna. Until then the antenna had been furled like an umbrella. It was supposed to open into a dish that would allow transmission to Earth of the vast amount of data that was expected. Each image was 800×800 pixels and the brightness at each pixel was encoded to 256 (or 2^8) levels of gray; so, 8 bits of information had to be transmitted for each pixel. The plan was to take about 100,000 images, and the real expectation was that we would take twice as much. Multiplying those numbers, the imaging alone was expected to send about 10^{12} bits of data back to Earth. The large antenna was essential, in order to focus enough of the transmitted radio power back to Earth from the great distance of Jupiter, providing the broad communications band needed to handle all the data.

The umbrella did not open. At least one of the struts must have remained stuck to the center pole. Efforts were made to shake it loose, but nothing worked. There were careful studies of what might have gone wrong, but nothing was conclusive. Presumably the long delays and three road trips across the U.S. had not been a big help for the mechanism. Ultimately, the only working antenna was a small one that had been serving for communications during the trip to Jupiter. At that distance, the tiny antenna could only send about 40 bits per second back to Earth. That data rate is closer to a 19th-century telegraph than to modern broadband communications. At that slow rate, the planned pictures would have taken hundreds of years to arrive. But, the spacecraft's life would be only a few years, limited by fuel for navigation, by the aging of the camera's detector, and by the money available to keep people working on the project. At best we could get only a few percent of the anticipated images.

The problem was addressed in a remarkable re-engineering project. The ancient computer on board *Galileo* was reprogrammed to encode the image data more efficiently using new image compression techniques. The computer was so old that the original programmer had to be lured out of retirement to make it work. Strategies were adopted for averaging over groups of pixels, so that more images could be received, albeit at the expense of resolution. And observing sequences were planned to take pictures quickly during the short fly-bys of each satellite, store the data on an onboard tape recorder, and then slowly radio them home during the weeks between satellite encounters.

The canonical evaluation of this effort is that it was completely successful. In a sense it was. The engineering teams at JPL had done a spectacular job of reconfiguring a crippled robot 100s of millions of miles away and getting the most out of it. The story showed NASA and its teams working at their finest, in their best classic form.

But, ultimately, the canonical dogma went too far. The official spin was that all of the objectives of the original mission had been met. This conclusion depends on your objectives. As rosy as the official summary may have been, as we try to understand Europa, we are stuck with only about 2% of the images that we had expected.[2]

Most of the surface of Europa has only been imaged at resolutions of a few kilometers per pixel, like Figures 2.2–2.7. These pictures do not show the details needed for a survey of geological features. We do have very high-resolution images, like Figure 2.9, which show great detail, but only at a few selected spots. They do not cover enough of the surface for a survey, and they show places that were selected because they seemed especially interesting to someone for some reason, so they cannot be considered representative. In fact, as discussed in Chapter 3, these artificial selection effects have skewed some people's impressions of the character of Europa.

A good compromise between the global coverage at low resolution and the selected sampling at very high resolution is the large set of images that show the surface at about 200 m/pixel. Such images cover somewhat less than 10% of Europa's surface. Several of the regions that were imaged at that resolution were selected for their special interest; for example, Figure 2.8 is a mosaic of 200-m/pixel images selected because Conamara had seemed so prominent and unusual on global-scale images. However, a set of "Regional Mapping" images had been taken at ~200 m/pixel to cover broad, unbiased expanses of the surface for the purpose of geological mapping.

---

[2] There was some benefit to having the more restricted data set, as Greg Hoppa often points out. We had to be creative, but at the same time we had a manageable data set (all the Europa image data fit on one CD) that even one person could go through over a short period of time. My small research group, consisting mostly of undergraduate and graduate students, was able to finish definitive surveys in a timely manner. I believe, for example, that Greg Hoppa, Randy Tufts, Paul Geissler, and I were familiar with the detailed content of nearly every image of Europa. Greg and Randy certainly were. Such comprehensive knowledge would have been impossible if the data set had been too large. It would have been harder in some ways to pull together a global story so quickly. In contrast, it must be almost paralyzing (and certainly challenging) to try to make sense of the vast amount of images of Venus from *Magellan* or of Mars from *Mars Global Surveyor*.

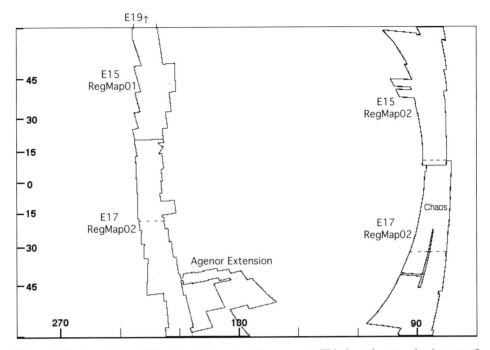

Figure 12.9. Locations of "Regional Mapping" image sets. This broad, non-selective set of data is ideal for surveys that avoid the biases of generalizations based on selected locales. Other surveys that used the Regional Mapping images are the survey of chaotic terrain (Chapter 16) and the survey of pits and uplifts (Chapter 19), as well as the survey of strike–slip in this chapter. Compare these locations with the map in Figure 9.1, and with the more complete set of regions imaged with resolution about 200 m/pixel mapped in Figure 16.10.

For our strike–slip survey we restricted ourselves to these relatively unbiased Regional Mapping data sets. Although we were not doing geological mapping in the traditional sense of marking and categorizing everything in the area, these images did suit the goals of our survey.

The Regional Mapping images cover two broad north–south swaths called RegMap 01 and RegMap 02. They were obtained with illumination at high enough incidence angles (inclined to the vertical), so topography and morphology are clearly shown. Both swaths range from within 20° of the north pole to within 20° of the south pole, and they are about 150° apart in longitude, as shown in Figure 12.9.

RegMap 01 covers a swath about 250 km wide around longitude 225°W (Figure 12.9), sampling a broad portion of the trailing hemisphere, located about half-way between the center of that hemisphere and the anti-jovian longitude (which was defined as 180°). The portion between 60°N and 20°N was taken during orbit E15, and the portion south of 20°N was taken during orbit E17. A northern extension was taken during orbit E19. Also, an extra loop of *Galileo* 200-m-resolution images extends from the southern end of RegMap 01 (near Astypalaea Linea) to encompass chaotic terrain features Thrace and Thera, and an important lineament called Agenor Linea (cf. Figure 9.1).

RegMap 02 samples the leading hemisphere, encompassing a swath around longitude 80°, of width similar to RegMap 01. I include as part of RegMap 02 its southern extension, which is sometimes called RegMap 03 in *Galileo* documentation. As with RegMap 01, the northern portion was imaged during orbit E15, and the southern during E17. The portion of RegMap 02 between 10°N and 30°S is dominated by a huge unit (1,300 km in diameter) of chaotic terrain (see Chapter 15) and therefore has no strike–slip tectonics to survey.

The completeness of the survey was limited in part by the data set and in part by the character of Europa itself. The resolution of the images limits recognizability to displacements greater then a few hundred meters. Moreover, strike–slip faults could only be identified if there were sufficient identifiable offset features marking the displacement. In many cases, especially for older faults, multiple cracking and rearrangements of the surface have left only relatively short segments of each fault, for which the probability is small that there exist usable indicators of displacement. Thus, older strike–slip faults are less likely to be recognizable. Hence, the results of the survey apply only to the more recent portion of the geological record.

Based on Alyssa Sarid's careful survey, we produced maps that showed the locations, orientations, and displacements of all strike–slip faults found in these regions. Samples of the maps are shown in Figure 12.10, showing the northern parts of RegMap 01, which include some of the most interesting cases of shear displacement. In these maps, each strike–slip fault is marked by a white line, with branching white lines showing some of the crossing lineaments that mark the displacement. Alyssa was fairly conservative, mapping only strike–slip faults in which the displacement is clear and unambiguous.

Readers interested in exploring these images for themselves should be sure to view the original, full-resolution versions which are in the public domain and available from NASA (via the Image Node of the Planetary Data System, which is online at JPL; JPL also produced a set of CD-ROMs with all of the image data). As reduced to fit in these pages, much of the detail is lost, but we can see nevertheless the locations and, in most cases, the evident displacements shown here.

12.5 PARTICULARLY STRIKING EXAMPLES

Several strike–slip faults are especially noteworthy in these maps because of their great length or large displacement or distinctive types of geometries.

12.5.1 The greatest displacement champion

In the far north of the trailing hemisphere, a fault ∼170 km long (labeled A in Figure 12.10a) displays a shear displacement of 83 km, nearly twice that of Astypalaea. All faults that far north should be left-lateral according to the tidal-walking theory, and

Sec. 12.5] Particularly striking examples 163

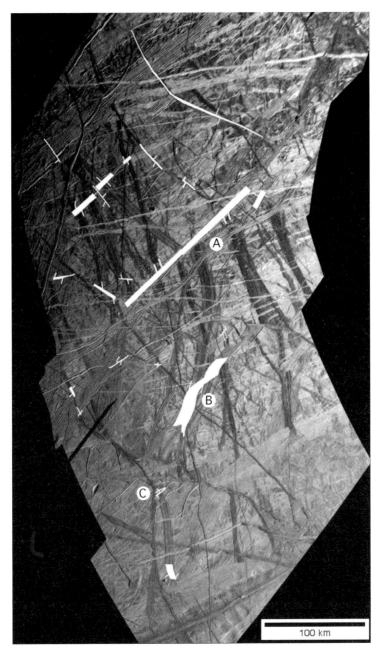

Figure 12.10a. Strike–slip faults in the far north (E19 portion) of RegMap 01, as mapped by Alyssa Sarid, are shown as white lines with tick marks that show the displacement-defining piercing points. Some interesting features are labeled: feature A is the most displaced known strike–slip fault (see Figure 12.11); B is an Astypalaea-like example of strike–slip-driven pull-aparts; and C is a narrow wedge between two strike–slip faults.

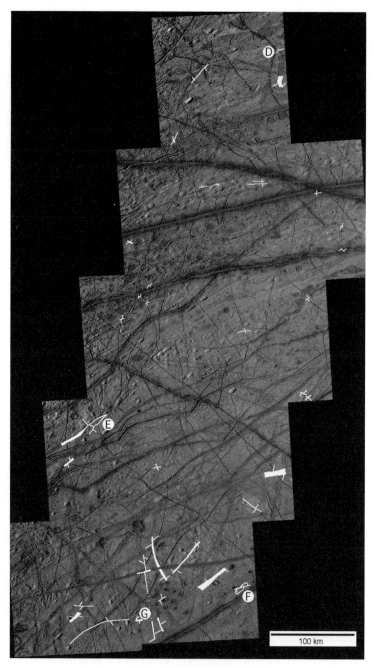

Figure 12.10b. Strike–slip faults in the E15 portion of RegMap 01, just south of Figure 12.10a. Labeled features: D and E are narrow wedges similar to A to the north in Figure 12.10a; F is an unusual sheared pit (see Figure 19.8); and G is the complex reconstructed in Figure 12.12.

Sec. 12.5]
Particularly striking examples 165

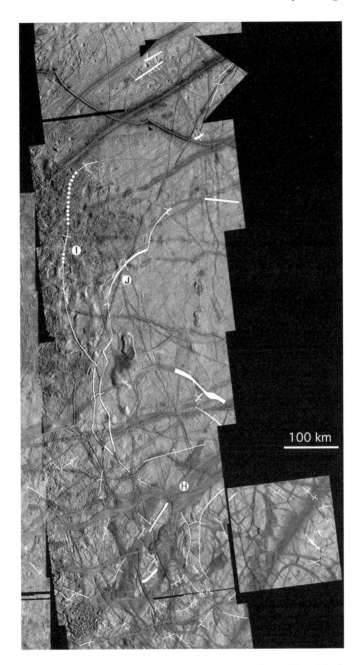

Figure 12.10c. Strike–slip faults in the equatorial regions of RegMap 01, from orbit E17, which overlaps Figure 12.10b, as shown in Figure 12.9. Labeled features: I and J are displaced cycloidal cracks; J is reconstructed in Figure 12.14; and H is a surface convergence zone, as revealed by the reconstruction in Figure 12.14 and as discussed in further detail in Chapter 17.

this large example conforms to that prediction, just as Astypalaea in the south conforms to the predicted right-lateral displacement. The 83-km offset is the greatest strike–slip displacement so far identified on Europa. This fault is displayed at larger scale in Figure 12.11a, and a reconstructed version is shown in Figure 12.11b. Even focusing on this particular feature, it is so long and its offset so great that we cannot display it here at full resolution.

In addition to its large offset, there has been relatively modest dilation of a few kilometers, as indicated by the width of the fault as marked in Figure 12.10a. The geometry is fairly simple, because this fault is so straight. Contrast that shape with the strike–slip fault about 100 km south, labeled B in Figure 12.10a. Fault B, like Astypalaea, evidently began as a curved crack, so its displacement had to involve substantial dilation as well as strike–slip, which led to the wavy, ribbon-shaped plan view shown in Figure 12.10a. Being in the far north, faults A, B, and all the others around them conform to the left-lateral displacement predicted there by the tidal-walking theory (Figure 12.8).

Figure 12.11a. The fault marked as A in Figure 12.10a is shown running diagonally from the upper right to lower left.

Sec. 12.5] Particularly striking examples 167

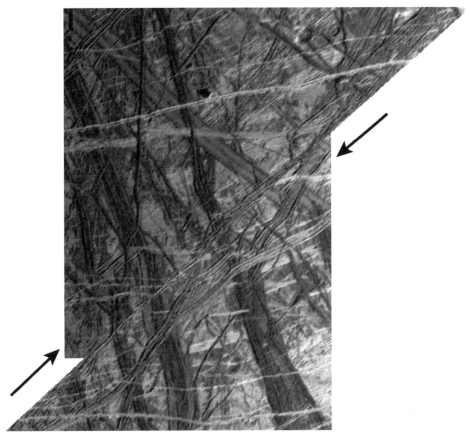

Figure 12.11b. Reconstruction of Fault E (Figure12.11a) demonstrates that there has been an offset of about 83 km in the left-lateral sense, consistent with predictions of the tidal-walking theory. The offset distance is the greatest so far identified on Europa. This reconstruction is based on 200-m/pixel images, which show more detail than can be reproduced here.

12.5.2 A time sequence of strike–slip

In nearly all the cases of strike–slip displacement that Alyssa mapped, the currently visible fault lines are fairly short and isolated, which is not surprising considering how active this surface has been. The tectonic record has been sliced and diced severely by the various resurfacing processes on Europa, including subsequent cracking and ridge building, dilation, and formation of chaotic terrain. Consequently, there are few examples of cross-cutting strike–slip fault traces in the surface record. The most complex case that we found is at the southern end of the E15 RegMap 01 region (Figure 12.10b), and extends slightly into the northern end of the E17 RegMap 01 (Figure 12.10c). A broad dilational band that cuts across the boundary between the E15 and E17 images in a NE–SW direction truncates this fault complex. (This band is marked by a dark "triple-band", which appears near the bottom of Figure 12.10b and runs across the top of Figure 12.10c.) None of the faults north of this band can be traced into the tectonic terrain south of it.

This strike–slip complex is shown in Figure 12.12a (see color section). In Figure 12.12b, the strike–slip lineaments in this area are labeled with capital letters and indicated by red lines, with colored lines showing some of the more prominent piercing features that are used to define the reconstruction. Several sets of strike–slip offset features in Figure 12.12b can be readily identified as part of a common, aligned, fault displacement, such as those labeled J, K, L, M, and N along a 100-km-long curved path.

Reconstruction of this area requires several steps because some realignments must be made before others. For example, going backward in time, we needed to realign the strike–slip displacement along faults H and B to straighten out the north–south crack ACG, before we could reconstruct the strike–slip on ACG. The sequence, going back in time is shown in Figures 12.12b–e (see color section).

Going farther back in time between steps c and d, we reconstructed the displacement along JKLMN, a 105-km-long arcuate fault. It appears that the adjacent terrain to the south rotated by about 1°, allowing strike–slip to occur uniformly along the entire curve. (If it had moved linearly, for example toward the east, part of the crack would have shown strike–slip, while other parts would have dilated.) A similar thing happened along DR, where the adjacent terrain rotated just enough to allow strike–slip, while avoiding a pull-apart. I cannot begin to explain this phenomenon, but it was fascinating to discover.

Reconstruction along fault P (shown between steps b and c) realigned an older band that crossed orthogonally under P. Fault P itself has the morphology of a dilational band, and the reconstruction of P includes closing the dilation as well as undoing the greater strike–slip displacement. This reconstruction also takes care of realigning the ridge complex that crosses F. In the process, going back in time, F opens substantially wider.

Think about that result: Lineament F was ~8 km wider in the past than it is now. Going forward in time, there was lateral convergence of the adjacent surface areas. With this discovery we began to resolve a major puzzle. Dilational bands, at which new surface is created, had been known to be common and widely distributed. Thus, we had known that somehow, somewhere surface had been removed to balance the surface area budget. Now, at last, a site of such removal of surface had been identified.

Lineament F provides us with an example of what convergence looks like on Europa (Figure 12.13). It is band-like, but unlike most dilational bands the opposite sides do not fit together. In addition, one side (to the north) has a slightly raised lip, which may have plowed over some of the adjoining terrain. The Himalayas of Europa are a few kilometers wide and 20 m tall.

12.5.3 A long, bent, equatorial cycloid in RegMap 01

Alyssa Sarid's survey showed strike–slip displacement all along an enormous lineament (labeled J in Figure 12.10c), composed of curved sectors joined at cusps, that

Sec. 12.5]   **Particularly striking examples**   169

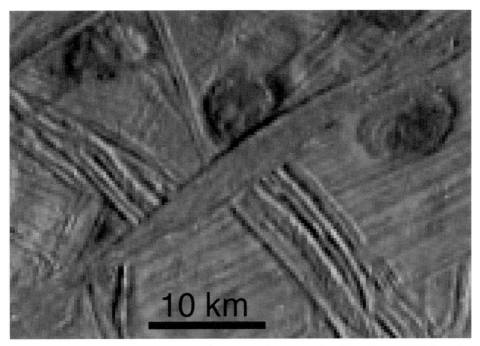

Figure 12.13. An enlargement of the convergence site at location F in Figure 12.12 (in color section).

bends across the equator for hundreds of kilometers. An image of this fault, and the region around it, is shown in Figure 12.14a, with the fault marked in Figure 12.14b. Note that for a good part of its length south of the equator, the strike–slip is divided along two alternative paths as shown. Figure 12.14b shows some of the prominent offset indicators of strike–slip displacement. Many more exist, so that reconstruction realigns a large number of linear features, including many that are not visible here, but appear clearly on the full-resolution images. Because this fault is neither straight nor a simple arc, geometry dictates that strike–slip had to be accompanied by dilational pull-aparts along sections of the fault, especially at the northeastern parts of the two arcs north of the equator, as clearly shown by the widening in Figures 12.14a and b.

The reconstruction is shown in Figure 12.14c. Going backward in time, a coherent plate over 400 km across had to be moved back north by about 8 km, with a rotation of about 1°. The rotation is necessary to realign right-lateral strike–slip displacement along the full length of this curving lineament.

When this plate is moved northward, as in Figure 12.14c, it pulls away from the terrain farther south, opening an 8-km-wide gap, well over 100 km long. Going forward in time, this geometry represents another lateral convergence zone, where surface area was eliminated. The convergence appears to have been accommodated within the complex of dark, band-like features (Figure 12.14d) running along the southern edge of the displaced plate.

170 Strike–slip [Ch. 12

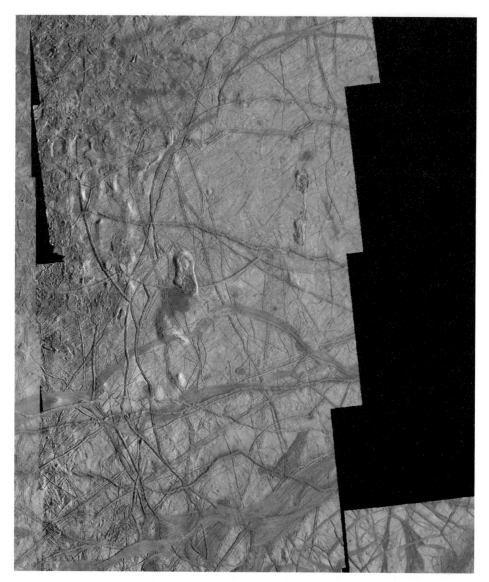

Figure 12.14a. The equatorial portion of RegMap 01 (cf. Figure 12.12c). Here the mosaic has been orthographically reprojected to minimize geometric distortion.

This complex of convergence features lies just south of a set of dilational bands, in fact some of the same ones that were discussed in Chapter 11 (see Figure 11.7). It may well be that this large convergence complex played a role in accommodating the dilation at those bands, but the strike–slip shown in Figure 12.14 occurred after those dilational bands had formed. Thus, the convergence complex in this zone (Figure 12.14d) may have been active long enough to absorb both the southward

Sec. 12.5] **Particularly striking examples** 171

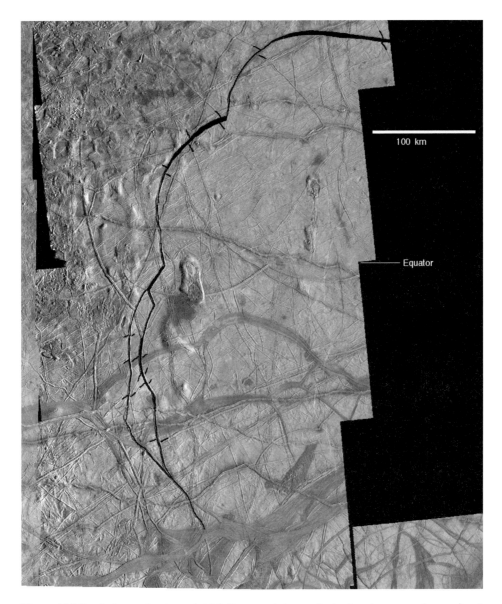

Figure 12.14b. A large, bent, cycloidal lineament is marked, showing several prominent indicators of strike–slip displacement, which is right-lateral along the entire fault. The displacement was accommodated along two adjacent faults in the southern portion.

motion of the large plate to the north and the earlier expansion at neighboring dilational bands.

The band-like complex at the convergence site consists of features very similar to the band identified as a site of convergence in Figure 12.13. Convergence bands here,

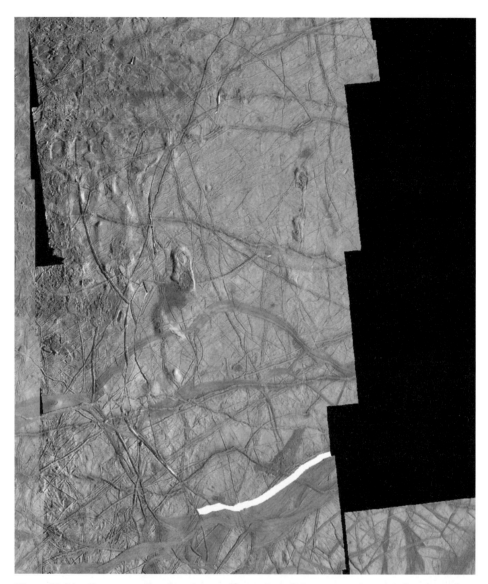

Figure 12.14c. Reconstruction (involving a 1° rotation of the crust to the east) shows excellent fit along the strike–slip fault, with an 8-km-wide gap at the south, identifying a zone where surface convergence appears to have occurred.

as there, consist of segments with fine internal striations, like dilational bands, but whose sides are not parallel or reconstructable. Like the convergence feature in Figure 12.13, each segment appears wider in the middle, and often has a slightly raised lip on one side.

The consistent appearance of terrain at both locations at which surface contrac-

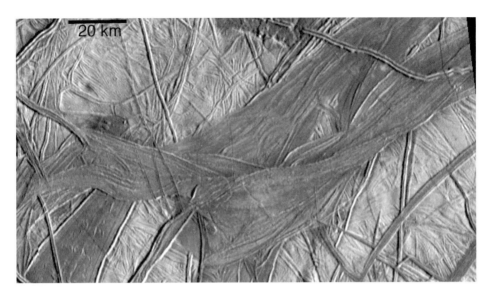

Figure 12.14d. The area of convergence, inferred from the reconstruction in Figure 12.14c. Note the similarity to the convergence site in Figure 12.13. Further discussion of convergence features is in Chapter 17.

tion seems to have occurred reinforced for us the likelihood that this type of feature is at least one type of manifestation of surface convergence. I return to the issue of surface area budget in Chapter 17 and discuss several lines of evidence that show that other, similar features may be convergence bands.

12.6 POLAR WANDER

The discovery of convergence zones on Europa, where crustal plates had evidently crushed together, helping to offset the addition of new area by dilation, had been serendipitous. Alyssa had mapped the strike–slip displacement, then together we reconstructed some of the interesting cases, and, because I had been worrying for a long time about surface area budget, the discovery of convergence zones jumped out at me.

In part, the original idea for assigning Alyssa to map strike–slip displacement came from a feeling that something unanticipated might come out of it, but we also had a specific goal. I wanted Alyssa to compare the distribution of left- vs. right-lateral offsets with the predictions of the tidal-walking theory. At least to me, that goal seemed specific. I had no idea what we might find, but, from years of experience, I knew that, if you compare a large new set of data with a solid theory, something interesting is likely to turn up. But, to Alyssa, who as an undergraduate was new to scientific research, the purpose of her survey seemed completely vague. After months of her mapping strike–slip faults, I told her to compare what she found with Greg's predictions (Figure 12.8).

As Alyssa wrote much later, as part of her application to graduate school, "I then compared the azimuth, location, and sense of strike to the expectations provided by the model. When they did not match up, I first thought that I must have done something wrong. But, I kept thinking about the problem and soon took my first step towards professional research science. I realized that the only way for the data and the theory to coincide was to conclude that Europa's ice shell had undergone polar wander."

Such a discovery is one of those great moments in a professor's life, when you realize that a student has become a real scientist, or perhaps had been a scientist all along. Either way, it was fun guiding her to that point of discovery.

Here was what Alyssa noticed when I insisted that she look for patterns in comparing her survey results with the theory. In the trailing hemisphere (RegMap 01), all strike–slip displacement in the far north (above about 50°N) was in the left-lateral sense, exactly as predicted by the theory of tidal walking. Similarly, in the far south, all strike–slip was right-lateral, again just as predicted. But then, she noticed something that did not fit. The predominance of right-lateral displacement in RegMap 01 continued much further north than it should have, all the way up to the equator, even for cracks that were oriented such that the theory predicted the opposite sense of shear. Moreover, the zone where the directions of shear offset were mixed, which should have spanned a wide band from about 30°S to 30°N according to the theory (Figure 12.8), actually ran from near the equator northward to at least 45°N. The entire length of the region spanned by RegMap 01 seemed to be too far north. It needed to be shifted south by about 30° to fit the theory.

One way to reconcile this discrepancy would be to throw out the theory. But the theory was elegant, and it explained the general trend (right-lateral in the far south, left in the far north, mixed in-between), so we were reluctant to throw it out.

If we trust the theory, it appears that the terrain in RegMap 01 (which extends from the far north to the far south near 225°W) may have moved roughly 30° northward, relative to the poles of the spin axis, since the time that most of the strike–slip occurred. Such a large portion of the crust was involved that a plausible explanation may be polar wander, in which the crust of Europa slipped as a single unit and became reoriented relative to the spin axis.

If the ice shell of Europa had moved northward on the trailing hemisphere, as the strike–slip record there had suggested, then it should have carried the terrain on the leading hemisphere southward. Later, when I needed to explain this geometry to a reporter, I made the sketch shown in Figure 12.15, which I later found posted on the Web at various space news sites. (Still later, I found Figure 5.3.) If such polar wander did occur, one would expect the strike–slip record in RegMap 02 (leading hemisphere) to show indications of a southward migration of terrain, similar to the northward shift that Alyssa had spotted in the RegMap 01 data.

For RegMap 02, the tidal-walking theory without polar wander had predicted that strike–slip displacement for this region, which is located very close to the center

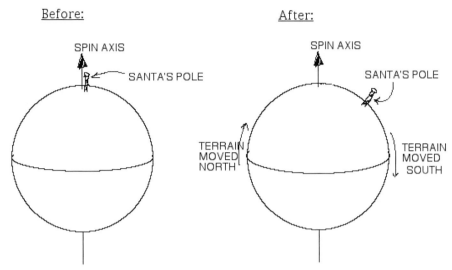

Figure 12.15. When the crust of Europa slips around so that the former pole site moves away from the spin axis, the terrain on one hemisphere moves northward, while it moves southward on the other side.

Artwork by R. Greenberg (cf. Figure 5.3).

of the leading hemisphere (longitude 90°), should be right-lateral throughout the southern hemisphere and left-lateral everywhere north of the equator (Figure 12.8). On the other hand, Greg Hoppa's early study of selected areas suggested that most strike–slip that formed on Europa has subsequently been carried from farther west to its current location by non-synchronous rotation. In that case, we would expect most of the strike–slip displacement in RegMap 02 to form according to the same general trend that we had expected for RegMap 01: right-lateral south of 30°S, left-lateral north of 30°N, and mixed in the equatorial belt between. In any case, without polar wander the distribution would have been symmetrical about the equator.

In RegMap 02 the data were less complete than in RegMap 01. First, there was no extension to the far north, unlike RegMap 01 with its E19 extension. Second, there is a very large equatorial chaos area (1,300 km across, see map in Figure 16.10) that extends south to 30°S, so there is no useful tectonic record in that zone. Nevertheless, there is enough data to tell us whether the terrain has shifted southward.

North of latitude 30°, Alyssa could find only three cases of strike–slip, of which two were left-lateral, but the statistics were sketchy. Between the equator and 30°N, the offsets are predominantly left-lateral, consistent with formation farther north. In the southern hemisphere, we have tectonics only south of 30°S. There are comparable amounts of left- and right-lateral offsets between 30°S and 60°S, where right-lateral shear would have dominated without polar wander. These results are consistent with the former equatorial zone having shifted southward by about 30°, again indicative of polar wander.

The distribution of strike–slip as surveyed by Alyssa Sarid provided evidence for polar wander, and the direction of polar migration appears to be consistent in independent evidence from each hemisphere: As the entire crust of Europa shifted as a unit, terrain formerly at the north pole of the spin axis has shifted by ~30° into what is now the northern part of the leading hemisphere.

We can estimate how recently this shift took place, and hence how rapidly the poles can wander, as follows. The age of the surface is probably less than about 50 million years, but the recognizable strike–slip features are among the more recent tectonic features created during that time. Most of the tectonic terrain is covered by older cracks, ridges, and bands, much deeper in the cross-cutting sequence of formation. Strike–slip is harder to recognize in these older features, which tend to remain only as short segments and which therefore are less likely to have identifiable piercing lineaments to define displacement. The strike–slip features described here are probably well within the most recent 10% of the surface age, and hence less than a few million years old. The polar wander inferred from those features must therefore have occurred within the past few million years.

Moreover, this shift must have happened fairly quickly compared with the period of non-synchronous rotation, and recently enough that not much strike–slip has taken place since. Otherwise, the distribution of strike–slip would have again become symmetrical relative to the equator and spin axis during subsequent rotation.

What could drive such polar wander? On a nearly spherical, spinning body, a very small local excess mass on the surface may cause such polar wander, as the site of the anomaly spins outward toward the equator. To make a similar point in a paper in the late 1960s, astrophysicists Peter Goldreich (Caltech) and Alar Toomre (MIT) considered an ant walking on a spherical planet. No matter where it walked, the land under its feet would stay at the equator. It is unlikely that the mass distribution of Europa's crust is sufficiently symmetric that a bug's mass would spin out to the equator and drag the icy shell along with it. However, the ice crust is very thin, has fairly low topography, and is probably uncoupled from the interior by the underlying global ocean. An anomalously-thick region might be pulled by centrifugal force out toward the equator, dragging the entire ice shell with it.

One plausible mechanism for driving such a process is thermal thinning of the equatorial ice crust in the leading and trailing hemispheres relative to the thicker ice at the colder poles, according to a model developed by Greg Ojakangas and Dave Stevenson at Caltech. Ojakangas and Stevenson had considered the diurnal tidal stress field (equivalent to Figure 6.2) and the equivalent strain. Real materials are never perfectly elastic, so some of the energy in working the crust must go into heat. The heating rate at any instant depends on the square of the "strain rate", which is the rate at which strain changes with a diurnally-varying tide. Ojakangas and Stevenson calculated that the greatest heating in the ice as it is stretched and squeezed in the diurnal tidal cycle would occur at the leading and trailing points (i.e., near the equator at longitudes 90° and 270°). The least heating would occur at the sub- and anti-jovian points (longitudes 0° and 180°). An intermediate amount of heating would occur at the north and south poles.

My student Dave O'Brien was mostly working on problems having to do with asteroids, but he could not resist getting involved with Europa as well. He compared the strain rates implied by Greg Hoppa's diurnal stress calculations with the heating rates that Ojakangas and Stevenson had computed and found perfect agreement.

Considering those heating rates, Ojakangas and Stevenson noted that the crust would become slightly thicker at the poles than at the leading and trailing points on the equator. They suggested that the crust would reorient itself about the Europa–Jupiter axis, with the thicker ice from the pole moving down to the equator, while the thinner ice at the equator moved to the poles. They did not need to invoke a hypothetical massive insect, like Goldreich and Toomre's. The non-uniform thickness of the ice might provide the excess regional mass to drive polar wander.

Polar wander had also been considered by other previous studies. For example, in 1996 a study by Bill McKinnon and his student Andrew Leith (Washington University in St. Louis) of whether there was any record in the orientations of tectonic lineaments on *Voyager* images was inconclusive, but hinted at a shift similar to what we found from the strike–slip survey.

Our result on polar wander was perfectly consistent with the prediction of Ojakangas and Stevenson. Because the model had the least heating at the sub- and anti-jovian points, the ice is thickest there and remains at the equator. The shift of the old north pole down into the leading hemisphere, as we found, agreed perfectly with their prediction that the flip would be about the Jupiter-aligned axis.

One difficulty with their theoretical explanation of polar wander is that it was predicated on an assumption that most of the tidal dissipation occurs within the crust. It is more plausible, instead, that tidal heat from the interior dominates the determination of ice thickness as discussed in Chapters 7 and 16; so, the poles might not become significantly thicker than anywhere else.

However, given the fairly smooth, uniform appearance of Europa's ice shell, it is probably susceptible to polar wander if any local mass anomaly occurs, whatever the cause. If the mechanism proposed by Ojakangas and Stevenson does not work, other causes might include local or regional variations in ice thickness due to thermal anomalies in the ocean or due to geological processes.

An important implication of polar wander is that the ice crust must be effectively decoupled from the interior, so that it can slip as a single unit. Substantial grounding of the ice layer (e.g., on subsurface rocky continents) seems to be ruled out if the poles have in fact wandered.

Corroborating evidence for polar wander comes from the distribution of chaotic terrain (Chapter 16) and of pits and uplifts (Chapter 19) in the same parts of the leading and trailing hemisphere investigated here. Each of those types of features has size and spatial distributions that are similar in northern-leading and southern-trailing regions, with another distinctly-different distribution found in both southern-leading and northern-trailing regions. Moreover, the qualitative character of the tectonic fabric in these regions displays a similar oblique antipodal symmetry. The symmetry axes for all these types of features seem to be similarly inclined by tens of degrees from the pole. Assuming that the character of the distribution of such features was related to distance from the equator (e.g., if it depended on ice thickness or heating rates), this tilted symmetry of the distributions could be explained by polar migration, consistent with the polar wander inferred here from the distribution of strike–slip faulting.

12.7 STRIKE–SLIP SUMMARY

The study of strike–slip faults has provided a powerful tool for investigating the geological, geophysical, and dynamical processes of Europa, as observed features are interpreted in the context of increasingly well-understood physical mechanisms. Strike–slip provided the first evidence for convergence bands that compensate for the dilation that has gone on elsewhere. Strike–slip provided further evidence for non-synchronous rotation. Strike–slip provided evidence for polar wander. Strike–slip demonstrates a direct link between diurnal tidal distortion of the satellite and tectonic displacement through the process of tidal walking.

The requirements of tidal walking place constraints on the structure of the crust. First, the process, like other displacement, implies a low-resistance decoupling layer, the ocean, below the crust. It also requires that the cracks must penetrate to that layer in order to allow the daily steps of tidal walking. Penetration of cracks means that the ice must be quite thin, probably less than 10 km, because tidal stress would have been insufficient to drive cracks through thicker ice.[3] Greg Hoppa likes to point out, too, that strike–slip is only found along cracks that have developed ridges, not along cracks without them. That observation is consistent with our models of ridge formation and of strike–slip walking. Both require cracks to penetrate to liquid. The crust of Europa seems most likely to be very thin ice over liquid water.

As studies of the tectonic record continue, in concert with increasingly complete theories of the underlying physical processes, there is the potential for significant improvement in our understanding of the history and structure of Europa. So far, only the most recent part of the tectonic record has been exploited, but, with these techniques as a model for investigation, future research should be able to penetrate even farther back in time to reveal more about the complex dynamics of Europa's crust.

The party line is that we cannot learn much more about Europa until we send another mission. Specifically, resolution of the debate about whether the crust is so thick that the ocean is isolated from the surface or is thin enough that the ocean is intimately linked to the surface is now said by the politically correct to require future spacecraft. That line is designed to sell space missions. But, the reality is that shortly after each major planetary mission, the data are set aside and ignored. We have already seen numerous examples of how old spacecraft data provide major discoveries decades after they were taken, when someone bothers to go back and take a look. Randy Tufts' discovery of the strike–slip fault Astypalaea in old *Voyager*

---

[3] In a debate at the Lunar and Planetary Science Conference several years ago, Bob Pappalardo argued that the ice is thicker than 20 km and that tidal walking is possible even where cracks go only partway down into such thick ice, but he never showed his calculations. At present, our model is all we have that is open and published.

images is just one example. If, instead of mothballing data shortly after mission press conferences, we are to pursue the lines of evidence that we have just begun to exploit, there is the possibility that we will continue to make major discoveries about Europa, even without waiting for the next mission.

13

Return to Astypalaea

Strike–slip faults had proven to be very important on Europa. They are common, and they provided a surprisingly large amount of information about the structure and processes on the little planet. After Randy Tufts discovered the archetypical example Astypalaea Linea ("The Fault") on *Voyager* images, he wanted to go back and look more closely.

Ordinarily I had little interest in the image-planning process. Partly, my feeling was it did not really matter where we looked, as long as it was on Europa. It would all be interesting, and we were likely to get a fairly good sample of every type of thing that there was to see. My only concerns were that we avoid isolated pictures by planning for contiguous frames, and that we obtain high-resolution images only where we had broader, lower resolution images for context. Another reason that I did not get engaged in the planning was that I was emphatically not welcome. Those team members who got the assignment to plan image sequences would have first rights to report what would be discovered on Europa, and politically powerful ones were preparing to control the discoveries.

Randy loved "The Fault" on Europa nearly as much as "The Cave" in Arizona, so he convinced me to push for high-resolution images during a later orbit. Other team members (and their students and assistants) had their own agendas and targets in mind. For example, Ron Greeley, a volcanologist was certain that some of the dark patches we had seen at low resolution must be volcanic lava flows (he called them "cryo-volcanic", because the material involved would be water rather than molten rock). Although similar dark patches had already proven to be chaotic terrain, Greeley was determined to continue his search for volcanism in higher resolution images. In the end, high-resolution images of Astypalaea would prove to be some of the most revealing images of Europa, while the search for volcanic flows only yielded more typical chaotic terrain. But first we needed to persuade the team to include Astypalaea in the high-resolution sequence.

Randy's long career in politics, social activism, and saving The Cave, made him a very astute and effective political operator. We also had Paul Geissler in our group, whose unassuming get-along demeanor was often our secret weapon in getting what we wanted.

The style of decision making on NASA space missions is confrontational and adversarial. The projects are divided into various scientific and engineering teams, whose jobs are to advocate for particular points of view. I suspect that this structure developed in response to engineering being a series of compromises between objectives and constraints, which involve understanding the best arguments, pro and con, for every issue. Whatever its origin, the structure and process yield decisions based on effective advocacy rather than on cooperation. The best space lawyer or politician gets his way. If the same amount of energy that went into the adversarial debate had been put into a cooperative approach to optimizing the mission, decision making would have been far more effective.

When the possibility to revisit The Fault at high resolution became available, Randy persuaded me that we would need to fight to get it into the imaging sequence, and that it would be worth the trouble. By this time in the mission, these decisions were being made via telephone conferences. Early in the *Galileo* project, back in the 1970s and 1980s, the Imaging Team had to meet every few months, usually at JPL in Pasadena. The plan had been that during the time that the spacecraft would be in orbit at Jupiter, we would spend most of our time in residence in Pasadena, so we would be able to work together continuously on updating the image sequence plans and on the image interpretation. By the time *Galileo* actually did get into orbit, the team members had gotten old and no one wanted to spend months on end together in California. Fortunately, thanks to the delays in the mission, modern telecommunications and data distribution were finally adopted, so people could butt heads at a distance from the comfort of their own offices.

For the teleconference for planning the images for orbit E17, my research group and I gathered around a speakerphone to go to battle for high-resolution images of The Fault. As always, the picture budget was limited. Also, we were up against the hobby horses of some very powerful operators. Ron Greeley needed to take pictures of his non-existent lava flows. Even worse, there was an unusually large range of competing interests in the planning process, because a computer glitch during orbit E16 meant none of the planned images of Europa had been obtained, and people wanted to make up during E17 for what they had lost. As if the challenge were not great enough, Randy also believed that an undercurrent of jealousy and animosity from the team's powers toward me undermined my effectiveness, so I kept a low profile. As head of our research group, early in the teleconference I did make a few formal broad points about the significance of Astypalaea Linea, but then let Randy with his deep understanding of The Fault and with his political sophistication carry the ball.

Ultimately, we did get a few frames that covered the region at about 200 m/pixel, and a limited set of high-resolution frames (42 m/pixel) that spanned about a quarter of the length of The Fault. The region imaged in this sequence proved to be remarkable in that it shows most of the major types of tectonic features found on Europa, including strike–slip, dilation zones, double-ridges and cycloids. It shows them at very high resolution, combined in interesting and unique ways.[1]

Figure 13.1 shows a mosaic of the high resolution images, superimposed on the 200-m images. Notice that the level of detail visible toward the middle (where we have 42-m/pixel data) is not available toward the sides (where we only have 200-m/pixel data). The bottom of this mosaic, where we have neither 42- nor 200-m/pixel data, is blacked out. Compare this image with Figure 12.1a, the *Voyager* image of the same region. In Figure 12.1a, the fault runs from the upper right to the lower left. Here, it runs vertically from the top to the bottom, and only covers 1/4 of the length. At the top of Figure 13.1 lies the parallelogram seen near the upper left end in Figure 12.1a, and near the bottom of Figure 13.1a Astypalaea is crossed by the cycloidal ridges that appear about 1/4 way down the fault in Figure 12.1a.

The mosaic of high-resolution images runs about 8,000 pixels from top to bottom. All the detail cannot appear on this single printed page, so we will examine portions of the region in larger blow-ups. First, though, let us consider the overall geometry exhibited in Figure 13.1. It becomes evident that this fault includes a set of parallelogram-like pull-aparts (as diagrammed in Figure 13.1b), not simply strike–slip with the one large parallelogram that was identifiable in the *Voyager* image (Figure 12.1a).

A simple strike slip with a pull-apart feature is shown schematically in Figure 13.2. At the top, shear is beginning along the horizontal sectors of the fault, but, at the jag in the fault, this displacement leads to dilation, as shown at the bottom. In the dilation zone, new material comes up, fills in, and freezes in place. On Europa, the slight diurnal working of the crack would cause parallel grooves within this pull-apart, as sketched in Figure 13.2, with a distinct center line where the newest material comes up. This picture seemed to explain the geometry of Astypalaea as seen at *Voyager* resolution.

At high resolution (Figure 13.1), we see much more detail. Astypalaea initiated as a somewhat cycloid-shaped fault, as shown schematically in Figure 13.3. Here, strike–slip opens a sequence of parallelograms. This geometry is evident on inspection of Figure 13.1, where it is marked explicitly.

If we look closely at any of the pull-aparts (e.g., Figure 13.4), we see all the usual features of dilational bands. The fine parallel furrows surround a central groove (marked by an arrow). At the edges are low ridges that developed as the gap was just beginning to open. More pronounced ridges are also evident along the lines of shear that separate this parallelogram from the ones just above and below it. Because these boundaries were in shear, rather than being pulled directly open, the crack remained closed long enough for diurnal working to build ridges. (The crack to the left of the arrow formed after Astypalaea, so it is not relevant to this discussion.) If

---

[1] Greg Hoppa recalls that one reason we were granted the high-resolution images of Astypalaea was that Pappalardo and Head wanted to see the cusps of the cycloidal ridges at high resolution. They believed that the morphology would prove their "linear diapirism" story for ridge formation (Chapter 10), and disprove my group's tidal-squeezing model. Once the images came down, we never heard any more from them on the subject. The appearance at the cusps is perfectly consistent with our ridge-building model, and very difficult to explain with linear diapirism.

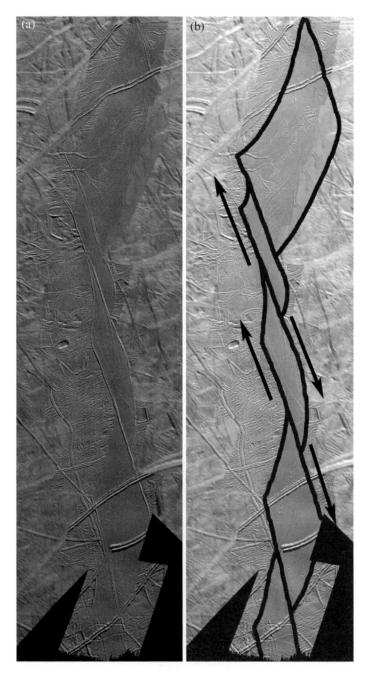

Figure 13.1. A mosaic of the high-resolution (42-m/pixel) images of Astypalaea covering about 1/4 of its length (200 km of 800 km), which was all we could get from the negotiations. The marked version at the right shows the displacement of the neighboring terrain, with lines of shear and corresponding pull-apart areas that look just like other dilation bands.

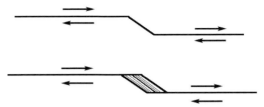

Figure 13.2. A schematic of the general geometric relationship between strike–slip displacement and a pull-apart zone.

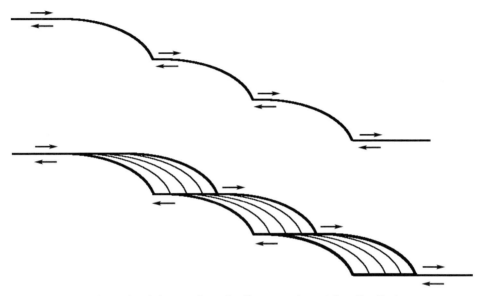

Figure 13.3. A schematic of the opening of pull-aparts when strike–slip displacement occurs along a cycloid-shaped crack. At the top is a cycloid-shaped crack; at the bottom is the geometry that can result from strike–slip (shear) displacement.

we closed up this parallelogram, the adjacent older terrain at the lower right would match the terrain at the upper left.

The double-ridges that developed along both sides of the shearing sections of the fault were eventually themselves sheared apart, like trains passing in the night. We can see this effect in, for example, Figure 13.5, which shows the shear zone between two parallelograms (near the bottom of Figure 13.4). The ridges have sheared apart so they are now double only in the middle of the shear zone. The central groove from Figure 13.4 cuts in between the two ridges, and then emerges from between them at the bottom to continue as the central groove in the next parallelogram. All of these images also show tiny "secondary" craters, formed by bits of ice that were ejected when other, larger craters, perhaps Pwyll, were formed.

The next parallelogram (Figure 13.6), near the southern end of the region imaged at high resolution, shows details of tectonic features that have not been

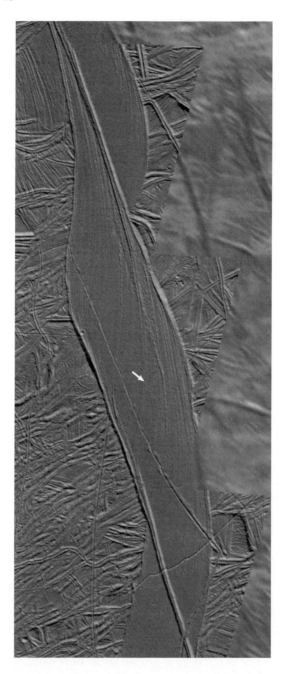

Figure 13.4. Enlärgement of part of the high-resolution mosaic of Astypalaea in Figure 13.1, showing detail of the pull-aparts and their similarity to other dilational bands, especially the fine furrows. The central groove is marked with an arrow. Its path can be traced down the center of the pull-aparts and between the offset double-ridges on the shear segments.

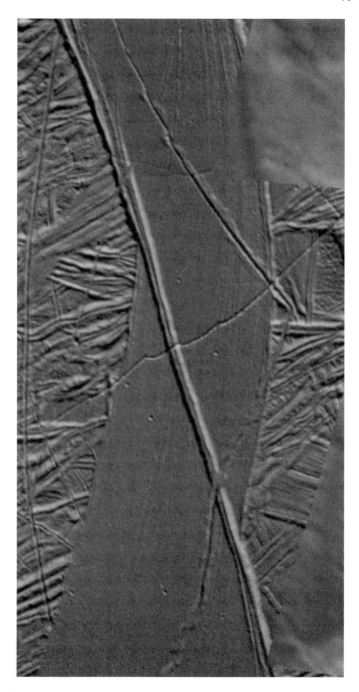

Figure 13.5. Enlargement of another part of the high-resolution mosaic of Astypalaea in Figure 13.1, showing detail of one of the shear segments, with ridges passing like two trains in the night.

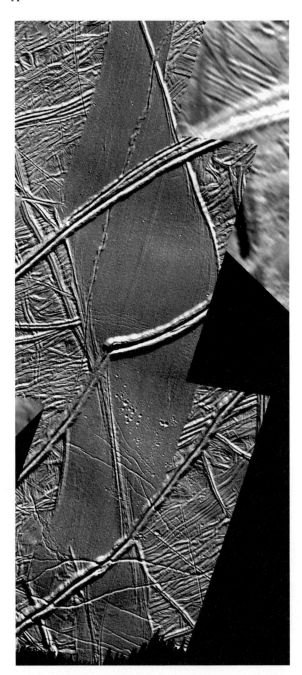

Figure 13.6. Enlargement of part of the high-resolution mosaic of Astypalaea in Figure 13.1, showing detail toward the southern end of the portion imaged in this sequence. This area includes parts of the cycloid-shaped ridges that cross Astypalaea, as well as the corrugations (25-km wavelength) in the pull-apart zone that run orthogonally to the fine furrows.

seen elsewhere. They happen to include (crossing Astypalaea) some of the best images of the details of cycloidal ridges (see Chapter 14 for discussion of these chains or arcuate ridges). Remember, the cycloids had been evident, and puzzling, ever since we obtained the *Voyager* images, and some of the most prominent ones imaged by *Voyager* cross directly over Astypalaea (Figure 12.1a). In the high-resolution views (Figure 13.6), we see that close up they are very typical double-ridges. These views of cycloids even show fortuitously the detailed appearance of a cusp of one of the cycloids. Note that just to the left of the cusp the double-ridge seems to fade away; that is an observational effect caused by the lighting which happens to run along the ridge there, so the morphology becomes nearly invisible. Another cycloid displays, at high resolution, a type of branching that seems to be fairly common. Only here can we see it in detail. Cycloids are discussed in detail in Chapter 14.

In Figure 13.6, I have increased the contrast somewhat to bring out a feature unique to this set of images, a set of corrugations indicated by three or four dark zones that appear to be subtle troughs running perpendicular to the furrows in the pull-apart zones. The spacing of these troughs is about 25 km (Figure 13.6 is about 75 km from top to bottom). The trough near the center of Figure 13.6 happens to have, by chance, a spray of secondary craters in it. There are also sets of fine cracks crossing Astypalaea, which may correlate with the rises between the troughs. If the lithosphere has been corrugated, it seems plausible that fine tensile cracks would form near the tops of the uplifted portions. One set of such cracks is seen just above the cusp of the cycloidal double-ridge. This set is clearly on one of the uplifted parts of the corrugation, although the center of the uplifted area seems further upward on this image, so the correlation is not perfect. The set of cracks just below the spray of ejecta craters does seem to be right at the peak of a rise, but the last set of cracks, at the bottom of Figure 13.6, seems to be at the bottom of a trough. Thus the cracks do not seem to be perfectly consistent with the interpretation that they formed in conjunction with the corrugations.

In principle, corrugation could be a way to reduce surface area and contribute to balancing the surface area budget. However, this type of feature may be unique to the Astypalaea region, or on the other hand it may be common and only happens to be visible with the high-resolution and smooth surface at Astypalaea. The issue of surface convergence, and how and where it balances the creation of new surface is discussed further in Chapter 17.

Astypalaea Linea consolidated everything that we had seen on Europa. The 40-km offset on an 800-km-long fault confirmed the mobility of the crust over a low-viscosity decoupling layer. Right-lateral displacement is consistent with the tidal walking theory. The pull-apart segments have exactly the same morphology as typical dilation bands, confirming our interpretation of those bands, because here we know precisely the pull-apart geometry and driving mechanism. The ridges along the shear zones, sheared apart like trains on opposite tracks, confirm the complete penetration to liquid that both produced the ridges and allowed strike–slip displacement. Everything we see at Astypalaea is consistent with thin ice over liquid water.

14

Cycloids

Among of the weirdest things revealed about Europa by *Voyager* images in 1979 were the prominent cycloid-shaped ridges, found especially near Astypalaea and points east, where fortuitous lighting made these features especially prominent. We have already seen (Figures 12.1a and 13.1a) that several of these ridges crossed over Astypalaea after crustal displacement had already occurred.

The International Astronomical Union (IAU) decided to call these features "flexi" and gave official names to some of them, but not others (Figure 14.1). Everyone involved in studying these features calls them "cycloids", in reference to the geometric shape that is a chain of arcs connected at cusps. As usual, the IAU nomenclature is mysterious and inexplicable, the inevitable result of a huge amount of discussion. At least in the case of the cycloids, I know of no major misconceptions that resulted.[1]

I had been thinking about these features for a long time when my daughter Leontine, then in junior high school, asked me to explain the regular chains of island arcs found on Earth. The island arcs mark sites of crustal convergence, where crustal area is removed, to balance the creation of new surface at dilation zones. At the moment, one of the most dramatic examples of dilation on Earth is the spreading of the Atlantic Ocean seabed, although for those who prefer to contemplate dilation where it is not covered by water, there is the East African Rift and its extension parting the Red Sea. Most of this expansion is accommodated by convergence of plates, where one rides up over another, especially at this moment around the periphery of the Pacific Ocean.

When one plate is bent down under another, it is forced into the hot mantle, where it begins to melt. This crustal material is the least dense type of rock. That is why it had floated up to the surface of the Earth and formed the crust. As it is melted

---

[1] Unlike the "lenticulae", for example, which have been the basis of much confusion and misunderstanding, as discussed in Chapter 19.

Figure 14.1. Cycloidal ridges that appear prominently in a *Voyager* image, near the far south of Europa. For location, compare this image with Figure 9.1, noting the familiar Astypalaea at the left (running under the "e" in Sidon Flexus and the "l" in Delphi) and the large dark chaos area (dubbed Thrace Macula) at the top, right of center.

again, this low-density rock tries to squeeze its way back up to the surface, through any cracks it can exploit in the overlying crust. When it reaches the surface, it forms volcanoes, building chains of mountains or islands, that mark the region of plate convergence. This type of volcanism, in which recooked crust floats back up to the surface is called "andesitic", named after the mountains that formed this way as the Pacific crust dipped under South America.

Usually, these chains of mountains form arcs at the Earth's surface. The current widening of the Atlantic is balanced by subduction especially around the Pacific, and that is where the arcuate chains of volcanism are found. The Bering chain of islands extends west from the mainland of Alaska in a great arcuate swoop. The west coast of South America, namesake for andesitic volcanism, has the form of a two-arc cycloid.

The question that Leontine asked in the late-1980s was: Why are these features arcuate in shape? The standard answer follows from simple three-dimensional geometry. If you try to bend a plate on the surface of a sphere down into the interior of the sphere, the crease will tend to follow a curve along the surface. Imagine making a cut in a hollow rubber ball and then bending down the surface. The crease would form a curve. Therefore, it has been assumed that when crustal plates on Earth dip down under others, they follow curves dictated by this geometry.

It occurred to me that the same effect might explain the cycloids on Europa. My student Mike Nolan and I worked out a theory of the optimally-efficient geometry for dipping one Europan plate under another. The theory was very elegant, but it never was published in detail. The main problem with that model was that it was difficult to understand why the initial cracking would predict the arcuate form that would be the most efficient folding pattern for a plate that would not bend downward until later, as it dove under the adjacent plate along the crack.

I always found it interesting that the same basic flaws in our model for arcs on Europa applied equally to the standard explanation for the shape of island arcs or of subduction boundaries in general on Earth. Yet, for the Earth, the model is accepted as part of canonical knowledge. Already, I was beginning to suspect that the canonical acceptance of scientific theory might not be correlated with whether the idea was well founded or true. In this case, once *Galileo* images were in hand, it became evident that, whatever the true mechanism may be on Earth, on Europa the model did not apply.

Galileo images showed that cycloidal lineaments are ubiquitous on Europa, and that they display all the same types of evolved morphologies as any other cracks. We have already seen several important examples in Chapter 13. Astypalaea Linea proved to have started as a cycloidal crack, albeit a somewhat irregular one, which then sheared along parallel portions of the crack and dilated along others. The several cycloids that formed later across Astypalaea (Figure 14.1) are long, beautiful examples that display typical forms and dimensions: Each arc is ~100 km long and chains often comprise a dozen or more connected arcs.

We also saw the remarkable cycloid in Figure 12.14, where the chain of arcs curves around a large region, hundreds of kilometers across, which has rotated slightly, yielding strike–slip displacement, and some dilation, along the crack. In Figures 11.3 and 11.6 we saw that most of the characteristic dilational bands in the Wedges region started out as cycloidal cracks.

A beautiful set of fairly-fresh cycloids that runs roughly north–south in the far north is shown in Figure 14.2. Along much of the length of these cracks, double-ridges have formed. Elsewhere the cracks have opened, forming dilational bands.

The double-ridge morphology that is typical of cracks that have not undergone displacement is common along cycloids. A pretty example that runs east–west across Figure 12.14 is enlarged in Figure 14.3, along with high-resolution details of several others.

If my earlier idea had been correct, that these features were sites of crustal convergence, where one plate had ridden up over another, the morphology would have been very different. We would have expected to see an asymmetry from one side to another, because one plate rides up, as the other is pushed down. On Earth, in front of island arcs there is a steep drop-off from the edge of the overriding plate, down to deep trenches in the floor of the ocean over the subducting plate. Even where subduction is only incipient, as in a thrust fault, where one plate rides up along an obliquely-inclined crack, there is a scarp, or cliff, where the surface drops from the higher plate down to the lower one.

Nothing like that is found along the Europan cycloids. Instead, the ridges are as symmetrical as along any other cracks. From this morphology, we have every reason to believe that these features began as simple tension cracks, like most of the other lineaments on Europa. In fact, we have also found cycloidal cracks that have not developed ridges. In general, these features are difficult to find because they are subtle, but Greg Hoppa and Randy Tufts have identified quite a few of them (e.g., Figure 14.4).

Figure 14.2. Cycloidal ridges (with some dilation) in the northern hemisphere (E15 RegMap 02) near 80°W, 60°N. The E15 RegMap 02 sequence was taken at about 230 m/pixel, but this image is somewhat foreshortened by the viewing angle toward the pole.

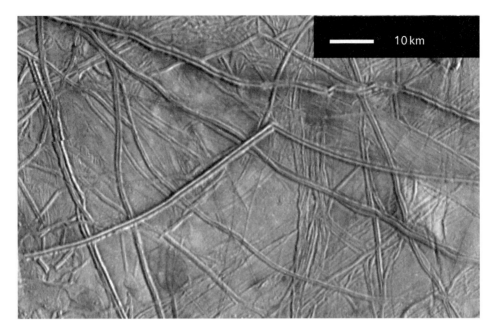

Figure 14.3. Examples of the double-ridge morphology of many cycloids. (a) An enlargement of part of the E17 RegMap 01 image sequence (also shown less enlarged in Figure 12.14), imaged at 230 m/pixel.

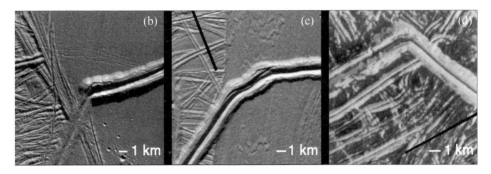

Figure 14.3. (b) A high-resolution view of the cusp of a cycloid that crosses a pull-apart portion of Astypalaea Linea, from the E17 high-resolution sequence (39 m/pixel) discussed in Chapter 13. Note that the ridges seem to disappear to the left of the cusp because the illumination is along the ridge line, illustrating a significant selection effect that can hide ridges. (c) Cycloid cusp in an E17 high-resolution image of Libya Linea at 40 m/pixel. (d) A cusp near the crater Cilix (Chapter 18) imaged during orbit E15 at 64 m/pixel.

The cycloids had impressed me as such striking features, so unique to Europa, and so ubiquitous there, that they seemed to be a key to the character and processes of the place. I tried to keep them as a high-priority issue for my research students. Of course, I was still thinking about ways to rescue my old idea about plate

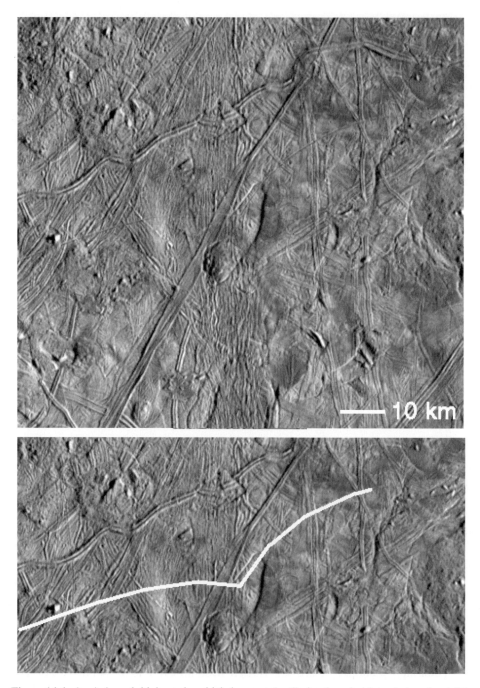

Figure 14.4. (top) A cycloidal crack, which has not (yet?) developed ridges, is barely visible crossing from east to west, in this image just southeast of Crater Manannán. The key at the bottom helps locate the crack, which is marked in white.

convergence. Fortunately, my students Randy Tufts and Greg Hoppa considered other possibilities.

Remember, in broad patterns, the lineaments of Europa seemed to correlate with tidal tension. And Randy, as the structural geologist in my research group, kept emphasizing that ice tends to fail in tension; it is much stronger in its resistance to shear or compressional failure. Randy attacked these problems of tensile cracking with an intuitive and graphic approach, and he was giving a lot of thought to the cycloids.

He shared a temporary cubicle in my lab with Greg Hoppa, and as *Galileo* images became available, he printed them on sheets of standard-size paper, scotch-taping them into huge mosaics covering every wall and partition in sight. Greg and Randy discussed these images constantly. At the same time, Greg was producing the global stress plots, which, in the format that I had designed, gave an intuitive, graphic picture of how tidal stress changed over time.

Randy often came to our weekly research group meetings with notebook sketches of how the tectonic structures of Europa might have evolved. We had seen similar sketches used by other geologists associated with the *Galileo* Imaging Team. We physicists in my own group called it "cartoon science". The geologists seemed to think that if you could draw a cartoon scenario, it might be what happened on Europa. These drawings formed the basis for many of their published interpretations. Most of the drawings reminded me of *Roadrunner* cartoons where the coyote runs off a cliff and does not fall until he realizes there is nothing holding him up. Just because you can draw it, does not mean that there is any real support. Randy assured me that the problem was not that geology is bad, just that we were seeing bad geology. In any case, with so many physics-types in my group, Randy was not likely to sell any cartoon physics, unless it could be supported by quantitative analysis.

One week Randy showed us a series of sketches of a crack propagating across the surface of Europa, as the direction and magnitude of the stress changed according to Greg's calculations. The result was a smoothly-curving crack trajectory. Naturally, I was skeptical, but Greg had been in on the development and the stresses were based on quantitative theory. Greg and Randy worked up the model in detail. To the delight and excitement of everyone in our group, they showed that the patterns that follow from propagation of tensile cracks over the course of several Europan days follow exactly the patterns that were observed on Europa.

Consider the crack propagation shown in Figure 14.5. Here, initial cracking occurs at a selected location in the far south near 69°S, 260°W, at the time (in this case 33 hours after orbital apocenter) when the tensile stress exceeds the strength of the ice. As the crack propagates eastward, the diurnal tidal stress changes direction and magnitude, so the crack gradually curves so as to be orthogonal to the tension. Eventually, in this case, about 35 h after the crack initiates, the tension falls to such a low value that even the active crack can no longer propagate and the crack goes to sleep. Then, roughly one Europan day (or orbital period) after the crack first began to propagate, the tension rises again to the critical value allowing continuation of crack propagation. At this time, the direction of the tension is similar to what it had

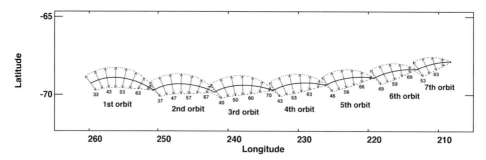

Figure 14.5. Propagation of a cycloid from an initial cracking site (at left) in the far south of Europa. At 33 h after apocenter, the tidal stress has become great enough to initiate tensile cracking. As time goes on, and the crack propagates to the east, tidal tensile stress (arrows) changes, resulting in curved cracking, between intervals when the stress is too low for crack propagation, which results in formation of cusps.

been a day earlier (although slightly different because the location is now farther east); the direction is quite different from what it had been when the crack had gone to sleep, so the new direction results in a cusp.

This sequence repeats itself, each day yielding a new arc and cusp. A typical cycloid of 10 or 12 arcs takes that many Europan days to form, or about one Earth month.

The details of the diurnal change in tensile stress vary with each day, as the crack propagates to locations hundreds of kilometers away from the starting point, so the chain may gradually change direction; the size and shapes of arcs may change along the chain as well. This effect is evident to a moderate degree in the case shown in Figure 14.5. Depending on the location of the crack, more extreme changes may occur along the chain. For example, the modified cycloid in Figure 14.6 also follows exactly the shape dictated by diurnal variations in the stress at its location on Europa.

This theoretical model has proven successful at explaining the shapes and orientations of cycloids all over Europa, and has been widely accepted as the basic explanation for these crack patterns. However, as one of the developers of the model, I cannot fully set aside some caveats and questions. One concern is that the geometry of any cycloid depends on the selected values of several unknown parameters. First, the strength of the ice (i.e., the critical value of the stress at which a crack starts to propagate) is very important. In the diurnally-varying stress field, it determines the initial azimuth direction of the crack. Second, the rate of propagation determines the curvature of each arcuate path. Very fast propagation will yield straight crack trajectories if the crack forms before the stress changes significantly; slow propagation would yield tight curves. Third, the critical stress below which a crack stops propagating determines how far each arcuate segment can propagate. Fourth, the elastic parameters of the ice crust as a whole determine the stress that is induced by the distortion of the shape of the satellite.

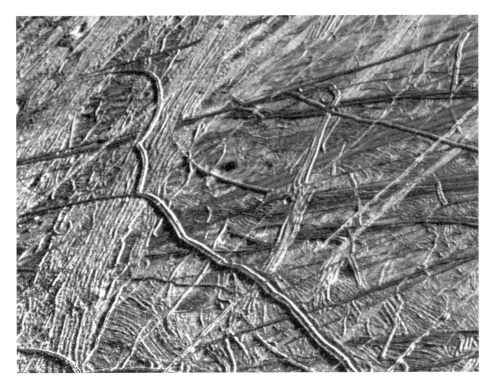

Figure 14.6. Cycloids can take on irregular geometries, depending on the diurnal variations in the stress field along their particular propagation trajectory. This one is in the far north near 70°N, 220°W, imaged during E19 in the northern extension of the RegMap 01 sequence at about 170 m/pixel.

The difficulty is that we do not know a priori the values of any of these parameters. In order to simulate the observed cycloid patterns, we have had to adopt values that yielded similar crack patterns. In that sense, the theory of cycloidal crack patterns is much less convincing than it would have been if we had been able to use known parameters based on understanding of the properties and behavior of ice. Unfortunately, we do not know enough about the strength or cracking processes of ice in general, nor of the conditions of ice on Europa, to have independent ways to determine the parameters for our theory.

However, we have found that most of the cyclical features on Europa require similar parameters in order to fit the model. In fitting the theory to the observations, we have found that the required value for the stress to initiate cracking at a cusp is typically comparable with the tensile strength of ice. The critical stress value at which propagation stops (setting up the next cusp) is somewhat smaller, but comparable. The required speed of propagation is a few kilometers an hour, comparable with a typical human walking speed. (The Discovery Channel illustrated this parameter on television by showing footage of me walking across the Arctic ice sheet next to an animated propagating crack.) Our calculations all assume the same elastic parameters for the ice crust as were used in our calculations of global tidal stress (Chapter 6). We do not know enough about the properties of Europan materials to know independently whether these parameter values are correct. Nevertheless, the consistency of inferred parameter values for most cycloidal features offers some assurance that the model is a good representation of the process that controls many of the important crack patterns on Europa.

It is possible to account for some diversity of patterns with modest changes in the parameters from one case to another. For example, if the propagation rate is made modestly dependent on the tension that drives the cracking, the radius of curvature will vary along each arc, producing a regular, slightly-skewed arc geometry that is observed in many cases.

While such variations are to be expected, they are only minor. The general apparent uniformity of the properties that control crack propagation is most remarkable. All cycloids, wherever they are, seem to have developed in similar material. If there had been differences in strengths, thickness, or structure, one might have expected significant effects on crack patterns. Moreover, many individual cycloids have propagated for hundreds of kilometers with extremely smooth regular cracking patterns following the figure predicted by diurnal stress. The icy crust of Europa must have been fairly uniform over long distances during the month that it took to create each of the long, regular cycloidal cracks. Moreover, the crust has been similar wherever and whenever other cracks form.

Not all cracks seem to follow cycloidal patterns, however. In part, this observation may be due to the fact that the surface has been so sliced and diced by tectonic displacement that older cycloids get cut into segments too short (<100 km) for us to recognize the cycloidal pattern. Most recognizable cycloids are fairly recent in the record of cross-cutting lineaments. Older tectonic lineaments are often slightly curved, probably because only a short segment of the original tide-driven crack pattern remains. Furthermore, we know from the existence of chaotic terrain that from time to time and place to place, the properties of the ice must have been atypical. Some areas must have been thermally modified at times, so crack patterns would be irregular. Full-fledged cycloids required the ice to have been fairly uniform over a large area for at least a month. Most significantly, the fact that all cycloids require similar parameters means that, whenever the surface has become uniform over a large region, it has returned to a similar condition to all other times and places where cycloids have formed.

Taking a global perspective, Figure 14.7 shows the crack patterns predicted to form due to diurnal stress variations according to our theory. Here, we have simulated the propagation from starting points spaced evenly in $10°$ intervals in latitude and longitude, with propagation starting toward the west. (Similar patterns are produced for eastward propagation, although the curvature of the arcs and the direction of the cusps is reversed.) The crack patterns depend on the rate of crack propagation, which controls the curvature of each arc, and on the stress levels required to start and to maintain crack propagation, which limits each arc and yields cusps. The crack patterns generated for Figure 14.7 are based on parameter

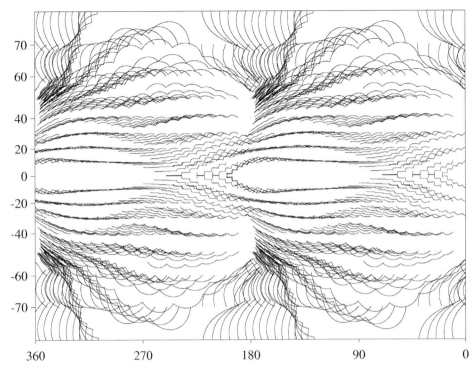

Figure 14.7. Theoretical patterns produced as cracks propagate at a finite speed through a diurnally-changing stress field from starting points spaced every 10° in latitude and longitude and propagating initially westward. The general shape is cycloidal, although rounded boxy patterns are produced as well, which may explain the crack patterns (though not the dilation) in the Wedges region and in the region diametrically opposite the Wedges. Patterns are shown in a Mercator projection. As always, longitude 0° faces toward Jupiter.

Diagram by Gwen Bart.

choices similar to those that yield reasonable fits to observed cycloids. These parameters are plausible for Europan ice, and they yield cycloids similar to those observed on Europa. However, the crack trajectories shown can only be considered qualitatively representative of the patterns to be expected at any location. A specific crack's path will depend on details of the local parameters.

For comparison, Figure 14.8 shows a map of cycloidal lineaments actually observed on Europa. (This map is affected by observational selection: The clustering in certain regions is due to the fact that imaging conditions were not uniform. Cycloids only seem rare in the locations where imaging was inadequate to reveal them.) Comparison of Figure 14.7 and 14.8 shows how well the model agrees with the observations.

The long east–west cycloids observed (Figure 14.8) in the far south around longitude 180° are the ones revealed by *Voyager* in the Astypalaea area (Figure 14.1). The theory predicts similar cycloids at the same latitudes (−50° to −60°),

Figure 14.8. A preliminary map of observed cycloidal lineaments on Europa, marked in black. The features are indicated on a background mosaic image in a Mercator projection, with longitude 90° and the equator running through the center, and with longitude 0° at the left and right sides. Most cycloids have been displaced from their original location of formation (cf. Figure 14.7) by non-synchronous rotation and/or polar wander. The map includes the extreme cases that form circular-to-boxy patterns, which are found in the Wedges region south of the equator and west of 180° longitude (Figure 2.4), and also found diametrically opposite in the sub-jovian hemisphere. These features lie just east of the locations where stress patterns suggest that cracks of this shape and size should form (Figure 14.7), probably displaced eastward by rotation of Europa.

Diagram by Gwen Bart.

but not at the same longitudes. We have interpreted these discrepancies as the result of an eastward shift, since their formation, due to non-synchronous rotation. In fact, Greg Hoppa has shown that a westward shift of the observed cycloids from that area would put their shapes in precise agreement with the theory. This procedure may provide a powerful constraint on rotation rates, by precisely specifying the longitude of formation of each cycloid, as discussed in Chapter 15.

A second similarity between observations and theory is the tendency for long east–west chains to run just north and south of the equator. This pattern is quite prominent on the theoretical plot (Figure 14.7). Such cycloids are observed on Europa at those latitudes, especially in the sub- and anti-jovian regions (around 0° and 180° longitude) where we have adequate image data (Figure 14.8). They cannot be seen at other longitudes, either because image resolution is inadequate, or because such tectonic features have been disrupted by the formation of extensive chaotic terrain (as in the splotchy region from 240° to 340° longitude).

A third point of agreement is that cycloids that run north–south on Europa (Figure 14.8) tend to be relatively short, and are usually far from the equator. The examples mapped in Figure 14.8 in the far north near longitude 90° are the same ones shown in Figure 14.2. This observation is also consistent with the patterns predicted by the tidal stress theory (Figure 14.7).

A fourth similarity between the arcuate cracks formed by theory and those observed involves the patterns in the Wedges region. We have included in the plot of observed cycloids a class of boxy or circular crack patterns that may not display the long chains of arcs typical of the classical (*Voyager*) cycloids, but which do show smooth regular curvature. Most of these cracks are seen in the anti-jovian hemisphere just south of the equator and west of longitude 180°, where they have dilated into the wedge-shaped openings which gave this area its name. At the far western edge of the Wedges region in Figure 14.8, we can recognize the arc of the Sickle dilation band from Figure 11.3.

The agreement between the boxy crack patterns in the Wedges region and theoretically-predicted crack patterns (Figure 14.7) is striking. The theory predicts cycloids that degenerate into stair-step patterns very consistent with the boxy shapes that characterize the Wedges.

These patterns are produced by the diurnal variation in tidal stress. They can only explain the shapes of the initial cracking, and do not directly address the issue of dilation, which is so prominent among the Wedges.

Dilation in the Wedges region has often been attributed to tidal stress due to non-synchronous rotation, but that proposed mechanism is based on a common misconception. The idea is that this region is currently just west of the anti-jovian longitude (180°), so that non-synchronous rotation is carrying it toward the top of one of the tidal bulges, where it is being stretched, causing dilation. In fact, non-synchronous rotation does produce tension in this region (Figure 3.7). However, the magnitude of the corresponding strain is very small. We have already discussed in Chapter 10 that a tensile stress of about 10^5 Pa, which would be produced by a degree of non-synchronous rotation, corresponds to a few meters of stretch per 100 km. Even if non-synchronous stress could accumulate indefinitely, without relaxing, which is implausible, the crust would be stressed at most about 100 m per 100 km, which would yield a negligible amount of dilation, compared with what is observed in the Wedges region, ~10 km per 100 km (see Figures 11.3, 11.6, or 11.7). Thus, what has driven dilation there remains unknown.

If the characteristic appearance of the Wedges were driven by tides, we would expect to find similar features 180° away, in the sub-jovian region. In fact, there does not appear to have been much dilation there, consistent with our conclusion that dilation has not been driven by tides. However, theory (Figure 14.7) would predict that initial cracking patterns should be the same in both hemispheres. In fact, it appears that the same type of tightly-curved, boxy crack patterns that are found in the Wedges are also located diametrically opposite the Wedges, just north of the equator and just west of 0° longitude (Figure 14.8, this locale is shown in Figure 17.3).

The similarity in shape and scale of the observed features to those predicted is striking, both on the sub-jovian and the Wedges region. In comparing the two regions it is important to remember that the similarity is in the original crack patterns, not in the current character of the cracks, which have dilated throughout the Wedges region, and not at all in the sub-jovian region.

In comparing the global patterns of cycloids, from theory and observation, we find that some features seem out of place, usually displaced in longitude. This displacement probably records rotation that has occurred since the cracks formed. Some cycloids seem out of place in latitude as well, probably due to polar wander. Because we seem to have a theory that gives tight constraints on where each crack occurred, we have a tool for unraveling the history of Europa's non-synchronous rotation and other changes in the orientation of its shell relative to Jupiter. This approach may be especially powerful where we have cross-cutting intersections with morphologies that allow us to determine the order in which these features formed.

In any case, most of the features with recognizable cycloidal patterns are very recent in the recorded geological history of Europa. Older faults have been sliced and diced by subsequent tectonics to the point that they are no longer recognizable. Because the recorded geological history of Europa only goes back a few tens of millions of years, due to continual resurfacing, the useful record of cycloids goes back only a few million years at most. Yet, even in that short time significant and complex reorientation of the ice shell has occurred, according to the evidence for polar wander and non-synchronous rotation.

Our understanding of the origins of cycloidal patterns provided the first definitive evidence that Europa did indeed have a global ocean. (At least, the *New York Times* and the BBC called it the "strongest evidence yet" and the "most convincing evidence yet", respectively.) The strong correlation between theory and observation showed that tides were driving the cracks, and only with a global ocean could there be adequate stress. Randy and Greg's discovery was front-page news across the U.S. My own feeling was that the general characteristics of the tectonic and chaotic terrains were already convincing enough evidence for the ocean. But after we explained the cycloids, even the most conservative scientists acknowledged that there had to have been an ocean.

But they still would not agree that there necessarily is an ocean at the present time. It seemed to me, given that there had still been an ocean during at least the past 1/100 of a percent of the age of the solar system while the cycloids were cracking, and given that the geological activity of Europa had been fairly constant over at least the past 1% of the age of the solar system for which there is a record, that it would be fairly strange to think that the ocean would have suddenly (on the cosmic scale) frozen just in time for *Galileo*. It seemed obvious to me from the geological record that the ocean must still be there.

Over the next couple of years, after we had developed the cycloid model, additional evidence for a currently-existing ocean accumulated from the magnetometer on *Galileo*. This instrument had measured changes in the magnetic field around Jupiter as *Galileo* swooped past Europa. These changes were consistent with what would be expected if Europa had a global shell of electrically-conducting material, such as salt water, just below its surface. According to the theory, as Europa orbited through Jupiter's magnetic field, electrical currents would be induced in the conducting shell, in turn generating a magnetic field that would add to the jovian field. The changes detected by *Galileo* near Europa fit that model. This result complements the strong evidence from cycloids that an ocean was present while the current surface was being created, indicating that the ocean still existed as recently as the late 20th century.

Cycloids remain a critical part of the evidence, however, for a couple of reasons. First, the magnetometer results depend on acceptance of the conducting ocean as the unique way of explaining variations in the detected field. That model is simple and elegant, so Ockham's razor gives it great credence. However, whether any other adequate models might work is an open question. The existence of cycloids, therefore, provides essential evidence. Second, the fact that cycloids formed so recently compared with the age of Europa would make it very implausible that the ocean would not still exist today. So, cycloids, as well as magnetometer data, do relate to the current state of Europa. Both the cycloids and the magnetic field model offer good evidence for the ocean. With both together the case is very strong.

15

Rotation revisited

15.1 CYCLOID CONSTRAINTS ON THE ROTATION RATE

Development of understanding of the formation of the distinctive crack patterns, the cycloid and related arcuate shapes, provided us with a tool for unraveling the history of the orientation of Europa's shell relative to the direction of Jupiter, going beyond what we already had found. We had theoretical reasons for expecting non-synchronous rotation, and we already had what seemed like good evidence that such rotation was significant, based on the systematic changes in orientation of cracks that Paul Geissler had noted in the Udaeus–Minos region (Chapter 9). These results had indicated that the rotation period was less than a few million years. We also knew from Greg Hoppa's comparison of *Voyager* and *Galileo* images that the rotation had to be slow, with a period greater than 12,000 years. With our understanding of cycloids, we could begin to narrow that range of uncertainty.

As a model for what may eventually be done with the complete set of known cycloids, Greg Hoppa showed how the original *Voyager* cycloids could be used to constrain the rotation period. While these features were prominent in an important *Voyager* image (Figure 14.1), they were also imaged under several lighting conditions and viewing geometries by the *Galileo* spacecraft.

We considered three of these cycloids that have cross-cutting relationships, which define a time sequence for their formation. The three are the group that cross Astypalaea in a tight formation, as shown toward the left center of Figure 14.1. They include Sidon Flexus and Delphi Flexus, labeled in Figure 14.1, and the unnamed cycloid (with relatively short arcs, concave northward) that crosses Astypalaea between the other two. Short segments of these three cycloids also were imaged at very high resolution where they cross Astypalaea (at the bottom end of Figure 13.1), including a cusp of the unnamed cycloid, which is also shown in Figure 14.3b.

This cycloid was never named by the International Astronomical Union (IAU), perhaps because it was somewhat foreshortened, and thus less obvious, in the *Voyager* viewing geometry. When my student Mike Nolan did a survey in the early 1990s of cycloid properties from the *Voyager* images, he labeled this one "G" in honor of his thesis advisor. Later, when Hoppa et al. published the paper showing how these cycloids constrained Europa's rotation period, that notation was retained, and to my students' amusement I became the only person to have had a feature on Europa named after himself.

Examination of the points where these cycloidal ridges cross over one another shows that "Greenberg Flexus", although not named after a site in Homeric literature, is older than both Sidon and Delphi. The point of intersection of Sidon with Delphi was imaged only where it was severely foreshortened, but reprojection and careful examination of the image shows that Delphi overlies Sidon. Thus, the order of formation of these three cycloids appears to have been G, followed by Sidon, followed by Delphi.

These features are all typical double-ridges, and the cross-cutting sequence gives the order in which they formed. We can reasonably assume that cracks between the ridges formed in the same order. As we discussed in Chapter 10, after a crack forms under tidal stress, diurnal working will continue to open and close it for some time, until non-synchronous rotation moves the crack into a different stress regime, or the crack anneals closed for some other reason. Ridge formation probably requires that the crack remain active and it would cease once the crack annealed. Another crack would not likely develop close to the earlier one until after the earlier one ceased to be active and thus ceased to be able to relieve tidal stress. Therefore, it is most plausible that ridges form at a crack prior to the next cracking event in the neighborhood. While one might concoct scenarios in which ridges form in a different order than their underlying cracks, and such things could conceivably have happened on Europa, we have adopted the more straightforward and physically plausible assumption.

We can also infer from tidal stress theory the longitudes at which each of these features formed, which may be different from the current location due to the rotation of Europa. We found that Delphi could have formed 75° farther west than its current location, Sidon 52° west, and G 50° west. Crack formation at those locations would give cycloidal geometries that match the observed features.

If these three cycloids had moved from west to east with non-synchronous rotation, the implied order of formation (Delphi, then Sidon, then G) would have been exactly opposite the order that is shown by the cross-cutting sequence. This conundrum is resolved by noting that tidal stress is symmetrical, so that, for example, if Sidon could have formed 52° west of where it is now, it could equally well have formed 1/2 of a rotation period earlier (180° farther west), or 1/2 of a period even earlier, etc. Thus, if Sidon formed 232° (i.e., 180° + 52°) west of its current position, its order of formation relative to Delphi can be reconciled with the cross-cutting record. And because the cross-cutting record shows that G formed earlier than Sidon, and even had time to build a ridge, it must have formed at least $50° + 360°$ (i.e., $50°$ plus one full rotation) west of its current location. At a minimum, it seems that these cracks formed over nearly a full period ($360°$) of non-synchronous rotation.

This result implied that no more than a couple of cracks form in a given region during a non-synchronous rotation period. Such a rate is reasonable if each crack relieves tidal stress in its region; tidal stress cannot begin to accumulate again within that region until the crack anneals. Globally, this crack production rate would correspond to ~50 new tectonic features forming during $180°$ of non-synchronous rotation. Even though the rate of crack and ridge formation is limited in that way, over 60% of Europa's surface has been covered with tectonic terrain (the other 40% is chaotic terrain), and in the region under consideration it is closer to 100% tectonic. These statistics can be used to estimate the number of rotational cycles that it has taken to do this tectonic resurfacing.

Resurfacing the region around Sidon and Delphi (an area roughly $800 \times 1,000$ km, shown in Figure 14.1) with tectonic features (ridges ~2 km wide along 1,000-km-long cracks) would require a minimum of ~400 cracks, corresponding to at least 200 complete cycles of non-synchronous rotation with one new feature forming every half-revolution with respect to Jupiter. (Recall that the symmetry of the tidal stress field means that there are two cycles of tidal stress in every rotation period.) Assuming an age for the surface of 50 Myr based on the crater record and assuming that the rate of tectonic resurfacing and the rate of rotation have been fairly uniform over the surface age, then the upper limit for Europa's non-synchronous rotation period would be 50 Myr/200 or about 250,000 yr.

This result allows tens of thousands of years for each new crack to remain actively worked by diurnal tides, before it anneals and lets stress build up again to create the next crack. This amount of time is more than adequate for the ridges to grow as described in Chapter 10. Most significantly, this line of evidence reduces the upper limit of the non-synchronous rotation period from millions of years to only 250,000 yr. With the rotation period previously constrained to be >12,000 yr from comparing orientations at the *Voyager* and *Galileo* epochs (Chapter 9) we conclude that the rotation period is approximately 50,000 yr, with an uncertainty of a factor of 5.

This analysis was based on only one region on Europa. Given how widely distributed cycloids are on Europa, and how many we have imaged, the work on cycloids in the Astypalaea region probably should be considered only a pilot study. It demonstrates, as a proof of concept, the potential for future studies of the tectonic record for unraveling the history of Europa. When future spacecraft obtain more complete global coverage at adequate resolution and with appropriate illumination, Europa's dynamical history will be laid out and ready to read. In the meantime, the data already in hand from the *Galileo* mission have not yet been fully exploited in this way, and probably still have a great deal to reveal while we wait to return to the Jupiter system.

15.2 CONTRADICTIONS WITH PREVIOUS WORK

One aspect of these results was somewhat disturbing to us. By itself, the idea that only one or two new cracks could form during each rotation relative to the direction of Jupiter was reasonable. That way each crack would have time to grow ridges before annealing and letting the stress build up again for the next crack. The problem was that this result contradicted our original evidence for non-synchronous rotation (Chapter 9), which had been based on the gradual, uniform changes in azimuthal orientation of cracks with time in the Udaeus–Minos region (Figure 9.2). There, we had suggested that numerous cracks of gradually-changing orientations had formed in one region during only a few tens of degrees of rotation, quite contrary to the idea that only one or two cracks could form in a given region during an entire rotation.

This earlier work had been very convincing and was universally accepted for good reason. The model elegantly explained the several sets of lineaments that were progressively and nearly continuously farther clockwise with decreasing age.

The original approach had been so appealing that various other researchers had begun to fit the geological record into a presumed sequence of gradual, continuous changes in crack azimuth. Geological mapping of various regions was interpreted as being consistent with rotations of lineament azimuths with time, and thus with non-synchronous rotation of Europa. Two of those investigations involved the *Galileo* regional mapping area "RegMap 02", where images at about 200 m/pixel cover the broad swath from the far north to far south across the center of the leading hemisphere (Figure 12.9).

Ron Greeley's graduate student Patricio Figueredo studied the northern portion of RegMap 02 (imaged during orbit E15), and constructed a hypothetical history in which various sets of tectonic lineaments were associated with geological episodes, yielding a proposed sequence of formation of lineaments in the region. Within the constraints of this sequence, and comparing the observed azimuths with our plots of tidal stress fields, Figueredo and Greeley arranged the orientations of the lineaments in a sequence consistent with nearly two non-synchronous rotation periods. However, in their proposed geological scenario the sequence of azimuth orientations is discontinuous and non-monotonic. The results by themselves do not suggest such rotation as compellingly as the nearly-continuous, monotonic variation that Paul Geissler had identified in the Udaeus–Minos region. Hence, while the Figueredo and Greeley scenario is not inconsistent with the stress orientations expected from non-synchronous rotation, it does not require it. Another problem was that Figueredo and Greeley attributed the orientation of many lineaments to tensile stresses from our theory that, while appropriately oriented, would have been too weak to crack the ice. Furthermore, the geological scenario that provides the basis for their proposed crack sequence is based on numerous implicit assumptions intrinsic to their geological mapping techniques, so the proposed crack sequence scenario is subjective and may not be unique.

In the southern-hemisphere portion of RegMap 02 (from images taken during orbit E17), Paul Geissler identified sets of lineaments, each set having a typical azimuth, that had formed in a time sequence based on cross-cutting relationships

(Figure 15.1). Various sets of parallel lineaments are evident in Figure 15.1. Most prominently, a large set of parallel ridges running with a 10 o'clock orientation (WNW to ESE) all formed at about the same time in the sequence determined by cross-cutting relationships. That is to say that, as far as we can tell from cross-cutting relationships, all other lineaments are either older or younger than the ones in this set. In a preliminary study shortly after we received the E17 data, Paul identified three such sets. Following up on his earlier work on the Udaeus–Minos region, he suggested that here, too, each of the three sets of parallel ridges was produced during a particular phase of a single rotational period.

However, as with the northern-hemisphere results reported by Figueredo and Greeley, the azimuth variation reported for the south was neither continuous nor obviously monotonic. Thus, neither of these studies of the leading hemisphere provided independent evidence for non-synchronous rotation, unlike the seemingly compelling relationships that Paul had discovered on the low-resolution images of the Udaeus–Minos region on the other side of Europa.

The high-resolution images of the densely-ridged patch of tectonic terrain just north of Conamara (part of which is shown in Figure 2.9) show sets of ridges of various orientations with a fairly clear sequence of formation, based on cross-cutting. In 2002 Simon Kattenhorn, a geophysicist at the University of Idaho, fit this sequence into a model of continuous azimuthal variation, based on our earlier interpretation of the Udaeus–Minos region and determined that the sequence of crack azimuths would require at least 2.5 non-synchronous rotation periods. At this higher resolution, in a locale where the surface has not been disrupted by formation of chaotic terrain, it seemed plausible that the visible tectonic record goes farther back in time, consistent with the greater number of cycles found in the record.

As Simon himself has noted, one problem with that simple interpretation was that at this low latitude (15°N), tidal theory shows that the surface would be in tension only during a small portion of the non-synchronous rotation cycle (Figure 6.3f), at which time it only runs north–south, contrary to the range of crack azimuths there. Thus, the stress history, in this particular location at least, must have involved more complex variations than simply the periodic changes in tidal stress during non-synchronous rotation.

Each of these efforts to fit an observed geological sequence of lineament formation into a systematic sequence of monotonically-varying azimuths yielded a scenario in which multiple cracks formed during a single rotation period, in several cases with many cracks forming within a short fraction of the rotation period at about the same orientation. These results were precisely contrary to what we had inferred from the cycloids in the Astypalaea region.

15.3 BACK TO UDAEUS–MINOS

This inconsistency was not widely appreciated, but Greg Hoppa kept reminding me about it. Because my own group was responsible for the evidence on both sides

Figure 15.1. In the southern-hemisphere portion of RegMap 02 (from orbit E15; for location see Figure 12.9), several sets of lineaments cross one another, each set having a distinct azimuthal orientation and a distinct position in the sequence of formation as indicated by cross-cutting relationships.

(azimuthal variation in the Udaeus–Minos region and cycloids in the Astypalaea region), it seemed especially important for us to resolve the contradiction. We decided to return to the Udaeus–Minos region, where the low-resolution images from orbit G1 (Figure 2.5) had provided the compelling and widely-accepted evidence for non-synchronous rotation. The earlier work had been solid research for its time, but it was based on the limited data from the first orbit, when *Galileo* had not even come very close to Europa. Strangely, even though this same region had been imaged at much higher resolution (200 m/pixel) as part of the RegMap 01 sequence (Figure 12.9) during orbit E15, as late as the summer of 2002 no one had used these images to confirm and improve what we had done with the G1 data. The more recent E15 images (Figure 2.5) were far superior to the G1 data (Figure 12.10) for exploring the cross-cutting sequence of cracks in that area.

Late that summer, Alyssa Sarid began her senior year at the University of Arizona, having already completed and interpreted her productive survey of strike–slip displacement in the RegMap images. Because she was already an expert on identifying and characterizing tectonic features in the RegMap images, I asked her to help me use the E15 RegMap data set to check the results from the G1 data. The first step was a survey of all intersections in the northern part of RegMap 01, where it overlapped the Udaeus–Minos region that we had studied in the G1 images.

We tabulated, for each lineament, all the other lineaments that it crosses where there is a recognizable cross-cutting order. From the cross-cutting, we tabulated which appears to be younger or older. We also listed the azimuthal orientation (or, in geological jargon, the "strike") of each lineament. Because lineaments often follow curved or irregular paths, this value was subject to interpretation but is probably meaningful to within several degrees. Most of the lineaments in the catalog are double-ridges, although other tectonic lineaments, including ridge complexes, dilational features, and other bands are also represented. The cross-cutting order for each intersection is generally based on the appearance of one feature clearly covering over another topographically. Very fine lines on the surface, where topography was unrecognizable, are probably cracks where ridges have not developed. In most cases the cross-cutting relationships of such cracks with other lineaments were not definitive, except in cases where a crack has breached an apparently earlier ridge or band. Where such cracks could be placed in sequence, they were nearly always found to be most recent, just as we had found in the low-resolution G1 data and consistent with the interpretation that ridges have not had time to form on the most recent cracks.

As in our sequencing of cycloids, we assumed that the observed cross-cutting relations among ridges reflect the order in which the underlying cracks formed. This assumption is consistent with the likelihood that ridge formation follows immediately after a given crack forms, continuing as long as diurnal tidal stresses keep the crack active; later, a different stress condition would allow the crack to anneal, stopping ridge growth. Only then could stress build up again to yield a new, differently-oriented crack that might cross the earlier one. With that rationale, a premise of our analysis was that the order of ridge formation corresponds to the order of crack formation.

One complication was that many of the lineaments in the region are ridge complexes, including, among many others, the "triple-bands" Udaeus and Minos. (Remember, most triple-bands have ridge complexes at their core, with diffuse darkening along their margins; other ridge complexes do not necessarily have the dark margins.) On close examination, we found that while some of the ridges in a given complex might go *under* a later, crossing ridge (i.e., one that is not part of the same complex), others in the same complex might cross *over* it. In that case, it appears that after some ridges in the complex were formed, a later crack and ridge formed, crossing it, presumably after considerable rotation of Europa had caused the tidal stress at this location to be reoriented. Then, the first crack system was reactivated, and additional ridges built along it, presumably after further rotation caused the stress to revert to its earlier orientation. If ridge complexes had formed according to our model for these features (Figure 10.11), then it would seem that various secondary ridges in the complexes might have formed during various rotational cycles of the satellite, while other cracks and ridges formed across the complex in-between those times.

The rich record of reactivated cracks in the *Galileo* regional mapping images and elsewhere has the potential for revealing a great deal about the history of Europa, about tidal stresses, and about the tectonic processes involved. So far, this aspect of the geological record has not been exploited, in part because it has been dauntingly challenging. Still, this source of information should be exploited more fully if we are to make the best use possible of the information we already have in hand.

For our rotation study, we avoided the complication of reactivated cracks by treating such intersections as places where we simply could not determine the cracking order from the morphology.

In an ideal data set, each and every crack in the selected area would cross both the next oldest and next youngest, and the morphology of the intersection would unambiguously display their order of formation. Assuming that the expected rotation of the tidal stress field (clockwise in the northern hemisphere) controlled cracking, the azimuth of the newer crack should be further clockwise than the older one. If a newer one is oriented farther counterclockwise, it must have formed in a subsequent rotational cycle.

In order to clarify this crucial point, let us consider an example from a locale that is rich in cross-cutting relationships, the neighborhood of the intersection of Udaeus and Minos, as seen in E15 RegMap data (Figure 15.2). Consider the two ridged lineaments located toward the eastern (right) side of this image that run roughly north–south. They have been cross-cut by a lineament that runs toward the northwest, so the impression is that the stress must have rotated 45° counterclockwise, based on the shortest direction between the two orientations. However, in order to fit the theory, we would have to assume that the stress actually rotated clockwise by about 135°. Similarly, Minos, and then Udaeus, formed even later in the sequence. A direct rotation from the north–south ridge orientation to Udaeus would involve a 70° counterclockwise rotation, but the requirement of clockwise rotation tells us that it must have been by 290° clockwise. Of course, the stress rotation between the north–south ridges and Udaeus might have been greater by

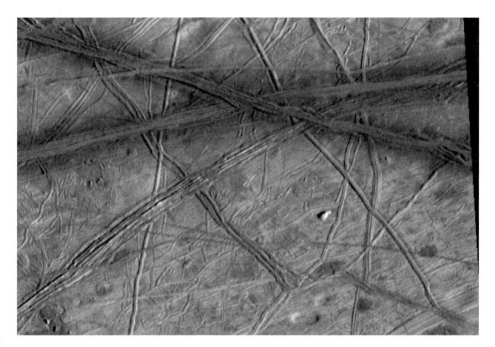

Figure 15.2. The intersection of the triple-bands Udaeus and Minos imaged at about 200 m/pixel as part of the Regional Mapping sequence during *Galileo*'s orbit E15. For context, compare this version with the same image in a mosaic with the other E15 RegMap 01 images (Figure 12.10b) and with the low-resolution view of the region from orbit G1 (Figure 2.5). Udaeus and Minos are typical global-scale "triple-bands", seen at this resolution as complexes of double-ridges, with dark margins extending several kilometers outward from their flanks. Using the navigational convention for azimuth (north = 0°, east = 90°), Udaeus is oriented at about 300° (or 120°) and Minos at about 260° (or 80°). This region is very rich in well-defined cross-cutting sequences.

any multiple of 180° (i.e., it could have been 290°, 470°, 650°, etc., clockwise, but not less that 290°).

We identified over 100 lineaments for which intersections placed constraints on their order of formation. If, for every crack, we could identify the next older and next younger crack, the sequence would be linear along a single track. Unfortunately, in practice, individual cracks only randomly cross other ones, and only a fraction of the intersections display morphologies in which the crack sequence (or more precisely the ridge sequence that we use as its proxy) is evident. Thus, the sequence information that connects between them is incomplete.

Alyssa developed a graphical visualization scheme which allowed a systematic search for sequences of crack formation that contained enough evidence to show the greatest amounts of azimuthal rotation. The longest formation sequence that she found includes 15 lineaments whose order of formation is well defined by intersections. If we fit this sequence to our model of continually clockwise-changing tidal stress azimuth, it must have required, at a minimum, about 1,000° of azimuthal rotation. Of course, it is possible that the sequence took place over even more azimuthal rotation, because the cross-cutting relationships do not preclude any number of extra 180° rotations between each crack in the sequence. (A rotation by any multiple of 180° would give a crack with the same orientation.)

At mid-latitudes in the northern hemisphere, we saw (Figure 6.3f) that, as non-synchronous rotation carries real estate eastward, maximum tidal stress cycles through 360° of azimuth during each period of Europa's rotation relative to the direction of Jupiter. Thus, if the long sequence identified by Alyssa actually represents the effect of non-synchronous rotation, we are seeing a tectonic record that goes back through almost three periods of rotation of Europa, at a minimum.

The sequence of crack orientations that we had inferred several years earlier from the low-resolution G1 data had spanned only a few tens of degrees of azimuth variation, apparently only a small fraction of one non-synchronous rotation period. The new result at first glance seems to be a extension back in time, a natural improvement made possible by the improved image quality. However, on closer inspection, the new results seem to pull the rug out from under the older interpretation. We took a look at the lineaments that we had identified in the G1 study and considered where they fit within the more complete sequence that we had derived from the E15 images. We found that several new lineaments fit between the G1 lineaments in the formation sequence, and they show that the latter actually formed over the course of nearly two rotation periods. Evidently, it was merely fortuitous that the selection of identifiable lineaments had appeared to represent a smooth continuous change in azimuth over a few tens of degrees.

This result meant that our study of the E15 data had not extended the G1 sequence farther back in time, but instead was calling into question our interpretation of the G1 data. Now we had to consider whether the new information on the azimuthal sequence supported non-synchronous rotation, even though the older work no longer seemed adequate. What had been so compelling about the G1 results was that they had seemed to show a nearly continuous record of varying azimuth. We asked ourselves whether the record from E15 offered an equally-compelling sequence, and I developed a test to see whether it did.

I developed my technique for testing the significance of the sequence by considering artificial cases like the following. Suppose that our study region contained a large population of cracks which developed with an increasingly clockwise azimuth through time, and that the geological record clearly showed the order in which they had formed. Further, suppose that the sampling in azimuth was so dense that there was no gap greater than 5°, and that the monotonic sequence in azimuth fit within, let us say, 180°. Such a tectonic record would be strong evidence for non-synchronous rotation, even better than our G1 story which had been universally accepted.

Now, suppose we tried to fit the same geological record to a model in which the azimuth varied counterclockwise with time. For example, if one crack was aligned due north, and the one just younger than it had formed 5° farther clockwise, then to fit a counterclockwise model we would need 175° of rotation between the two. In that case, it would take many cycles to fit the entire sequence into this wrong direction.

Conversely, if a tectonic record required several azimuth cycles to fit a clockwise sequence and could also be fit into a similar number of cycles in a clockwise sense, then the sequence could hardly be construed as evidence for non-synchronous rotation. In fact, when Alyssa fit the formation order to an artificial model in which we assumed the cracks formed sequentially in the counterclockwise sense, she found that nearly the same number of cycles was required either way.

We conclude that there is no evidence in the cross-cutting record described here that supports non-synchronous rotation. Similarly, when Alyssa considered the record nearly diametrically opposite, in the E17 RegMap 02 area (see Figure 15.1), she concluded that, although there is a rich record of cross-cutting relationships, there is no compelling evidence in that record for non-synchronous rotation.

On the other hand, none of the analysis here is inconsistent with non-synchronous rotation either. In fact, these results were perfectly consistent with our results based on cycloids in the Astypalaea region, which had indicated that only a couple of cracks could form in any region per rotation period. This result would explain why there would be a lack of any record of recognizable azimuthal variation, even if Europa has been rotating non-synchronously.

When Alyssa presented these results at a couple of scientific conferences, I was surprised at how aggressively she was attacked. First, I expected that a young undergraduate would have been treated with at least some sensitivity. Second, the news she was bringing was that my group's own earlier interpretations were now in question. We were comfortable with that result, so why should anyone else get upset about it? The problem was that our earlier interpretation had been accepted into the canonical set of *Galileo* discoveries. It had been used by others as the basis of their own geological interpretations, and with that investment they had taken ownership of the idea. Now, in reversing our earlier result, we were pulling the rug out from under them. My approach to science has been to follow the evidence where it takes us, which is good science, but bad politics. Ironically, in this case, the part of the canon we were questioning had been our own discovery.

While our 1998 interpretation of the G1 data was reasonable given the limitations of the G1 data, with the more complete imagery available with higher resolution and more advantageous lighting from orbit E15, we found that the case for non-synchronous rotation from lineament azimuths is hardly compelling. Similarly, continuing investigation of the large set of data acquired over the entire *Galileo* mission can provide deeper understanding, and can correct misconceptions based on the earlier limited data. It seems likely that Europa does rotate non-synchronously, but not based on the evidence we perceived at first. Images already in hand are likely to provide greater insight regarding many of the issues under discussion regarding Europa. We need to exploit all that information, and follow its implications. If we change our minds about what Europa is like, it will not be a sign of failure, but of success in pushing forward. To a large extent, the images and other data from the *Galileo* mission remain to be explored and interpreted fully. That work needs to be done if we are to justify the effort that went into the *Galileo* mission, and to prepare for subsequent spacecraft exploration of Europa.

As invested as I am in modeling of Europan tectonics in terms of the effects of tides and of non-synchronous rotation, I am comfortable with the possibility that we may have it all wrong, or that things may be much more complicated than we have assumed. For example, in the E17 RegMap 02 images there is a clustering of parallel crack orientations running roughly NW–SE (evident in Figure 15.1). Alyssa found that, in the context of the non-synchronous rotation model, these cracks formed in clusters and repeated at the same orientation during at least two successive azimuthal rotation cycles, and probably during many cycles. Perhaps the formation of a set of parallel lineaments during one period might have left a weakened structure that favored the same azimuth for cracking in a subsequent cycle. Alternatively, it is possible that all the cracks near a single azimuth formed during one rotation cycle (e.g., due polar wander), but that the ridges that we see now were formed at various later times due to reactivation. In other words, while we have been assuming that the ridge formation sequence is a valid proxy for the corresponding crack formation sequence, it is not necessarily so.

Additional information may come from more complete consideration of the global context. Support for this idea comes from, for example, the fact that the parallel sets of cracks in the E17 RegMap 02 region are aligned with a set of globe-encircling (near-great circle) lineaments. In that case, their orientation may not have been dominated by the regional orientation of tidal stress, in an area where they happen to have been well imaged, but rather by the global-scale processes that produced such lineaments.

One candidate effect might have been a reorientation of the global crust relative to the spin axis (i.e., polar wander). We have seen, from the distribution of strike–slip faults (Chapter 12), evidence for such a dislocation within a fraction of a rotation period. If this process happened once, and recently, in the geological record, then it probably happened frequently over the entire history of the surface. We cannot rule out the possibility that such slipping of the whole crust around the planet has happened repeatedly, either at random or in some systematic way that we do not yet understand. In that case, it may be that a dominant driver of tectonic cracking is the stretching of the crust that occurs as it re-conforms to the tidally-elongated figure of Europa. From the point of view of computing tidal stress, polar wander would be a sort of non-synchronous rotation in arbitrary directions. If it happens suddenly, tremendous stress would ensue, possibly dominating the tectonic record. We have only begun to consider the range of possibilities. Whatever it is, the story has been written in the tectonic record, and is there to be read.

16

Chaos

16.1 CHARACTERISTIC APPEARANCE

On Earth, when the icy surface of a northern lake or the Arctic ice cap breaks up, rafts of ice float around on patches of open water. When such openings in the ice occur (e.g., by local heating), the rafts may be portions of the ice crust that are thick enough to resist melting, or they may break off from the edge and drift toward the interior of the melted zone.[1] If the weather turns cold again, the rafts freeze in place, with a matrix of refrozen water between them, often surrounded by icy crust that had not melted through.

When we obtained our first moderate-resolution images of Conamara Chaos, taken during orbit E6 in February 1997 at about 180 m/pixel, with favorable lighting, we saw surface that is indistinguishable from such terrestrial melt-through sites (Figure 16.1a). Large rafts of ice display portions of the previous surface, which in this case had been predominantly densely-ridged tectonic terrain, like the terrain that surrounds Conamara.

Many of the rafts can be pieced together like pieces of a jigsaw puzzle,[2] allowing reconstruction of their original locations relative to the surrounding terrain and to one another. Toward the westward side, the rafts have moved very little away from the banks of the opening, so the continuity of their surface features is obvious. Along the southeastern edge of the opening, cracks in the adjacent terrain indicate incipient rafts that have not moved significantly into the opening. Aside from the fact that they have not moved, there may be no other difference between them and rafts that have drifted into the opening. Other rafts

---

[1] Rafts or not, such openings can be important to marine life in the Arctic (e.g., by opening breathing holes for migrating whales and seals).
[2] That is a jigsaw puzzle where many of the pieces have been chewed up by a toddler. Greg Hoppa's son Wyatt spent a good part of his toddler years in our lab, so the metaphor has some basis in reality. More about the relevance, or lack thereof, of terrestrial sub-oceanic volcanic vents to Europa is discussed in Chapter 21.

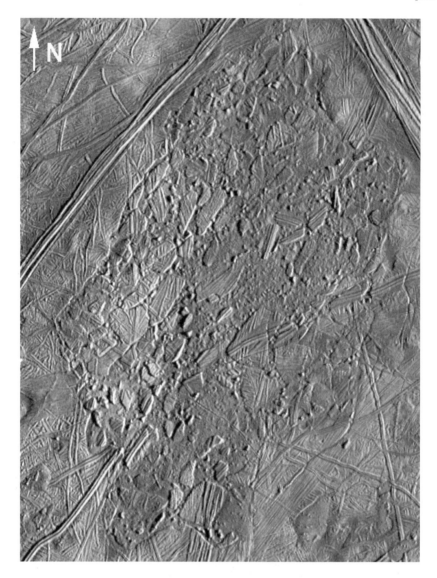

Figure 16.1a. Conamara Chaos imaged during *Galileo*'s E6 orbit (enlarged from Figure 2.8). The chaos area is about 80 km wide.

can be reassembled into chains that link back together the pieces of ridges that had formerly crossed this area. For example, the prominent double-ridge that approaches Conamara on adjacent terrain from the southwest can be identified on numerous rafts that can be reconstructed linking the ridge to its continuation on the east side (Figure 16.1b).

The picture is even clearer in Figure 16.2, where details of the rafts and their relationships are clear. Most of the pieces needed for reconstruction of the ridge that

Sec. 16.1]	Characteristic appearance 221

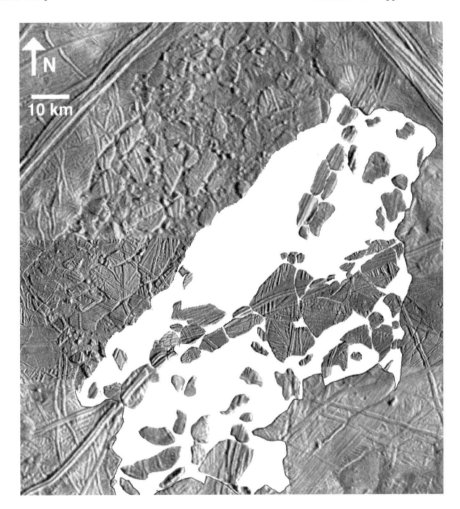

Figure 16.1b. Reconstruction of many of Conamara's rafts shows them to comprise the portions of the previous tectonic terrain where major ridge systems crossed. Chains of rafts reconstruct nearly all of the double-ridge that runs from the lower left across to the middle right. The rafts to the northwest have not been reconstructed here, but are obviously broken from the edge of the chaos area, because they have not moved very far from their original locations.

crossed this area are still intact. The lumpy, bumpy matrix between the rafts appears to consist of small bits of previous, unmelted crust too small to show pieces of the previous terrain, imbedded in the refrozen water between the rafts. As noted in Chapter 2, new cracks and ridges have already begun to form across this refrozen terrain, beginning the transformation of the chaotic region back to tectonic terrain.

The east end of the mosaic in Figure 16.2 should be viewed with its context in Figure 16.1a in mind. Here, we see distinct patches of irregularly-shaped chaotic

Figure 16.2. A mosaic of high-resolution images spanning Conamara at 54 m/pixel, also from orbit E6, spanning Conamara. Compare with Figure 16.1a. Note that the high-resolution mosaic is displayed within the surrounding context from the lower resolution image in Figure 16.1a, which appears relatively fuzzy when compared with the higher resolution.

terrain, each with the usual characteristics, including small rafts and the lumpy, bumpy matrix. On the other hand, the terrain between these small patches of chaos and the main part of Conamara had been broken somewhat by cracks. Aside from having not been displaced, this terrain could be considered to be a set of large rafts lying between the exposed matrix in Conamara and in these small patches to the east. Viewed that way, the small patches are actually part of the Conamara structure, especially because they likely formed as part of the same disruption event.

In December 1997, during orbit E12, a set of several ultra-high-resolution images (about 10 m/pixel) were taken of an area across the middle of Figure 16.2a. These images were laid out and reprojected so they could be laid over the background of the earlier images (Figure 16.3). They show, in fantastic detail, the rafts and matrix between them. Toward the west end of this mosaic is a raft covered by parallel ridges which has cracked in two. An oblique view, looking into the crack as if from an airliner's window is shown in Figure 16.4, which is part of the same set of ultra-high-resolution images, but which has not been reprojected from its original viewing geometry. It shows the detail of the cliffs dropping from the sides of the rafts down to the refrozen matrix below.

Another telling high-resolution view of chaotic terrain is shown in Figure 16.5. This area lies within the broad RegMap 02 region in the northern part of the leading hemisphere. RegMap 02 images provide a context for Figure 16.5, and show that it lies at the western edge of a large expanse of chaotic terrain. In this image, at the right we see chaotic terrain with large rafts that have broken away from the older tectonic terrain in the center. However, even that older terrain has been heavily cracked and differs from the rafts only in that they have not floated eastward into the main portion of the chaos. Several patches of chaos are scattered throughout the region, and appear to be locales where melt has extended to the surface, leaving a typical lumpy, bumpy matrix, with some small, and some tilted, rafts.

Not all patches of chaotic terrain are large. The patch shown in Figure 16.6 appears in low-resolution images (those with >1 km/pixel) as one of the dark spots (e.g., it is visible in Figure 2.5), for which the name "lenticulae" was invented. At 230 m/pixel some structure can be seen (on the left side of Figure 16.6 or printed smaller in Figure 12.10b). The identifying details of chaotic terrain can be difficult to recognize if the patch of chaos is small or the resolution is marginal. On the right side of Figure 16.6, we see that with much better resolution all of the identifying characteristics of chaotic terrain are visible, just as on Conamara: rafts displaying parts of the previous tectonic surface, especially where the crust was thicker with ridges; the lumpy, bumpy matrix; and cracks and ridges formed across the chaos since the matrix refroze.

In general, the identifying characteristics of chaos scale down with the size of the patch of chaotic terrain. For example, the chaos in Figure 16.6 contains the same features as Conamara, but the rafts are smaller and the matrix texture is finer in proportion to the size of the chaos (about 1/5 as wide as Conamara).

As a result, patches of chaos reach a limit of "recognizability" if they get too small relative to the pixel size in a given image. To some extent, the recognizability of

Figure 16.3. A mosaic of very-high-resolution images (9 m/pixel), reprojected to fit into a background from Figure 16.2a. What appeared to be sharp and detailed in Figure 16.2 compared with Figure 16.1, here appears as a relatively fuzzy background compared with the super-high-resolution view. Like most of the mosaics used in this book, this one was assembled by student assistants at the University of Arizona's Planetary Imaging Research Lab. Comparing with Figure 16.2a, we see that this image spans about 30 km from east to west.

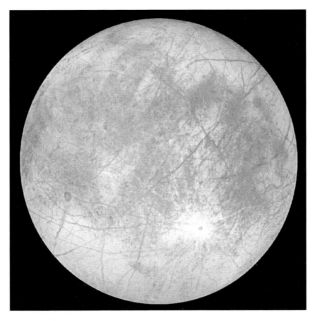

Figure 2.2. A full-disk view of Europa taken early in the *Galileo* mission, during orbit G2. Only orbits with prefix E passed close to Europa; higher-resolution, close-up views were not possible during this orbit. Here, the center of the disk is near the equator at longitude 290°W. The prime meridian (longitude 0°) is by definition in the center of the Jupiter-facing ("sub-Jupiter") hemisphere, so Jupiter would be toward the left of this view, and the satellite's orbital motion is approximately away from the viewer. This image shows most of the trailing hemisphere, which is centered at 270°W.

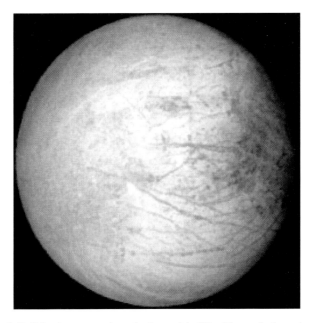

Figure 2.3. This full-disk view was taken during orbit C9 with resolution about 12 km/pixel. The center of the disk is near the equator at longitude 40°W. The sub-jovian point is in the dark splotchy area to the right.

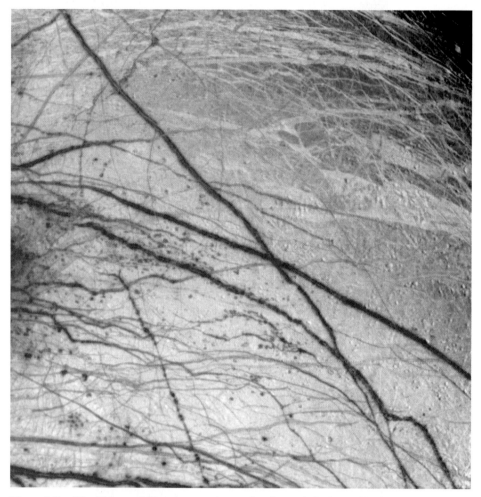

Figure 2.5a. The Udaeus–Minos intersection region imaged during orbit G1 at 1.6 km/pixel. At this resolution, the global-scale dark lines have bright centers, so they were called "triple-bands", and the spots were dubbed *lenticulae*, Latin for freckles. The terminologies defined to describe appearances at this resolution have caused major misunderstandings when carelessly applied to high-resolution images. This color composite was produced by Paul Geissler (LPL, Univ. Arizona), using separate black-and-white images (Figures 2.5b, c, and d) taken through three filters, 0.559 μm (green), 0.756 μm (a wavelength slightly longer than red), and 0.986 μm ("near-infrared"), and combining them as if they were red, green, and blue for display to the human eye.

Figure 2.6. The region from Crater Pwyll (near the bottom) to Conamara Chaos (near the top), about 1,000 km from bottom to top, imaged during orbit E4. Global-scale lineaments are visible with the appearance of "triple-bands". The ejecta rays from Pwyll extend to, and across, Conamara. Note that the color here is even less real than that in Figure 2.5; here very-low-resolution global color data have been overlaid on the higher resolution E4 images.

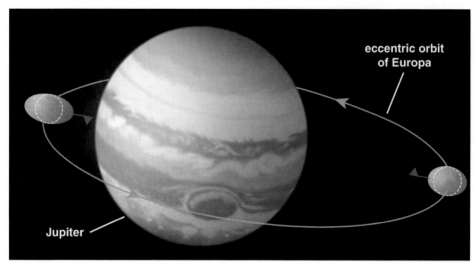

Figure 4.2b. The change in Europa's tidal figure between perijove (closest point in the orbit to Jupiter) and apojove (farthest point). The orbital geometry is less abstract than in the Europa-fixed reference frame of Figure 4.2a, and the scale is less egregiously wrong: At least here Jupiter is larger than Europa, but still not to scale.

Illustration by Barbara Aulicino/*American Scientist*, based on my sketch.

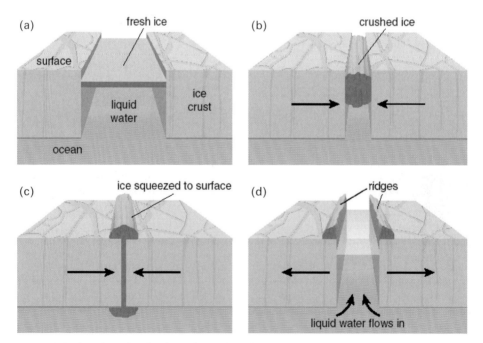

Figure 10.3. Another visualization of the ridge formation process (cf. Figure 10.2).

Barbara Aulicino, *American Scientist* magazine.

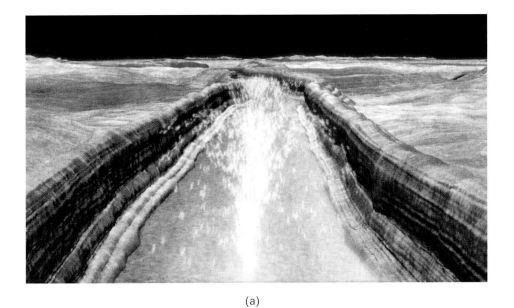

(a)

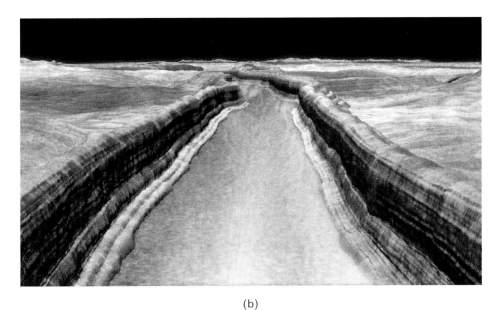

(b)

Figure 10.4. (a) Newly-exposed water in an opening crack is boiling as it freezes in this illustration from the NHK (Japan) TV production *Space Millennium*. Animations this good make it easy to forget that the process is only theoretical and has never actually been observed. (b) A couple of hours after the liquid is exposed, a thin layer of ice has formed.

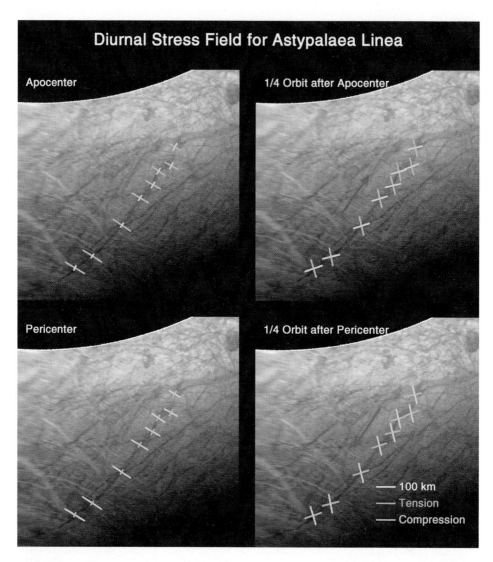

Figure 12.2. At apocenter, stress at Astypalaea is tension across the fault; a 1/4-orbit later, diagonal tension and compression yield right-lateral shear (orange arrows); a 1/4 orbit later at pericenter the fault is squeezed shut by compression; a 1/4 orbit after pericenter the shear is reversed, but friction in the closed fault prevents slippage. During each orbit, the fault takes a step in right-lateral displacement.

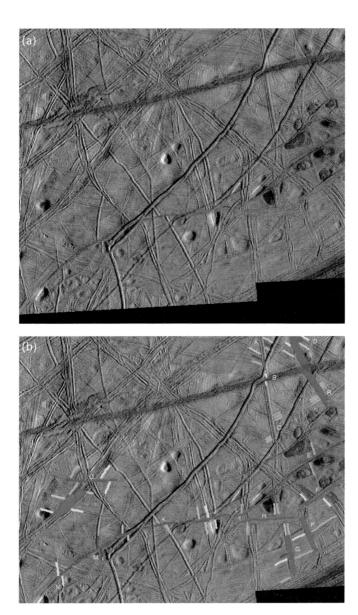

Figure 12.12. A time sequence reconstruction of the southern end of the E15 RegMap 01 area (Figure 12.10b). Faults are marked in red, with prominent crossing features marked in other colors. The steps go progressively backwards in time from the present (a and b) to the earliest time (e). Reconstruction of the arcuate crack involves considerable rotation of the adjacent plates. The gap that opens (going back in time) at the location marked F here represents a site where surface convergence has apparently occurred (see Chapter 17).

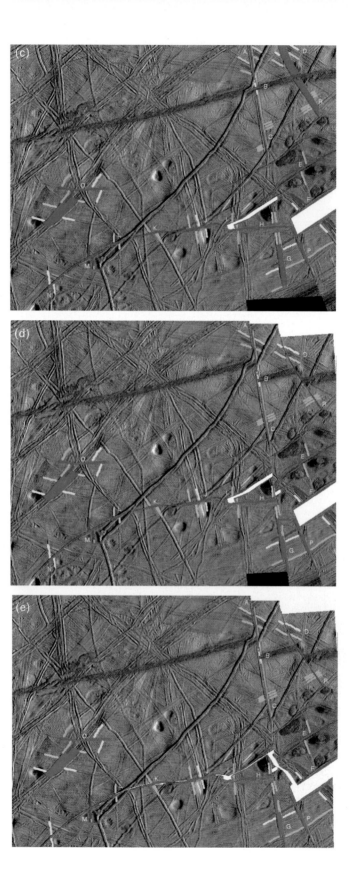

Figure 16.10. A Mercator map of chaos in areas imaged by *Galileo* at ~200 m/pixel, which are outlined in white. In our investigation we avoided the terrain marked by dark stripes, which was dominated by major impact features or illuminated from too high above the local horizon (in the E14 Wedges region). Areas marked in red and green are fresh and relatively modified chaotic terrain. Features marked in blue are the suspected remains of earlier chaotic terrain.

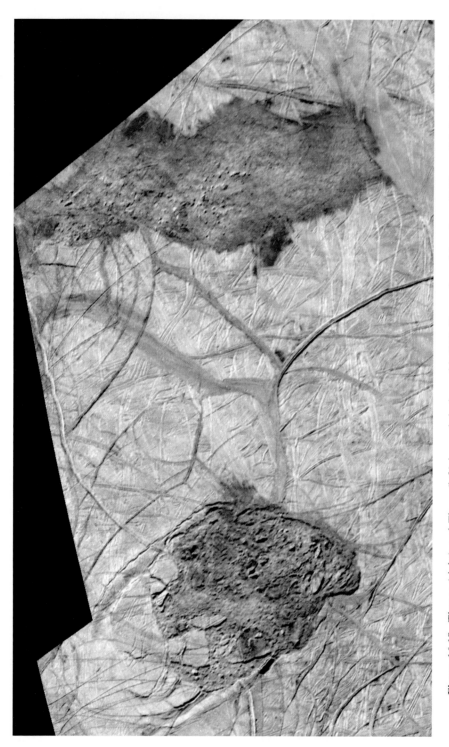

Figure 16.18. Thrace (right) and Thera (left) imaged during orbit E17. For scale and location, cf. Figure 9.1 (lower right).

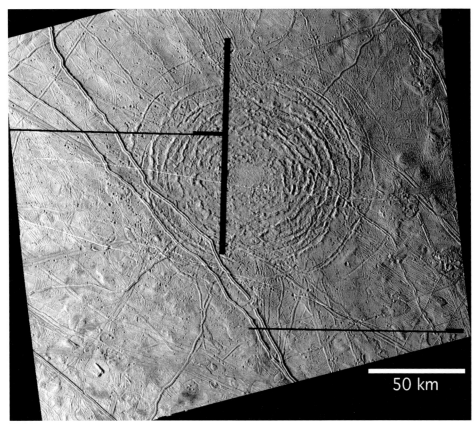

Figure 18.1. A color composite of the impact feature Tyre Macula, from images taken during *Galileo* orbit E14 at about 200 m/pixel.

Planetary Imaging Research Laboratory, University of Arizona.

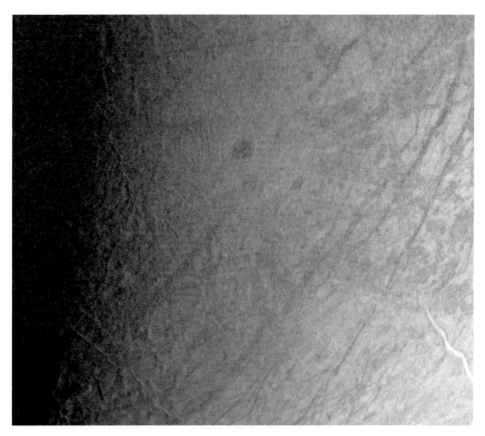

Figure 18.8. Pwyll appears in this regional view from orbit E12 as a dark orange splotch (to upper left of center), which extends several kilometers beyond the crater itself. The other orange splotches are patches of chaotic terrain. At this phase angle, the rays of fine ejecta, which were so prominent in Figures 2.2 and 2.6, are not visible. This color composite uses images taken though filters at 0.756 µm, in green, and in violet, and combines them as if they were red, green, and blue. Note the west end of Agenor Linea at the lower right.

Color composite by Paul Geissler.

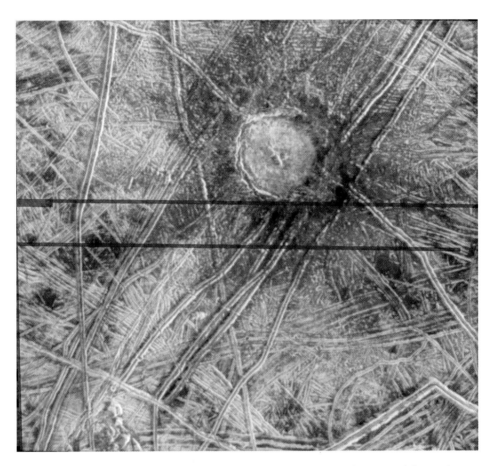

Figure 18.10. Color mosaic (combining image frames taken through green, violet, and near-IR (0.968 µm) filters) of Crater Cilix taken during orbit E15, at 110 m/pixel resolution. Gores of missing data in individual frames in the color sequence yielded the strips of inconsistent color. Note the chaos at the southwest (lower left) corner, and the cusp of a major cycloidal ridge pair.

Figure 20.1. Color-coded elevations near Crater Cilix superimposed on the color mosaic of the region from Figure 18.10. In the area color-coded for elevation, reds represent 650–750 m, yellows 450–550 m, and blues are 100–200 m, as computed by Nimmo et al. (2003). Black diagonal lines show where Nimmo et al. constructed elevation profiles.

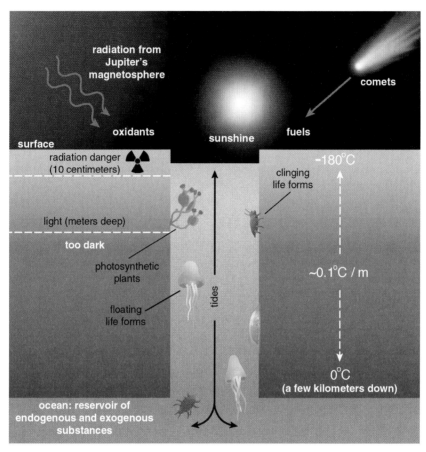

Figure 21.1. Tidal flow though a working crack provides a potentially-habitable setting, linking the surface (with its low temperature, radiation-produced oxidants, cometary organic fuels, and sunlight) with the ocean (with its brew of endo- and exogenic substances and relative warmth). Photosynthetic organisms (represented here by the tulip icon) might anchor themselves to exploit the zone between the surface radiation danger and the deeper darkness. Other organisms (the tick icon) might hold onto the side to exploit the flow of water and the disequilibrium chemistry. The hold would be difficult, with melting releasing some of these creatures into the flow, and with freezing plating others into the wall of the crack. Other organisms (jellyfish icon) might exploit the tides by riding with the flow. This setting would turn hostile after a few thousand years as Europa rotates relative to Jupiter, the local tidal stress changes, and the crack freezes shut. Organisms would need to have evolved strategies for survival by hibernating in the ice or moving elsewhere through the ocean.

Artwork, based on my earlier drawing, is by Barabara Aulicino/*American Scientist*.

Figure 22.1. What is wrong with this picture? According to JPL, "This artist rendering shows a proposed ice-penetrating cryobot and a submersible hydrobot that could be used to explore the ice-covered ocean on Jupiter's large satellite, Europa ... The cryobot would melt its way through the ice cover and then deploy a hydrobot, a self-propelled underwater vehicle that would analyze the chemical composition of the ice and water in a search for signs of life." Whether a device could actually drill down through several kilometers of ice is uncertain. In the picture, the surprisingly-jagged base of the ice is very close to the bottom of the ocean, so here the ice must be over 100 km thick. There is no evidence that Europa's ocean floor has volcanic vents, but with ice so thick that it isolates the ocean, such sites would be the only hope for life. In fact, the ice is probably thin and permeable, so life might prosper near the surface, where it would be readily accessible. If this submarine is named, it might be called the *Red Herring*. Selection of a name for the cryobot sticking down from the ice is left to the reader. Image courtesy NASA/JPL-Caltech.

Sec. 16.1] **Characteristic appearance** 225

Figure 16.4. Part of the high-resolution image set shown in Figure 16.3, but here in its original viewing geometry, an oblique view as if from the window of an airliner. Fans of debris have spread below the cliffs, a fine crack has formed across the matrix, and tiny craters are visible.

the "Mini-Mitten" in Figure 16.6 is already degraded in the 230 m/pixel view at the left, where the chaos is about 70 pixels wide. For another example, consider the high resolution (34 m/pixel) view of small patches of chaotic terrain in Figure 16.5, part of which is enlarged in Figure 16.7 (left). Here several patches of chaos, each a few kilometers wide, are surrounded by fairly intact tectonic terrain. At 34 m/pixel, the characteristics of chaotic terrain are clearly visible. In the regional mapping image of the same area at 230 m/pixel (on the right side of Figure 16.7), the features are nearly unrecognizable.

There is considerable diversity in the numbers and sizes of rafts among various areas of chaotic terrain. In many cases, especially among smaller patches of chaos,

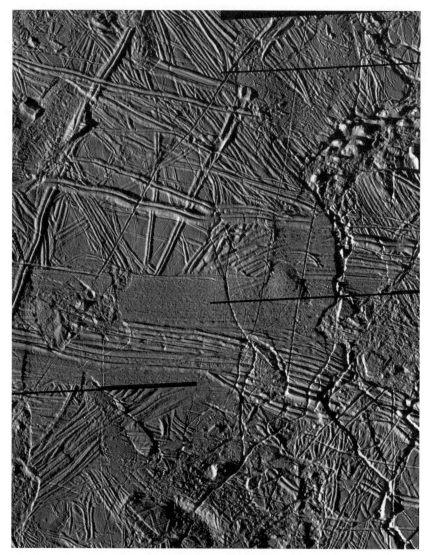

Figure 16.5. A high-resolution image (about 30 m/pixel) from orbit E11 of a locale in the northern RegMap 02 region, which is at the western edge of a large area of chaotic terrain. The area shown here is about 40 km across.

there are hardly any rafts. For example, within the chaos shown in Figure 16.6 you could select a circle 2 km wide that would contain lumpy matrix, but no rafts. Often such small patches of chaos do not have rafts, but the characteristic matrix unambiguously identifies the chaos. A good example of a small patch of chaos (about 15 km across) with no rafts was shown near the top of Figure 2.11, although it is possible that at higher resolution some of the lumps in the matrix texture might be

Sec. 16.2]   **Three hypotheses for formation of chaos** 227

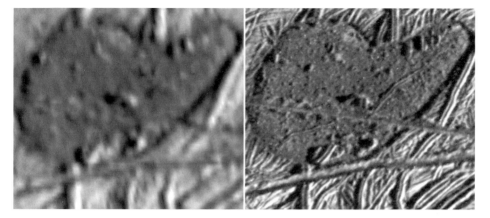

Figure 16.6. The "Mini-Mitten", a small patch of chaotic terrain in the RegMap 02 region near 36.1°N, 227°W. At left is a view at 230 m/pixel blown up from the Regional Mapping image sequence. The area shown is about 18 km across. The data are the same as in Figure 12.10b, where this feature appears at the left center as a small dark spot, but are printed much larger here. At the right is the same feature at 64 m/pixel.

revealed as rafts with recognizable bits of the former surface. It is important to bear in mind the types of observational biases that are introduced by resolution limits.

Another fairly common, but hardly ubiquitous, raft configuration is the tilted raft. A good example is prominent in Figure 16.7. No systematic study of the distribution of these features has been done, but they might potentially provide insight into the dynamics of chaos formation and evolution. Any study would have to account carefully for selection effects, because tipped rafts are most obvious, and thus most often spotted, near the terminator where long shadows bring out their tilts, sometimes making them look like shark fins.

16.2 THREE HYPOTHESES FOR FORMATION OF CHAOS

The general appearance of chaotic terrain is similar to sites on Earth where surface ice melts through, exposing liquid water from below, and then refreezes around blocks and rafts of drifted ice. When the *Galileo* Imaging Team saw the first images of Conamara that revealed its morphology, this interpretation was prevalent. As plans were made for preparing a set of papers that would report the mission's preliminary results in the magazine *Nature*, Mike Carr was put in charge of a paper that would follow this line of thinking, using the structure of Conamara as the then strongest evidence for an ocean on Europa.

At that time, in early 1997, as the team's expert in tidal processes, I was deeply engaged in investigation of the tectonic record. While Conamara was interesting to me and the melt-through explanation seemed obvious, understanding chaotic terrain was somewhat tangential to the specific work that I was doing. However, the human

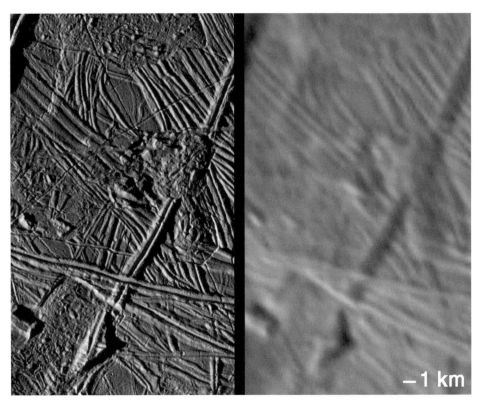

Figure 16.7. Small patches of chaotic terrain, at left enlarged from the 34-m/pixel image shown in Figure 16.5, and at right shown at 230 m/pixel from a Regional Mapping image sequence (E15 RegMap 02).

dynamics of the preparation of the Carr et al. paper were too fascinating to miss. During this time Jim Head and his protégé Bob Pappalardo lobbied hard for their view that the ice was too thick for such oceanic exposure. Ultimately, the paper was a somewhat baffling compromise, with its title ("Evidence for a subsurface ocean on Europa") and much of its description of Conamara seeming to lead toward an oceanic exposure, while the conclusion was that solid-state convection processes, rather independent of the ocean, were at least as plausible as the cause of the chaotic disruption.

Pappalardo had only recently completed his graduate work on interpretation of the geology of other icy satellites in terms of solid-state geological processes analogous to those on Earth, so it was natural for him to look at chaos on Europa from that perspective. He pointed to the several examples of small round patches of chaotic terrain around Conamara (like the one seen in Figure 2.11) as evidence for solid-state convection. The interpretation was predicated on a claim that these chaos features, as well as a putative family of related features called "pits, spots, and domes", were all about the same size ($\sim$10 km) with fairly uniform spacing. In that view, these features represented the tops of convection cells, and their size and spacing indicated an ice thickness of about 20 km.

In fact, it was arranged that Pappalardo would tell this story as first author of one of the team's articles in the same issue of *Nature*. In the same time frame, with his strong political backing, Pappalardo was designated as the team's spokesperson, being assigned to present the early results and interpretations of the *Galileo* Imaging Team at various scientific conferences and policy forums. With such strong political traction, the thickness of the ice and the convection within it became entrenched canonical knowledge, independent of the fact that the putative evidence for it had never been quantitatively developed and was incorrect.

If you did buy the idea that the 10-km-wide chaos patches were the tops of convection cells, then the model proposed that larger areas of chaos, like Conamara, were somehow the result of coalesced small patches of chaos. How that coalescing would have worked was always somewhat vague, but evidently it sounded convincing. The solid-state explanation for chaos and for other features became a consistent theme of Jim Head and Bob Pappalardo's group at Brown University, extending, for example, to their explanation for double-ridges that I described in Chapter 10.

People tend to see on Europa whatever they are used to seeing. Ron Greeley has studied volcanism on the Earth and other planets. To him and his team at Arizona State University, chaos, especially the small round patches, looked like volcanic upwelling. Because ice flows and melts at much lower temperatures than rock, this process came to be called cryovolcanism.

A number of chaoses (as I like to call distinct patches of chaotic terrain) have appearances that are indeed suggestive of upwelling flow consistent with cryovolcanism. The chaos in Figure 2.11 is typical of this type. These chaoses appear to be bulged upward, to have lobate edges suggestive of viscous flow spreading over the adjacent crust, and sometimes have cracks in the adjacent crust suggestive of loading by such flow out over the surface.

Perhaps the best, and the biggest, example of this type of chaos is the Mitten in the northern RegMap 02 region (Figure 16.8). Compared with the similar-sized Conamara, the Mitten has relatively few large rafts, although some can be seen in the thumb, for example. There are many small rafts and lumps and bumps within the matrix. The matrix appears to be updomed somewhat, a perception based on the variations in brightness under this illumination. Greeley's student Patricio Figueredo, who studied the E15 RegMap area for his dissertation, argues persuasively (even if ultimately incorrectly) that the morphology of the Mitten suggests viscous upwelling. Along the western and especially the southwestern edges, the matrix seems to have ridden up and over the adjacent crust. The crust there is cracked and tilted, as if weighed down by the overlying material.

At first, I listened to the initial geological interpretations of the chaotic terrain with some detachment as I worked on the problems of tidal stress and tectonics. But, as that work progressed, the tectonic evidence was strongly arguing in favor of fairly thin ice. Our model for strike–slip displacement needed the cracks to penetrate to liquid, the tidal stress seemed insufficient to crack down more than a few kilometers,

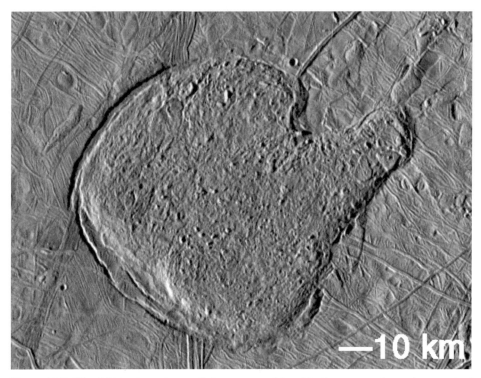

Figure 16.8. The Mitten is a chaos of size comparable with Conamara, imaged as part of the E15 Regional Mapping sequence at about 200 m/pixel. It is easy to inadvertently perceive the topography in this image inverted, with high places looking low, etc. In order to perceive the topography correctly, note that here the light is coming from the left; the two craters (about 2 and 2.5 km wide) on the surrounding terrain should look like holes with rims; and the double ridges should not look like grooves. The lumpy matrix appears to be bulged upward.

and only cracks with double-ridges were displaced by strike–slip, consistent with formation of ridges by tidal pumping of oceanic water to the surface. A consistent picture of thin ice was emerging from our studies of tidal tectonics, at least thin enough that cracks could link the ocean to the surface.

Such thin ice was inconsistent with the emerging canonical geological interpretations of chaotic terrain. The convection model seemed to require much thicker ice. Although the exact requirement was uncertain, the model's advocates usually cited 20 km as a minimum. And cryovolcanism required deep source chambers for the upwelling flow. Having established several lines of evidence for thin ice, I needed to understand and evaluate the basis for each of those competing models.

The convection model was based on the putative "pits, spots, and domes" that supposedly were the surface expression of convection cells. For 4 years nearly every presentation reviewing what *Galileo* had learned about Europa included a required mantra stating that "Europa-is-covered-with-pits-spots-and-domes-that-all-have-about-the-same-size-and-spacing-and-demonstrate-that-there-was-solid-state-convection-and-ice-thicker-than-about-20-km." Yet, the only documentation for that established fact was a couple of paragraphs in Pappalardo's *Nature* paper. Eventually, we showed that the taxonomy and generalizations about pits, spots, and domes were premature and not supported by a more complete survey (Chapter 19).

The volcanism model seemed weak as well. Buoyancy presents a challenge for models of cryovolcanism because the liquid water or slush must rise through, and then over, less dense solid ice. In contrast, on terrestrial planets volcanism can be conceptualized in global terms as part of the differentiation process, with the least dense materials, which constitute magmas and lavas, rising to the surface (albeit with important local variations in the dynamics). On Europa, volcanic flow requires inexplicable pressure chambers to force the liquid upward or *ad hoc* additives to lower the density of the water. In either case, the watery magma cannot be part of the global ocean: the ocean could not be pressurized enough to reach the surface; and if the density-reducing additives were widespread, the crust would sink into the ocean rather than the ocean squeezing onto the surface.

Our thin-ice results from tidal tectonics were much more consistent with the original interpretation that chaos was simply the manifestation of melt-through of the ice from the ocean below. If the ice is thin enough, fairly modest local heating anomalies could cause occasional melt-through.

Nevertheless, convective upwelling and volcanism had become the canonical models, even though they seemed to be predicated on only superficial, preliminary, and highly-questionable evidence, and they seemed to contradict what we were learning from tidal tectonics. In contrast, melt-through was consistent with the ice being thin, and it provided a simple explanation of the character of chaos. I had not planned on getting involved in interpreting chaotic terrain. The ghost of Ockham made me do it.

16.3 OUR SURVEY

I needed to roll up my sleeves and examine chaotic terrain more closely. More precisely, I needed to roll up my students' sleeves and get them to examine chaotic terrain. Greg Hoppa and I spent a good part of the summer of 1998 looking over every available image in preparation for a systematic survey of chaoses. The most extensive set of images were those taken at about 200 m/pixel with lighting appropriate for revealing geological morphologies. At that time we already had received the data from E15 Regional Mapping, which covered large areas in the northern hemisphere (the northern parts of RegMap 01 and 02, see Figure 12.9) that represented a relatively unbiased sample of the surface, because they had not been targeted to specific features. We also had other sets of similar images that covered fairly-broad regions around various targeted sites.

Greg and I pored over these images, comparing them wherever possible with the few high-resolution images (as in the comparisons shown in Figures 16.6 and 16.7), training ourselves to recognize chaos even where it appeared at the limits of recognizability (e.g., on the right side of Figure 16.7).

At the end of the summer, undergraduate student Jeannie Riley was assigned by the NASA/Arizona Space Grant program to work with us. Jeannie was a smart and careful geology student. She quickly became skilled at recognizing chaos, and we set her to work mapping all the chaoses that appeared in the available data sets (see Riley et al., 2000).

As we surveyed these features, we realized that they spanned a continuum of degrees of degradation by subsequent tectonic processes, especially cracking and ridge building. We had seen that Conamara, one of the freshest and most recent features on Europa, already displayed signs of incipient degradation as cracks and ridges had already begun to cross over it. In other places, hints of older chaos were barely recognizable. Figure 16.9 shows an example of a locale where cracks and ridges have nearly obscured older chaotic terrain, and chaos has disrupted tectonics, in a sequence of discernable episodes.

This realization added a new dimension to our disagreement with the canonical description of Europa. The geologists had been creating maps that identified various terrains and features. They used those maps to create stratigraphic sequences, that

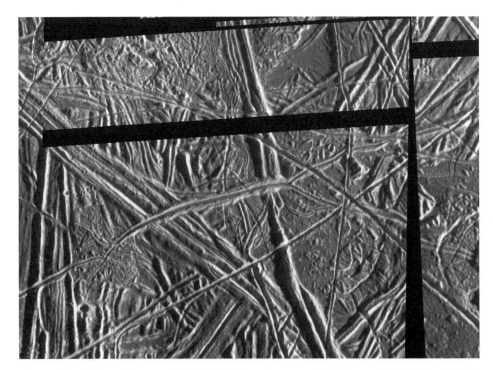

Figure 16.9. This region near 11°N, 328°W (contiguous with Figures 10.5, 10.6, and 11.1) taken at 26 m/pixel during orbit E4 shows details of sequential degradation of chaotic terrain by tectonics, and disruption of tectonic terrain by chaos formation. Unfortunately, no images exist showing the regional context of this high-resolution view.

were supposed to represent long-term change in the character of the processes on Europa. These maps and sequences seemed to us to be quite fanciful. One problem was that the notion of stratigraphy as developed in terrestrial geology made no sense on Europa.

Terrestrial geological mapping was developed on the basis of an understanding that the time sequence of geological history is recorded in a series of layers, or *strata*, that have been crunched and tilted, and then sliced off at the surface, where observations can be made. Thus, a geological map characterizes a slice that intersects those layers and can be converted into a record of changes with time. On Europa there are no strata, only a series of things that have modified the surface at different times and places. When traditional mapping protocols were followed outside the appropriate context, and with inadequate thought about how maps should be used in this new context, strange results were bound to come out. Geological mapping can be done meaningfully on other planets, but only by taking great care to review the fundamental concepts. Vicki Hansen (University of Minnesota, Duluth), a terrestrial geologist who has examined planetary geology as well, has emphasized this point. It is no coincidence that she had gotten into considerable political trouble with the same players who were now controlling *Galileo* geological interpretation.

The geological mappers for *Galileo* decided that chaos must be a relatively-recent phenomenon, because the chaos that appeared on their maps was very fresh. They interpreted this observation in the context of my decade-earlier work on the tidal evolution of the orbits. I had found that tidal heating on the Galilean satellites might have decreased significantly over the past 100 million years, which would have allowed the ice on Europa to thicken relative to the past. The geologists suggested that those changes led to the conditions for formation of chaotic terrain, which in their way of thinking required fairly thick ice.

As much as I liked the idea that my earlier work on celestial mechanics might be supported by what we saw on Europa's surface, as my group looked at the images, we saw no evidence that chaos was an especially recent phenomenon. The impression that it began to form only recently comes from the obvious fact that the newer, fresher chaos was much more easy to spot than the older examples which were obscured by subsequent processing of the surface.

I come from a background in astronomy, where observational selection biases are an everyday issue. If a picture doesn't show asteroids that are small or hard to see, we don't take it as evidence that they don't exist. In fact, we usually assume that they do exist, and observers try to find ways to see them. With chaos on Europa, the *Galileo* Imaging Team's canonical result was that most patches of chaos are larger than about 10 km and that most chaos is fresh and recent. But that result ignored the observational selection effect that small and old chaoses are difficult to see.

The situation seemed analogous to the impact-cratering record. Surely enormous numbers of craters of every possible size have been created by impacts and removed by resurfacing, even as impact cratering continues as an ongoing process. Yet, all the craters and other impact structures recognized on Europa are fairly fresh, with minimal indications of degradation. If we applied the same poor logic to the impact record as the geologists were applying to chaos, we would conclude that cratering is a recently-begun process. Obviously, that conclusion would be incorrect. Europa has been continually bombarded throughout its existence. The only reason that craters look fresh is that the older ones have been degraded beyond recognition by the ongoing resurfacing of the satellite. It seemed to me that chaos, too, might only seem to be recent, even if it had been created continually throughout the history of Europa.

Jeannie Riley went to work and mapped chaotic terrain wherever she could find it. We asked her to mark fresh-looking chaos in red, and older, modified chaos in green. This system was a crude way to represent the continuum of ages, but it was about as accurate as we could be, especially because near the limits of recognizability (whether due to age or size) the results depend on subtle differences in lighting conditions among images. Figure 16.10 (see color section) shows the distribution over the globe in all the regions where adequate 200-m/pixel-resolution images were available (cf. Figures 9.1 and 12.9). By this time we had the E17 portions of the RegMap images as well, extending into the southern hemisphere, and we mapped chaos in a variety of high-resolution images as well.

As shown in Figure 16.10 (see color section), we did not attempt to map the western 2/3 of the E14 images of the Wedges region because illumination is too nearly vertical and morphology is difficult to see. It would be very difficult to spot chaos under such conditions, and especially older chaos that has been modified or covered. It is no wonder that mapping of precisely this region by the Brown University group had contributed in large part to the spurious result that most chaotic terrain is recent.

The largest patch of chaos that we mapped is the 1,300-km-wide splotch in the center of the leading hemisphere (centered near longitude 90°W on the equator) that we marked with green, indicating that it appears to be morphologically old and modified. This chaotic terrain, which we mapped using RegMap 02 data, is part of what appeared on the global scale in Figure 2.3 as the huge yellowish brown splotch to the left side of that hemisphere. The fact that this splotch is somewhat paler than the low-resolution (global-scale) appearance of other patches of chaos (e.g., Conamara) may also be a sign that this terrain is relatively old, because bombardment by energetic charged particles (discussed in Chapter 21) would likely have just such an effect.[3] Thus, low-resolution coloring is consistent with our geological interpretation of this terrain as old and modified.

Although this giant chaos is relatively old, within it are patches of much fresher chaos, indicated by the red spots surrounded by green in Figure 16.10. An example of such a site is shown in Figure 16.11, which shows a portion of the "green" chaos around 10°S, 80°W. Here, the older chaotic terrain, whose formation had included disruption of tectonic terrain, had subsequently been itself disrupted by formation of fresh new chaotic terrain. The fresh chaos (located in Figure 16.11 by brackets) appears as a red spot embedded in the large green area in Figure 16.10.

---

[3] This giant chaos occupies a large portion of the leading hemisphere. Paul Geissler and student Matt Tiscareno showed that this hemisphere may be subject to preferential net redeposition of water molecules that are sputtered off Europa by energetic ions in Jupiter's magnetosphere. Thus, the paler appearance of this chaos may be due to its location as well as its age.

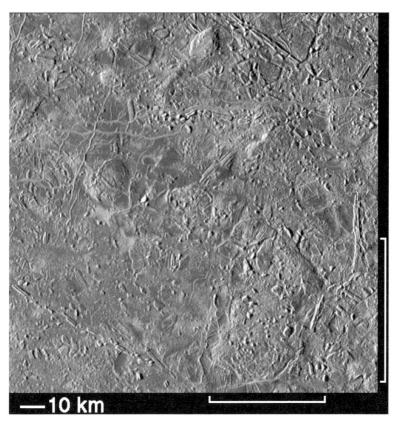

Figure 16.11. A sample showing the "modified" chaotic terrain that we mapped as green in Figure 16.10. This class of chaotic terrain appears to have been modified, subsequent to emplacement, by tectonic processes that have produced fine-scale lineaments, such as the dendritic ridges in the upper left quadrant of this figure. Even more recently, more fresh chaos (mapped as red in Figure 16.10) has broken through the older surface over a 30-km × 50-km area indicated by the brackets. The region shown is near 80°W, 10°S, within the largest chaos area on Europa, seen in Figure 16.10. The fresh chaos (shown here between brackets) appears in Figure 16.10 as a red spot embedded in the large green chaos area.

Formation of chaotic terrain disrupts whatever was there before. If the previous surface was tectonic, then cracks and ridges will appear on rafts. If the previous surface was chaotic, then pieces of the older chaotic terrain may appear on the new rafts. The sequence of tectonic and chaotic processes that have played out within Figure 16.10 appears to be very complex. However, here, as elsewhere on Europa, the basic processes are quite simple. Tectonic processes break up, cover, and destroy older terrain, and formation of chaos disrupts older terrain. In either case the older terrain might be chaotic or tectonic. In this way, Europa's surface has undergone the continual renewal that makes its surface so young compared with most other bodies in the solar system.

Jeannie had marked the chaoses on the digital images using image-processing software, so it was straightforward to obtain statistical information. I was particularly interested in the statistics of the sizes of patches of chaos, because the canonical model of convecting thick ice was founded on Pappalardo's claim that there was a characteristic, roughly 10-km, size for these features, corresponding to the scale of the convection. As far as I could tell the only supporting evidence for that notion was a qualitative impression from images (Figure 2.5) of the region immediately around Conamara (Figure 2.8) and the low-resolution freckles (the *lenticulae* in Figure 2.5). From Jeannie's maps, we had quantitative information about size distribution.

My previous experience investigating the size distributions of asteroids suggested using a similar format to plot the statistics of chaoses (Figure 16.12). We graphed a histogram of the numbers at each size, using logarithmic scales because, as with asteroids, the numbers increase so rapidly with decreasing size.

When we plotted the numbers from the moderate-resolution images (the Regional Mapping images and other images taken at $\sim$200 m/pixel), we obtained the curve shown by the solid line. There is a steep increase in numbers from the largest chaoses (hundreds of kilometers across, equivalent to areas of many thousands of square kilometers) down to those near 10 km across. For even smaller sizes, the plot shows a steep drop-off in numbers. The distinct peak in numbers of chaoses, with an area near 60 km^2 (equivalent to a radius of about 10 km), demonstrated quantitatively what Pappalardo and the Brown group had perceived in those images: There was clearly a characteristic size of about 10 km in these images.

I had worked with statistics like this before. The size distribution of known asteroids follows a similar curve, with a similar drop-off at small sizes. Astronomers understand that there is a limit to the sizes of bodies that they can see, so that drop-off reflects our observational limitations, not the statistics of what is actually out in space. The peak in numbers of observed asteroids shows the size at which asteroids become too small to recognize. It is a measure of our observational limits.

The drop-off in numbers of chaoses at sizes below 10 km is similarly an observational selection effect. We have already seen (e.g., Figure 16.7) that, at about that size and below, chaotic terrain becomes increasingly difficult to recognize in 200-m/pixel images. The peak in numbers of chaos at 10 km is a direct result of the resolution of the images. The initial impression of a prevalence of 10-km sizes should never have been accepted as an indication of reality on the ground, let alone become the basis of a canonical mantra.

From the 200-m/pixel data it was incorrect to conclude that there is a peak in the actual numbers at 10 km. The evidence simply is not there. On the other hand, we could not rule out such a peak. We simply cannot say, from those images, what the size distribution is below 10 km.

However, we can explore the size distribution below 10 km using the high-resolution images. We had already seen that chaoses that were marginally recognizable at 200 m/pixel were perfectly clear at 30 m/pixel. The problem is that we do not have much surface coverage at that higher resolution, but we did have enough in hand to extend our survey down to much smaller sizes. Counts from the high-resolution images are shown by the dashed line in Figure 16.12.

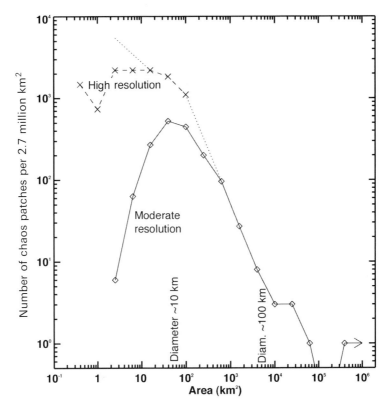

Figure 16.12. Incremental size distribution (as a log-log histogram) of all the chaos patches (Figure 16.10). The population is binned in log area, with the bin width corresponding to a factor $10^{2/5}$, and the plotted numbers are at the center of each bin. The solid line and dashed lines represent statistics from the moderate- and high-resolution images, respectively.

These results show the effect of more complete recognizability at sizes below 10 km. Starting at the large sizes, the numbers increase with decreasing size. Rather than leveling off and dropping below 10 km, the high-resolution data show that they continue to climb in numbers down to considerably smaller sizes. In fact, the histogram of the larger sizes, for which chaos is completely recognizable in the 200-m/pixel images, could be extrapolated to smaller sizes (the dotted line between 10^3 and 10^2 km^2 in Figure 16.12) and it exactly matched the numbers revealed by the high-resolution pictures.

It is clear that the peak at 10 km was entirely an observational artifact and the numbers increase at smaller sizes. The putative observational evidence for the convective thick-ice model was non-existent.

The high-resolution data tell us a bit more about the smallest chaoses. The observed numbers flatten out and then drop a bit below 10 km^2, in part because of recognizability limits. If chaoses become hard to find in 200-m/pixel images, then they likely become hard to find if they are smaller than a couple of kilometers in the high-resolution images. Thus, observational counts would be consistent with extending the dotted line between 10^2 and 10^3 km on to ever-smaller sizes.

We would quickly run into trouble with that extrapolation. There would be so many small chaoses between $1\,\text{km}^2$ and $100\,\text{km}^2$ that the sum of their areas would be as great as the total surface area of Europa. The entire surface would be saturated with tiny patches of chaos, with no room for tectonic terrain. Clearly, there are not that many tiny chaoses.

There is a more plausible way to use the dotted line connection between the high-resolution and low-resolution data to infer the numbers of tiny chaoses. This line shows that 200-m/pixel images reveal most of the 20-km-wide chaoses, but only 1/3 of the 12-km chaoses. If we assume a proportionate reduction in recognizability for the 30-m/pixel images, then we were counting most of the 3-km ones but only 1/3 of the 2-km ones. This line of thinking allowed us to estimate the actual numbers of small chaoses as shown by the dotted line extending to sizes below about $10\,\text{km}^2$ in Figure 16.12.

This extension to small sizes rises slowly enough that, even if we extend it to arbitrarily-small sizes, there is no problem exceeding the total area of Europa. However, if we add up all the chaoses from tiny sizes on up, we find that at least 30% of the surface, and more likely about 40%, is covered by chaotic terrain.

When the Imaging Team was initially focusing on the very fresh and obvious example, Conamara, the impression was that chaos was recent and unusual. Our survey showed that chaotic terrain has been forming, as far as we can tell, throughout the geological history of the surface, and it covers nearly as much of the surface as tectonics, the other dominant type of terrain. Moreover, there is no evidence for a typical size of 10 km.

16.4 MELT-THROUGH

The foundation of the convective thick-ice model was shaky, and the tidal–tectonic work suggested that the ice was thin enough for cracks to penetrate to the ocean. The idea that chaos simply represented sites of melt-through of the ice crust seemed increasingly attractive. Either an area of the surface is temporarily replaced by oceanic liquid exposed to space, or the ice becomes so thin that, from the point of view of its mechanical effects, it might as well be liquid. Such a process could create the observed characteristics of chaotic terrain.

We assume that tidal heating is not perfectly uniform, but is able to heat various regions of the crust preferentially at various times. This heating might be from below (e.g., if the dissipation is in the solid rock beneath a liquid ocean), or in the crust itself. Regional or local concentration of heat at the bottom of the ice shell might result from currents in the liquid or from anisotropic heating in the rock or ice. If tidal heating is predominantly in the ice crust, unless it is perfectly uniform, regional or local concentration might produce melt-through.

Consider the process of thinning a surface crust of ice over liquid water. In Figure 16.13a, the ice has begun to thin from below. Because the ice is supported

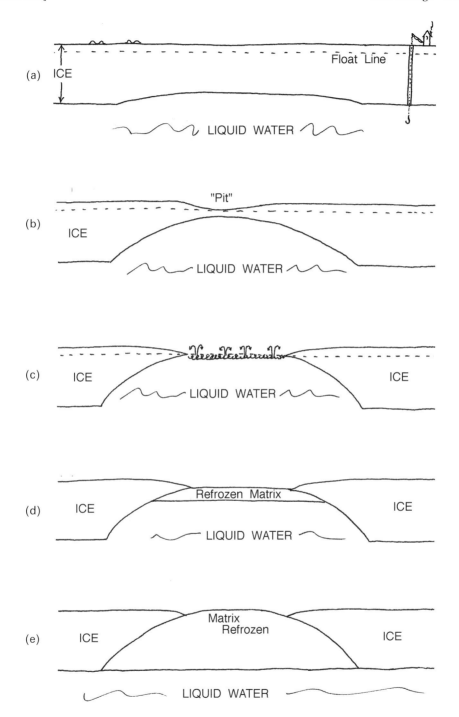

Figure 16.13. A schematic of the melt-through model of chaos formation.

buoyantly, it has a float line about 10% of the way from the surface to the base of the ice, which is the level to which water would rise through a fishing hole.

With more heat, as the ice thins further, buoyancy decreases, so the surface sags downward isostatically (Figure 16.13b). If melting does not continue beyond this point, but the crust remains thin, the surface may be left with a depression. This isostatic sagging may provide a genetic connection between the many "pits" seen on Europa and chaos areas.

Eventually, the ice thins so that liquid is either exposed at the surface or covered by a fragile, thin skim of ice (Figure 16.13c). The surface tapers to the edge of the opening where buoyancy supports it at the level of the float line. Thus, the ramped beaches characteristic of many of the shorelines of chaos terrain are formed. The liquid rises to the float line. Where liquid is exposed, its surface is agitated by boiling and freezing on exposure to space. The active surface might spray material, including salts, organics, or whatever other impurities are contained in the liquid. The activity continues (perhaps intermittently) as long as adequate heat continues to warm the liquid. Then, as the heat diminishes, the surface refreezes and a new crust forms and thickens (Figure 16.13d).

The remaining surface would be roughened to some degree by the agitation undergone during exposure of liquid and during the initiation of refreezing. The lumpiness of the texture would be enhanced by any floating chunks of unmelted crust, but, just as the numbers and sizes of rafts vary from case to case, the degree of lumpiness varies as well, so that some chaotic terrain can have a relatively fine matrix texture.

As refreezing continues, the crustal thickness under the chaos area increases until it becomes similar to the surrounding crust. Isostatic compensation due to the buoyancy of the refrozen ice raises the chaos area, with the topographic discontinuity at the edge forming an encircling moat (Figure 16.13e). This process explains the common domed shape of many chaos areas rising from a gradually-ramped shoreline as in Figures 16.8 or 2.11.

In some cases, downwarping might also result in cracking parallel to the shoreline (Figure 16.14a). If the cracking penetrates through the crust (which is quite thin at this location) rafts may be launched (Figure 16.14b) and float away from the shore (Figure 16.14c). Crustal rafts from the immediate edge of the chaos have tapered thickness, so that, as they float, their original surfaces will be tilted down to the float line, as is the case for many observed rafts. In other cases, melting of portions of the rafts may account for re-equilibration of the float angles, as observed. In any case, where crustal material has broken along or within the chaos area, flotsam of a wide range of shapes and sizes is likely to be created (as sketched in Figure 16.14c), ranging from small pieces that contribute to the texture of the matrix to large pieces that are part of the identifiable raft population.

Those chaos areas that are bordered by cliffs, rather than by gradually-sloping beaches, are simply the result of such separation of shoreline rafts. It is not necessary to invoke any separate formation mechanisms to distinguish between the two types of chaos appearance. The difference, from the point of view of the chaos formation process, is trivial.

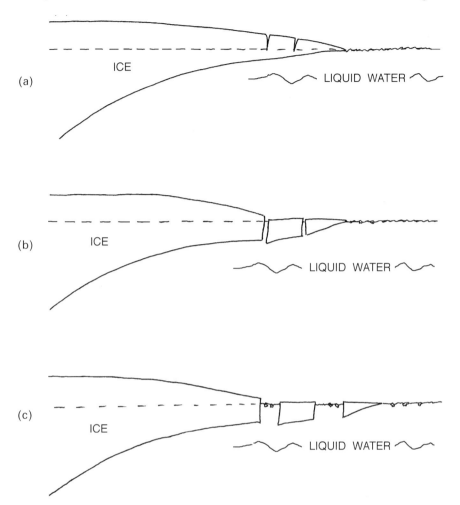

Figure 16.14. (a) Production of lateral cracking (e.g., near the Mitten). (b) Edge rafts in place. (c) Rafts migrate away from the edge. In cases where edge rafts do not separate from the surrounding surface, the bulged-up matrix yields the impression of a raised surface; in cases where edge rafts have moved away, the remaining coast has cliffs that give the impression that the chaos area is lower than surrounding terrain.

The thickness of rafts, even if we take into account the portion below the float line, is not necessarily indicative of the general thickness of the crust prior to the melting that created chaos, because the rafts may have broken free after significant melting had already taken place. However, the thickness does provide useful constraints. Based on the freeboard (the height of their surfaces above the surrounding matrix) of the rafts in Conamara, and assuming that like terrestrial icebergs they were floating with about 10 times as much ice below the surface, the rafts ranged in thickness from a couple of hundred meters up to about 3 km. Thinner rafts are probably indistinguishable from the other lumps and bumps in the matrix. The thickest rafts probably broke from near the edge of the melt-through and may have been nearly as thick as the surrounding ice. Thus, raft thicknesses in Conamara are consistent with the values of ice thickness that we were inferring from tidal tectonics.

The melt-through model also suggests that parts of the crust surrounding large chaos areas might have become quite thin during the process. Such thinning in the neighborhood might explain why large chaoses are often surrounded by numerous smaller ones; melt-through on a small scale is easier where the ice is thin. This effect explains why so many small chaos features surround Conamara, for example.

This model also has as a natural consequence the preferential survival of crust at pre-existing ridges, as observed in numerous places where the extent of chaos has been bounded by ridges, where peninsulas or causeways of older ridges jut across chaotic terrain, or where rafts preferentially preserve ridges (as shown in Figure 16.1b). At a pre-existing major ridge the crust is probably locally thicker than elsewhere, either because the ridge formation process creates a "keel" on the bottom of the crust, or because the ridge topography is supported isostatically, or simply because the topography itself is a thickening of the crust. As the crust is heated from below (Figure 16.15), it resists melting somewhat under the ridge sites, so the ridges and the area immediately around them have a preferential tendency to survive.

The thermal melt-through model also fits well with one of the most promising explanations for the darkening that borders mature ridge complexes (the "triple-bands") and surrounds most tiny patches of chaos. Sarah Fagents, a postdoctoral researcher at Arizona State University, suggested that warming near the surface results in sublimation of the ice, producing local enrichment of the other substances in the ice. Even slight enrichment could significantly darken the ice in this way. In the case of mature ridge complexes, repeated passage of warm tidal water up through the crust would provide such warming. Similarly, chaos formed by melt-through and warming of the adjacent terrain would have the same effect. In both cases, the terrain is not only darkened, but usually has topography that seems somewhat softened and subdued, consistent with Sarah's model.

What may be an extreme example of this effect is a feature seen in the isolated set of high-resolution images a few hundred kilometers west of Conamara (an image sequence also used in Figures 10.5, 10.6, 11.1, and 16.9). This feature, the so-called "Dark Pool" shown in Figure 16.16, is unique in our images, and frustratingly appears in this locale where we have no context images to supplement this high-resolution view. The dark pool lies in tectonic terrain, probably the fine furrows of an old dilation band, just west of the area shown in Figure 16.9, where considerable chaotic terrain has been formed during several episodes over time. In fact, included in Figure 16.16, on the right, is a patch of chaotic terrain almost the same size (3 km wide) as the dark pool. If Sarah Fagents is correct, and heating from below causes the darkening of the surface, then the dark pool might be an incipient patch of chaos, where melt-through nearly occurred (as in Figure 16.13b), leaving only a smoothed, darkened depression.

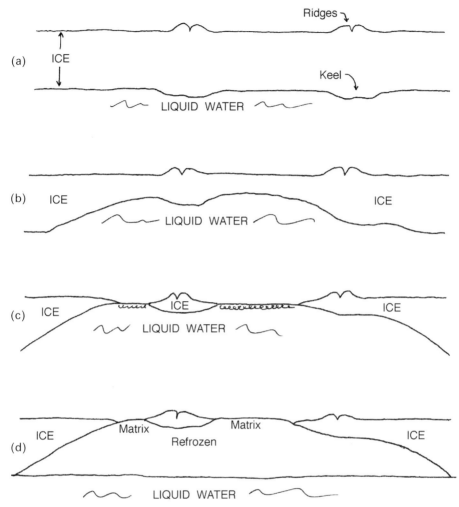

Figure 16.15. The melt-through model explains why pre-existing ridges can bound the edges of chaotic areas, and/or leave causeways, peninsulas, and chains of rafts across chaotic terrain.

16.5 VOLCANISM, NOT

The characteristics of chaotic terrain fit quite well with the melt-through model for its formation, while what we had learned about the size distribution was at odds with the putative underpinnings of the convective thick-ice model. As another alternative, volcanism did not seem plausible on theoretical grounds. Volcanism usually requires that the lava be less dense than the surrounding solid. On terrestrial planets, where molten lava is less dense than rock, volcanism is a major resurfacing process. It is not so easy when water is involved, because cryo-lava (water or slush) is denser than cryo-rock (ice) and unlikely to raise up and over the surface.

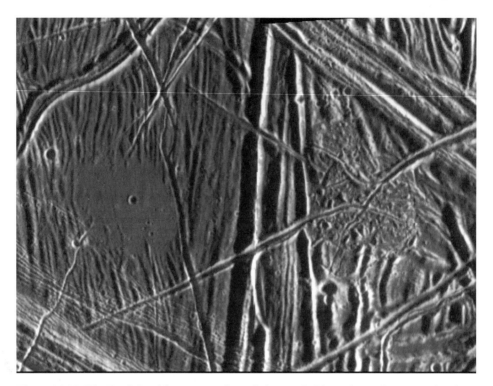

Figure 16.16. The Dark Pool is not a pool at all, but probably a place where warming from below has smoothed and darkened the surface. The craters are genetically unrelated to the Pool. Ejecta from larger craters probably just happened to crash here.

Liquid water tends to stay below the solid, whether in an iced drink, a frozen lake, or the Arctic ice cap.

On Europa, volcanism would require another physical driver to force the fluid to the surface. We have already seen how diurnal tides may work cracks, squeezing water and slush up and onto the surface to produce ridges. In a sense, that process could be called volcanism. However, it is difficult to envision a theoretical mechanism that could create chaotic terrain with volcanism, unless *ad hoc* conditions are invented.

We have seen, though, that some examples of chaotic terrain are suggestive of lava-like flow. Many small rounded chaoses, like the Mitten (Figure 16.8), or the one in Figure 2.11, or some of the other small ones around Conamara, do suggest that upwelling material has flowed out onto the surface, as in Figure 16.17. Such upwelling is consistent with the bulged-up appearance of the lumpy matrix and the slope of the surrounding terrain downward toward the edge of the matrix. It would also be consistent with the cracking of the surrounding terrain along the western and southwest edges of the Mitten, where the crust might be downwarped by the load.

On the other hand, these features are equally well explained by the melt-through model. The topography created by melt-through, followed by refreezing (Figure

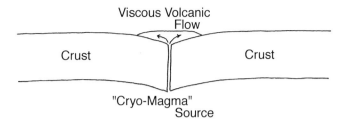

Figure 16.17. Volcanic flow of a viscous fluid up and over the crust might explain the topography of many small round chaoses or of the Mitten. However, exactly the same topography could be created in the context of the melt-through model for chaos formation by isostatic relaxation of the crust after refreezing (Figure 16.13e).

16.13e), would have the same appearance. The topography in Figure 16.13e is the same as in Figure 16.17. Thus, while in certain cases, patches of chaotic terrain might give the impression that the cross-section may look like Figure 16.17, with one type of material flowing over another, we cannot see below the surface. The melt-through process shown in Figure 16.13 leads to the same surface appearance. Moreover, the cracking along the edges of the Mitten might simply be the incipient formation of rafts along the edge of the melted region (Figure 16.14). Although some chaoses do indeed suggest volcanic flow up and over the surface, their appearance can equally well be explained by melt-through.

One potential way to discriminate between the volcanism cross-section (Figure 16.17) and the refrozen melt-through cross-section (Figure 16.13e) would be to measure the surface topography. If the updomed matrices of chaoses proved to be generally significantly higher than the surrounding terrain, the topography would suggest that something more than a buoyant, isostatic rebound pushed the material upward. Some claims have been made for such excess updoming in some cases, but the uncertainty in such measurements is great.

In principle, *Galileo* data can be used to infer the topography of chaotic terrain in two ways. One is by using stereographic analysis, where the two views of an area taken from two directions are compared, just as our brains construct three-dimensional views of the world by comparing the images made with both of our eyes. The other is by photoclinometry, in which subtle variations in shading are used to infer slopes, just as our brains interpret the morphologies of terrains by the variations in brightness in an image. The problem with stereo-analysis is that where we have more than one image of a feature, they are usually taken under very different conditions of illumination and resolution. The problem with photoclinometry is the darkening of the surface in and around chaotic terrain, which in a given image cannot be separated from the effects of changes in the inclination of the surface.

Several claims have been made that the heights of chaos matrices have been measured and that they are updomed relative to the surroundings. In order for such claims to be credible, they would need to describe in detail the quantitative methods, the limitations of the data, and the range of uncertainty of the results. The uncertainty should be displayed with error bars plotted directly on any topographic profiles that are presented as results. So far, such details have been lacking. My suspicion is that the limitations of the data currently in hand preclude any definitive results from topographic analysis. Though claims have been made to the contrary, the case has not been made adequately that topography supports models involving volcanic or convective upwelling.

What makes the melt-through model more attractive is not that it offers a better explanation of the appearance of the Mitten than volcanism, but rather that it explains all chaos with a single process. Observations show us that chaoses have a wide range of sizes and numbers of rafts. We can explain the difference between cases with abrupt cliff-like edges and those with ramped-down beaches simply by whether rafts have broken away from the edge or not. We do not need to invoke an entirely different mechanism like volcanism to explain some of them. Ockham's razor implores us to avoid multiple explanations for similar phenomena, and that logic compelled us toward the melt-through model. Melt-through provides a unified explanation for the formation of all chaos terrain, rather than requiring separate models for various sub-classes.

Even as I began to make the case for melt-through, the heavy-hitters on the *Galileo* Imaging Team were describing Europa's surface in terms of volcanism. There was considerable attention and image sequence planning devoted to getting high-resolution images of the volcanic flows that Greeley and Head thought they were seeing based on low-resolution images, an interpretation that was generally unquestioned. Within my research group, however, Greg Hoppa and Randy Tufts found their talk of volcanic flows to be inconsistent with the evidence. Whenever we had looked closely at dark splotches on Europa, they had turned out to be chaotic terrain. Applying Ockham's razor we expected that the other dark splotches would be similar chaos.

At high resolution that is what they proved to be. For example, one of the targeted areas included the two dark areas known as Thrace and Thera as likely lava flows. These dark areas, to the southeast of the Wedges region (visible in Figure 2.4 and labeled in Figure 9.1) were imaged during orbit E17 at moderate resolution (Figure 16.18, see color section). Thrace and Thera are both large, classic examples of chaotic terrain.

During the same orbit, high-resolution images were taken of the southwestern margins of Thrace (Figure 16.19), because Jim Head wanted to get a good look at his lava flows. That is exactly what he saw: Dark margins around the chaos area that he interpreted as flows of darkened material that seem to have filled in-between higher ridges.

Ockham's razor insisted on a different interpretation. We had already seen similar darkened terrain: the margins of mature ridge complexes, which appear in low resolution as the dark components of the low-resolution triple-bands; and the dark surroundings of chaotic terrain, which for the smallest chaoses appear at low resolution as the dark spots named lenticulae. Wherever we saw these darkened terrains (e.g., Figure 2.8 shows darkened ridge margins that appear as triple-bands and darkening around small chaoses that appear as lenticulae, Figure 1.10 shows the same blown up, and Figure 2.9 shows darkening along a mature ridge), they

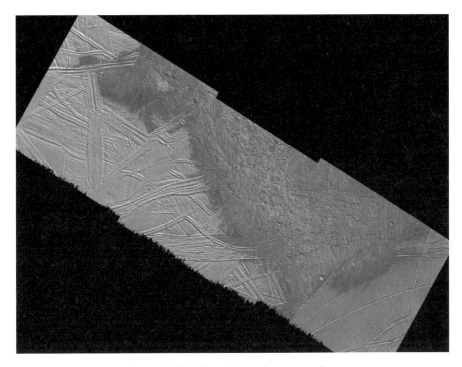

Figure 16.19. Thrace's southwest margins.

presented that same appearance: slight softening of the topography, and darkening in the lower flattened portions. This common appearance in those locations is consistent with Fagents' model in which heating has both softened the surface and darkened it where the most sublimation occurred.

There is no reason to attribute this similar appearance, in a similar context, at Thrace to have an entirely different explanation. The appearance at Thrace is most simply explained as a result of the same warming that likely creates the same appearance along cracks and other chaos. And those locations are where we would expect such heating, whether because of tidal working and oceanic flow through cracks, or the melting that leads to chaos.

The model of an ice crust thin enough to be linked to the ocean through cracks and melt events can explain all these features without resorting to a variety of *ad hoc* processes. Until we get more information to the contrary, William of Ockham tells us again that the simple model is best.

16.6 HEAT FOR MELT-THROUGH

The melt-through model assumes that adequate amounts of tidal heat can be concentrated under locales or regions of the crust. If the ice is only a few kilometers thick, those requirements are much less stringent than if the ice were greater than 20 km thick, as in the convecting thick-ice model.

The advocates for the thick-ice model fought vigorously against the idea of melt-through, using political advantage to promote the canonical model of thick ice. It was arranged that Pappalardo would put together a paper for the *Journal of Geophysical Research* defining the *Galileo* Imaging Team's position on Europa's geology in the context of the key issue of the time: "Does Europa have a subsurface ocean: Evaluation of the geological evidence". The subtitle gave the impression of a balanced, authoritative assessment. Unfortunately, the paper was an advocacy piece for the thick-ice model, concluding under the mantle of the authority of the Imaging Team that there was no geological evidence for any direct linkage between a liquid water ocean and the surface.

In order to make that case, the paper needed to show that the melt-through model was physically impossible, and it used the following argument. If you adopt an internal tidal-heating rate that gives a heat flux through the surface of a few tens of milliwatts per square meter, then it would take all of the total heat output from all over Europa concentrated under Conamara (0.024% of Europa's surface area, according to them) for hundreds of years to melt through a 2-km-thick crust. Obviously, such a concentration of all of the tidal heat under one location was out of the question.

I found this argument to be specious. There is no particular significance to the timescale hundreds of years, because there is no evidence that Conamara formed that quickly. A similar calculation with the same heat flux would show that it would take only 1% of the total tidal heat if the melting occurred over tens of thousands of years. Or, with a higher, but still plausible, heating rate of a couple of hundred milliwatts per square meter, only a very modest extra 0.02% of Europa's heat concentrated under Conamara (about 0.02% of Europa's surface area) for 100,000 yr would provide enough heat to melt open Conamara. As presented in the *JGR* paper, melt-through seemed impossible, but with appropriate logic and arithmetic we could see that it was well within the range of plausibility.

At the Lunar and Planetary Science Conference in the spring of 2000, David Stevenson, a planetary geophysicist at Caltech, pointed out a more legitimate obstacle to melt-through. He noted that the ice at the bottom of the crust, just above the liquid water ocean, would be so close to the melting temperature that, even though a solid, it would have a fairly low viscosity. Now, if slightly enhanced local heating melted a spot on the bottom of the crust, the warm ice from adjacent to it would tend to flow up and into the thin spot. Just as a fluid on the surface of a planet tends to flow down into a depression due to gravity, the fluid ice on the bottom of the crust would flow up into the thin zone due to buoyancy. On those grounds Stevenson proclaimed with great authority that melt-through was impossible.

Such pronouncements proved to be premature. My graduate student Dave O'Brien, who had been working with me on some challenging problems related to the size distribution of asteroids, naturally attended the weekly meetings in my lab, and followed the issues raised by our Europa work. Dave decided to set up a straightforward computer simulation of melt-through. He used the common finite element approach, in which the ice is conceptually divided into many small, finite cubical cells, and the computer tracks the heat flowing into and out of each cell from and to its neighbors. The thermal history of each cell is calculated, taking into account heat flow from below, radiation of heat outward from the top of the crust, the thermal conductivity of the ice, and the latent heat needed for melting.

We found that the ice can be melted through with only a very modest concentration of tidal heat flowing out from the interior. Obviously, it was easiest to melt through if the ice were only a couple of kilometers thick, or less, but in order to strengthen our conclusions we considered thicker ice as well. For example, if the heat flux is $0.1\,W/m^2$ the steady-state ice thickness would be about 6 km. Then, if a few percent of the total heat flux is concentrated over an area at the base of the ice about 200 km in diameter, within 40,000 yr a Conamara-size opening forms at the surface.

The cross-sections of the crust during the melting process looked uncannily like my earlier sketches of the melt-through process (Figure 16.13). A wide area of thinned ice surrounded the melt-through exposure. Numerical experiments with various amounts of heat concentration and varied parameters showed that a wide range of sizes of chaos could be formed. As we had determined with earlier back-of-an-envelope estimates, there is no problem melting through the ice with modest concentrations of plausible amounts of tidal heat.

Dave O'Brien's simulations showed that the bottom of the ice melts away so fast that the low-viscosity ice envisioned by David Stevenson is never more than a very thin layer near the bottom. It is not thick enough to allow significant inflow from the edges of the melt-zone up into the thinning area. Stevenson had raised an interesting issue, but Dave O'Brien had shown that it was not an impediment to melting holes in the crust.

Dave O'Brien presented his results at the autumn 2000 meeting of the DPS (the Division of Planetary Sciences of the American Astronomical Society). DPS presentations are limited to a few minutes' length, and the public discussion after each is even shorter. Although student presenters are usually treated relatively gently, Stevenson proclaimed before the large audience that O'Brien's work must be all wrong, but there was not enough time for any explanation. The audience had heard the definitive judgment by a renowned authority that this graduate student's work (which bucked the *Galileo* party line) was worthless. Later, during a private conversation, Stevenson agreed that Dave O'Brien's work was fine. But the DPS audience never heard that, and the damage was done.

Formation of chaotic terrain requires modest concentration of heat from below at various times and places. Is such concentration plausible? On Io, ~20% of the total tidally-generated heat flux escapes through the single volcano, Loki, which is a much greater portion of the heat concentrated into a much smaller area than we have suggested for Europa.

On Europa, the modest, occasional concentrations of heat adequate to melt through the ice crust might originate at volcanic vents at the base of the ocean. There is no direct evidence for undersea volcanism, but the plausibility has been accepted in the context of considerations of the possibility of life on Europa.

Astrobiological theorists were constrained by the canonical wisdom about Europa, which they had received through the authorized *Galileo* channels, that the ocean was isolated from the surface. They thought they needed to invoke volcanism as a source of the energy and chemistry for life. More likely, the ocean is linked to the surface, so life is much less constrained, but the plausible volcanoes would play a role in localizing the heat output from Europa's deep interior.

As *Galileo* imaging was providing evidence for an ocean, *Galileo*'s Project Scientist Torrence Johnson recruited John Delaney, a prominent geophysicist from the University of Washington, to participate in considerations of the role of volcanism in supporting oceanic life. Delaney had led important explorations of sub-oceanic volcanic hot springs on Earth.[4] Delaney in turn got Richard Thomson, an oceanographic hydrodynamicist from the University of Victoria, British Columbia, involved.

Thomson and Delaney examined the critical issue of whether the concentration of heat at a volcanic vent under the ocean could remain concentrated as it moved outward through 100 km or more of liquid water. As a Canadian and an oceanographer and a researcher of integrity, Thomson had no particular interest in supporting the *Galileo* party line, nor did Delaney. They concluded that thermal plumes within the ocean could keep the heat localized as it rises up to the ice, supporting our picture of thermal melt-through as the original of chaotic terrain. Constructing dynamical models of an ocean that we barely know exists has strong components of speculation, of course; so, whether the heat is transported as described by Thomson and Delaney remains uncertain, but at least within our range of uncertainty, melt-through remains a viable and perfectly-plausible process.

Moreover, heat concentration could be enhanced still further if the ocean is shallower than generally assumed. For example, if the low-density layer of Europa includes hydrated silicates, the liquid water layer may be only a few tens of kilometers thick, and volcanic sub-oceanic mountains could reach up nearly to the bottom of the ice. In that case, volcanic heat could be directly deposited in concentrated sites at the base of the ice. It is quite plausible that on Europa heat is concentrated more than in the conservative examples that we had investigated using Dave O'Brien's simulations.

The study of chaotic terrain implies that melt-through occurs at various times and places. It follows that the thickness of ice is quite variable with time and place. Chaos formation has been common, currently covering nearly half the surface, and continual, displaying a wide range of degrees of degradation by subsequent cracking and ridge formation. The sizes of chaos patches range from over 1,000 km across down to at least as small as recognizable on available images, ~1 km across. Early reports that chaos is rare and recent, or that there is a characteristic size to the small patches, were an observational artifact: Old or small examples are harder to see but they are most common. Over the surface age, formation of areas of chaotic terrain has been frequent and continually interleaved with tectonics, so at any given location melt-through has occurred at least every few million years.

---

[4] More about the relevance, or lack thereof, of terrestrial sub-oceanic volcanic vents to Europa is discussed in Chapter 21.

17

Crust convergence

17.1 BALANCING THE SURFACE AREA BUDGET

Europa's crust is highly mobile. The evidence for considerable movement of plates of a wide variety of shapes and sizes is clear. Tectonics not only provided the cracks that allow this relative displacement, but also created the lineaments (usually ridges) that we can use as markers to reconstruct the past positions of plates that have moved with respect to one another. Fortunately, too, there are no trees or other annoying manifestations of life to get in the way of our views.

Even the limited *Voyager* images showed clear evidence of both strike–slip displacement and dilation along cracks, although Schenk and McKinnon had encountered resistance to the seemingly-radical notion that a surface could be so mobile. Study of *Galileo* images confirmed that dilation has been common and widely distributed (Chapter 11). Large crustal plates (often hundreds of kilometers in scale) have separated, with the space between them filled by new surface material creating dilational bands. Similarly, new surface has also been created along strike–slip faults, where bends in the fault lines have resulted in pull-apart zones (Chapter 12), indistinguishable in appearance from dilational bands. The only thing that distinguishes strike–slip pull-aparts from other dilational bands is the particular mechanism that drove them open.

The dilation of cracks in Europa's crust creates new surface area. In contrast, the other resurfacing processes on Europa do not add new area. Consider ridge formation, in which tidal pumping and extrusion of slush from a crack over the adjacent terrain creates new surface for up to a kilometer on either side of the crack. In that process, new surface is created (often hundreds of square kilometers) by covering old surface. In the process, exactly as much old surface area is destroyed as new surface area is created. Similarly, consider chaos formation, in which the previous terrain is disrupted and melted and the new surface (matrix and displaced rafts) forms entirely within the area of the destroyed terrain. There, too, precisely as much old surface area is destroyed as new surface is created. However, with dilation, as a crack opens, the space between represents new surface area, but there is no local, intrinsic removal of surface area.

Because new surface area has been continually created on Europa in this way, someplace and somehow crust must be removed to compensate for it. The alternative, a planet with excess skin, would not be a pretty sight. In some way, the surface area budget must be balanced as new surface is created at dilation zones. A key question for several years has been: How and where has crust been removed for the total surface area to be conserved?

On Earth, we have the same issue, with rapid dilation occurring on a global scale, including the opening of the Atlantic Ocean, the separation along the East African Rift, and the parting of the Red Sea. With so much dilation between crustal plates creating new area, several processes help keep the surface area budget balanced. Most common in large-scale plate tectonics is plate subduction, where one plate of the Earth's crust rides down under another. The topography that marks the locations of subduction is very dramatic and distinct, although usually hidden by an ocean. Between a subducting plate and the edge of the plate that overrides it, the seafloor drops into deep narrow trenches, the deepest spots in the ocean. No such trenches have been identified in the icy crust of Europa.

When plates of continental crust, as opposed to oceanic crust, converge on Earth, they are too thick for one to slide under the other, so the material must pile up, notably when India crashed into southern Asia, creating the Himalayas; but, no such mountain ranges have been seen on Europa.

On a more regional scale, surface area can be taken up by thrust faulting, where an obliquely-dipped fault plane allows one plate to ride slightly up over another, but not nearly as much as in the case of subduction. Thrust faults are a common way that horizontal compression is relieved on Earth, and they are common and large enough on Mercury that they may be part of a global contraction process. There is no evidence for thrust faulting on Europa.

Europa's surface area budget has an enormous income, but where does it all get spent? We already saw evidence in Chapter 12 (e.g., Figure 12.13) for a type of band where convergence of the surface has occurred, as inferred from our study of strike–slip displacement. This type of terrain may be adequate to explain where the excess area has gone. Let's consider some other mechanisms and types of terrain that may play a role.

17.2 SURFACE CORRUGATIONS

In principle, horizontal compression could be accommodated by corrugation of the lithosphere in a series of crests and troughs. Such corrugations have been observed in one site, which lies within the pull-apart zone of Astypalaea. Figure 13.6 shows that there are three troughs that run nearly perpendicular to the trend of the fine furrows in the Astypalaea pull-apart. There are also three sets of fine cracks that also cross parallel to the troughs and which may correlate roughly with crests between the troughs. Actually, the correlation between these cracks and the crests of the corrugations is not very accurate, but it has been interpreted as the result of the down-bending of the lithosphere along the crests. The evidence for corrugations on Europa consists of these three wavelengths in this single, rather special location.

Louise Prockter and Bob Pappalardo, from the Brown University group, assumed that these corrugations were created by horizontal compression of the crust. They calculated the minimum horizontal pressure required to create such corrugations in an elastic lithosphere about a kilometer thick and found it to be $>2 \times 10^7$ Pa. They noted that that number is very large compared with the stress expected from tides, and they concluded that such folding could not have occurred if the ice crust is a thin elastic lithosphere. Instead, they proposed that folding of the brittle surface resulted from the response of a thick layer of viscous ice below it, and that this could be driven by only 3×10^6 Pa of pressure. They considered a plausible source of such horizontal compression to be tidal stress due to non-synchronous rotation, and they concluded that they had found the mechanism for accommodating horizontal compression.

Prockter and Pappalardo's theory of folding was predicated on their thick-ice model, with its isolated ocean, so they took the discovery of the folds to be evidence for thick ice. In their word, it "discounts" the view of Europa that had been emerging from my research group's work on tidal tectonics in which the ice is thin enough that the ocean is linked to the surface. Their entire paper on the subject was only two pages long (a report in *Science* magazine) and the proposed theoretical model was described in four sentences. No details of the analysis were presented in any follow-up paper. Nevertheless, the story was fast-tracked into the canon.

Their story is an interesting interpretation, but even if they had published the details, there would be little reason to take it very seriously. Consider the observations. Only three wavelengths in a very unusual locale do not compellingly indicate a globally-important phenomenon. These features are interesting and provocative, but ordinarily such oddities would not be considered the basis for such far-reaching implications. Prockter and Pappalardo pointed out two other examples of cracks or troughs, perhaps similar to those seen in Astypalaea, but examples with less than one wavelength are hardly convincing signs of corrugation. The observational evidence supporting the model is marginal.

Now consider the theoretical part of their story, at least to the extent we can, given that hardly anything about it is published. The key to its purported success is that the amount of stress required could supposedly be provided by tides. Figures 6.1–6.3 show that typical tidal stress levels are less than $\sim 10^5$ Pa (about the same as the atmospheric pressure on the surface of the Earth), 30 times less than the theory required. Prockter and Pappalardo invoked several tens of degrees of non-synchronous rotation to get the required stress. However, recall from Chapter 6 that it is unlikely that so much stress could build up. The lithosphere could not remain purely elastic during the slow non-synchronous rotation, so the stress could not build up. Even if it did, the crust would have cracked long before the necessary non-synchronous stress could have accumulated.

Another problem is that the displacement distances that correspond to tidal stress are very small, as we have already discussed in the context of ridge building and dilation. For example, the tidal stress invoked in the corrugation model corresponds to a strain of about 100 m over the surface of Europa. How such small amounts of displacement could have played a significant role in addressing the surface area budget is unclear. The forces that drove dilation acted over broad regions and caused long distances of displacement as discussed in Chapter 11. These same forces must have driven comparable amounts of convergence elsewhere. The relationship among these processes and the tidal forces was not adequately described in the 2-page *Science* paper to support adoption of the corrugation mechanism as a significant component of the surface area budget.

Finally, the effectiveness of corrugations at reducing surface area suffers from the geometry of the "cosine effect": The slopes of the corrugations are too low to produce much horizontal shortening. The troughs in Astypalaea are probably less than a few tens of meters deep, and they slope up toward the crests over about 10 km, so the slopes are probably less than about 1%. The shortening is given by the cosine of the slope, in this case 0.9999. If such corrugations extended for 1,000 km, they would account for about 100 m of surface contraction.

If corrugation plays a significant role in the surface-area budget, the case for it has not yet been made.

17.3 CHAOTIC TERRAIN AS A SURFACE AREA SINK

Nearly half of Europa's crust is chaotic terrain, and most of that terrain is in very small patches. The brittle, elastic lithosphere has been continually perforated by these openings, which could readily allow for accommodation of considerable amounts of surface contraction. Consider the analogy of perforated sheet metal; its area can be contracted leaving minimal evidence of distortion. In the same way chaotic terrain could have accommodated considerable surface contraction without creating obvious signs of distortion.

In this process, each patch of chaotic terrain would have its shape distorted slightly while the hole in the lithosphere was still open at that location, but, because each already had an irregular shape, we cannot detect any systematic observable effect. If every chaos was perfectly round, these processes would have created regions where all the chaoses would be ovals with parallel orientations. Unfortunately, chaoses are not shaped so regularly; so, even if the process has been the dominant way of accommodating lateral compression of the surface, it would have happened without leaving a trace.

Several of the people assigned to denigrate the thin, permeable ice model have mocked this explanation. Their line has been that if we cannot point to the specific effects of the horizontal compression, this mechanism has no merit. I find that logic baffling. We know that a huge fraction of the area of the lithosphere has been perforated through by chaos, and we know that such openings can relieve stress and allow an area to contract. Whether or not we found direct observational evidence for sites of surface convergence, chaos would provide a perfectly-plausible explanation for how surface area is reduced to compensate for dilation elsewhere.

Given that we know so much of the surface has been continually perforated by chaos, it is reasonable to assume that much of the surface contraction needed to balance the formation of new surface area due to dilation has been accommodated by these openings. Whatever process has been operating over broad regional scales to pull apart dilation bands (Chapter 11) has probably also been squeezing chaos openings closed somewhat at the same time. Study of the shapes and morphology of chaos alone cannot tell us whether this effect has been enough to accommodate dilation. However, it might be interesting some day to sort out from the geological record whether dilational bands tended to form preferentially at times when chaos openings happened to be particularly common. Eventually, we may be able to determine how much of a role chaos had in allowing regions of the surface to contract under horizontal compression.

17.4 CONVERGENCE BANDS

In Chapter 12 we saw how reconstructions of strike–slip displacement revealed two examples of linear surface features that appeared to have accommodated surface convergence (Figures 12.13 and 12.14d). (I use the term "convergence" to represent displacement in which crustal plates have moved together as surface area at their boundary has disappeared. I try to avoid the word *compression* in describing displacement because it would imply something about force, and hence about process. Here, we are simply describing the geometry of displacement: convergence is essentially negative dilation.) Convergence was inferred to have occurred at these sites because it was necessary to accommodate the observed strike–slip displacement.

At these locations, we found band-like structures that have an appearance distinctly different from dilational bands. Dilational bands are usually characterized by parallel boundaries of identical shape, with intervening material covered by fine grooves parallel to the edges, as in Figures 11.3 or 11.7. The pre-dilation configuration can be reconstructed by removing the intervening material and fitting together the adjacent crustal plates. In contrast, the convergence bands have curved boundaries that do not fit together. The example in Figure 12.13 has a lens shape, with one side forming a slight lip. The other case (Figure 12.14d) has internal fine grooves, which are not parallel, but rather form intricate shapes similar to microscopic views of animal muscle tissue. (While the convergence band in Figure 12.14d seems to comprise dark material, Figure 12.13 shows that albedo is not a reliable indicator of convergence.) Bands with this general appearance are fairly common on Europa, and it is plausible that they may generally mark convergence sites.

Another linear feature on Europa with similar characteristics has also been suggested to be a convergence feature on the basis of completely-independent, albeit circumstantial, evidence. Agenor Linea is nearly 1,400 km long, running roughly east–west about 40° south of the equator from longitude 180° to 250° (Figure 17.1). The similar, smaller Katreus Linea, whose location near and

Figure 17.1. Indicated by arrows, Agenor Linea is the bright band running across the lower portion of this region, and Katreus Linea is the smaller, similar bright band just north of Agenor. This region is shown as a portion of a Mercator-projected mosaic by the U.S. Geological Survey, which extends from about 7°N (at the top) to about 45°S, and from longitude 170° to 250°W (here 10° ~ 300 km). Most of the northern half of this image contains the Wedges region, which lies between the mottled-appearing chaotic terrain shown to the left and right. See also Figure 9.1 for the location and global context.

parallel to Agenor indicates that it formed in response to the same tectonic forces, is likely a related structure. As a long, bright, wide band, Agenor (with Katreus) is unique on the relatively heavily-imaged anti-jovian hemisphere. Three characteristics suggest that Agenor may be a convergence feature.

First, Agenor has resisted efforts at reconstruction. For dilational bands reconstruction has been easy and often obvious; neighboring plates fit together unambiguously. On the other hand, by definition, at a convergence site the adjacent terrain from the neighboring plates is missing. Hence, the inability to match the crust on opposite sides of Agenor could be explained by convergence.

Second, the proximity of the Wedges region (Figure 17.1), where considerable surface dilation has been a source of new surface, is consistent with Agenor being a site of corresponding surface removal by convergence, a point that was made very early on by Schenk and McKinnon in their study of crust displacement from the *Voyager* images.

Sec. 17.4]   Convergence bands   257

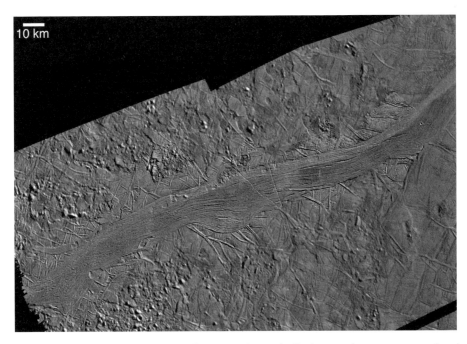

Figure 17.2. High-resolution image of Agenor shows similarity to other convergence bands.

Third, Agenor's appearance at higher resolution (Figure 17.2) is similar to the morphology of the convergence zones we found from the strike–slip reconstructions (Figures 12.13 and 12.14d), with curved boundaries and the non-parallel, "muscle tissue" striations. Unlike those cases, Agenor is bright compared with surrounding terrain, but we already saw in those previous cases that albedo is not necessarily a distinguishing characteristic of convergence.

The Brown group had interpreted the high-resolution images (Figure 17.2) completely differently. Based on qualitative impressions, they interpreted the morphology as evidence for strike–slip displacement along the lineament, and concluded therefore that Agenor is not a "contractile feature". Why strike–slip should preclude convergence is unclear. Certainly we have seen plenty of examples of strike–slip with accompanying dilation on Europa. Another difficulty with the Brown interpretation is that no one has been able to reconstruct the strike–slip by aligning features from the adjacent terrain, which has been the usual way to demonstrate such displacement. In fact, the lack of identifiable linkable features across Agenor actually supports the idea of convergence, in which the formerly adjacent terrain has been consumed in the convergence process.

The argument rests entirely on a qualitative comparison with terrestrial experience, where appearances that result from tectonic displacement may be quite different from morphological expressions on Europa. Agenor has a similar appearance to the known convergence bands on Europa, which may be a more relevant comparison than tectonic morphologies on a solid rocky planet.

While several pieces of evidence collectively point to Agenor (along with the associated Katreus) being a convergence feature, they do not prove it. However, further evidence comes from the Evil Twin of Agenor.

17.5 THE EVIL TWIN OF AGENOR

Imaging of the Jupiter-facing hemisphere of Europa has been limited. Aside from global low-resolution color images (Figure 2.3), and a few very-high-resolution images of isolated locales, the best broad coverage was obtained during *Galileo*'s 25th orbit as a mosaic of a dozen images at about 1 km/pixel. A number of important features have been identified on this image sequence that show similarity to the anti-jovian hemisphere 180° away, including considerable chaotic terrain and distinctive crack patterns, which are probably extreme cycloids as discussed in Chapter 14. Specifically, tightly-curved cracks that form circular or boxy patterns in the sub-jovian area are similar to those in the Wedges region, although unlike the Wedges these cracks exhibit little dilation.

Within these images of the sub-jovian region lies a linear band (Figure 17.3) that closely resembles Agenor (cf. Figure 17.1), but lies nearly diametrically opposite it (Figure 17.4). While Agenor runs westward from longitude 180° in the southern hemisphere, its twin runs westward from longitude 0° in the northern hemisphere. Agenor is somewhat farther from the equator, but not much. The appearance of Agenor and this nearly diametrically-opposite "twin" are quite similar. Both are bright bands, with occasional narrow dark edges abruptly cut off by the bright zone. The twin contains a large "island" of undisturbed older terrain. While Agenor itself does not include such an island, its companion Katreus does have one very similar in shape to that of Agenor's twin. While quite similar to one another, Agenor (with Katreus) and its twin are quite different from other linear features on Europa.

Given this similarity alone, Agenor's twin became another candidate convergence feature. Naturally, I could not resist referring to it as the Evil Twin of Agenor. Apparently, as soon as we had presented the significance of this feature at a conference, the nomenklatura of nomenclature decided they had better give the thing a more respectable name. But they were too late. By the time I learned its official name, *Corick Linea*, I had already published it as the Evil Twin.

What makes the Evil Twin special is that, unlike at the other convergence candidates, the plates bordering the convergence have prominent markings that allow reconstruction, even though substantial crustal surface has been eliminated at the interface. In contrast, Agenor and other convergence features have resisted reconstruction, because the terrain that previously existed between the current surface plates is now gone, so it is not available to show the past positional relationship between the two sides. There, the problem is analogous to finding the correct relative placement of two puzzle pieces without having the pieces that would fit between them. However, at Agenor's Evil Twin the plates bordering

Sec. 17.5]   **The Evil Twin of Agenor**   259

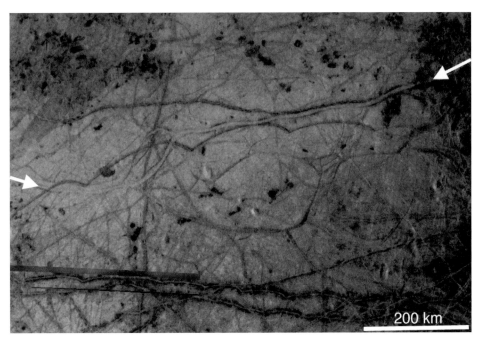

Figure 17.3. The Evil Twin of Agenor runs between the arrows. Compare its appearance with Agenor in Figure 17.2. The twin includes an "island" (left of center) similar in shape and scale to that in Katreus (Figure 17.2). The resolution here is about 1 km/ pixel, and the region shown is about 900 km wide. Note the circular and boxy crack patterns south of Agenor's twin, similar in shape and appearance to the crack patterns in the Wedges region, which are nearly diametrically opposite on the globe.

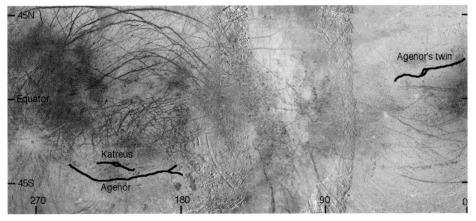

Figure 17.4. Locations of Agenor and its diametrically-opposite twin on a USGS global mosaic (Mercator projection). Agenor runs westward from longitude 180° in the southern hemisphere, while its twin runs westward from 0°. Note the location of the Wedges region north of Agenor (cf. Figures 9.1 and 17.3).

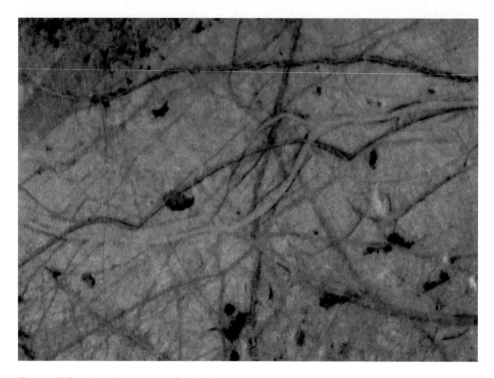

Figure 17.5a. An enlargement of a 440-km-wide portion of Agenor's Twin (from Figure 17.3). Reconstruction of Agenor's Twin is possible because of pre-existing dark lines (actually so-called "triple-bands") that cross obliquely and intersect on the island, near the center of this figure. A dark cycloid (seen in Figure 17.3 running across the figure from just below the left arrow) crosses the Evil Twin twice, and can be seen to post-date it, as shown by its appearance on the bright portion of the current convergence band.

the convergence have prominent linear markings that allow reconstruction by aligning the marks, even though substantial crustal surface has been eliminated at the interface.

Specifically, the reconstruction is largely guided by thick dark lineaments that obliquely cross Agenor's Evil Twin, and also cross one another on the "island". They define a particular displacement of the neighboring terrain needed in order to bring each of these lineaments into alignment. These dark lines are apparent in Figure 17.5a, which is an enlargement of the critical portion of the twin (a 440-km-wide portion of Figure 17.3). That these lines pre-date the formation of the Evil Twin is evident from the fact that they are not seen on the band that composes the twin, which we assume is terrain that has been created along the convergence zone by the convergence process.

The reconstruction is shown in Figure 17.5b. Going back in time, a gap of 25 km

Sec. 17.5] The Evil Twin of Agenor 261

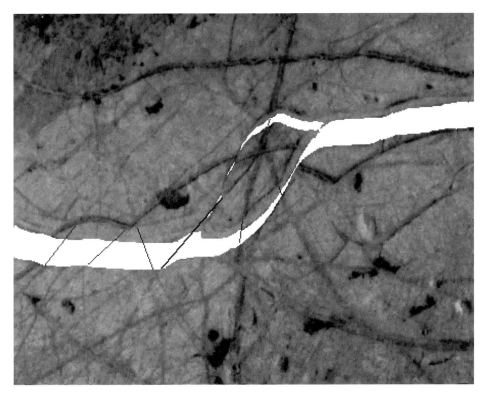

Figure 17.5b. Going back in time, a gap (shown in white) opens about 25 km wide, showing that that much surface has been removed during convergence along Agenor's Evil Twin. This gap is in addition to the approximately 5-km-wide current bright band, which represents area that has been reprocessed, but not removed, in association with the convergence displacement. Here the current bright band has been arbitrarily kept attached to the plate to its north. Minor lines that are realigned corroborate the reconstruction, as marked. Dark lines that formed after the convergence (such as the prominent cycloid) are broken up by the reconstruction.

opens, showing that this much surface has been removed. Over the length of this feature (~400 km) a total of 6,000 km^2 of previous surface has been eliminated, a significant component of the surface area budget.

The Evil Twin of Agenor provides the first direct reconstruction of a convergence band on Europa, as well as strong supporting evidence for the hypothesis that the type of morphology shown in Figures 12.13, 12.14d, and 17.2 is indicative of convergence features in general. This hypothesis was based on multiple lines of strong circumstantial evidence even before I reconstructed the Evil Twin: such terrain is observed where we had identified convergence based on neighboring strike–slip reconstructions, and at Agenor where convergence had been inferred from its location relative to the dilational Wedges. The Evil Twin strengthens that hypothesis further, because it is so similar to Agenor and is demonstrably a convergence feature.

In terms of topography and structure, the morphology of these convergence bands is fairly subtle and subdued, just the opposite of what would be expected if the ice were very thick. On solid planets, crustal plates either pass above and below one another (as in subduction or thrust faulting) or pile up together, as in continental collisions (e.g., building of the Himalayas). On Europa, such structures are not evident, even though substantial convergence has occurred. At convergence bands on Europa (e.g., Agenor's twin), a zone of crust more than 25 km wide has been compressed into a feature only a few kilometers wide.

If the surface ice layer were thick, it is unclear how the substantial amounts of ice displaced in the process got out of the way without leaving much topographic structure. One might speculate that the thick ice either had a low enough viscosity to avoid building topography during convergence, or the topography relaxed away later on. The identified convergence bands are among the most recently-formed features on Europa, so such relaxation would have had to have occurred relatively quickly and completely.

On the other hand, the lack of structure at convergence sites is perfectly consistent with there simply not having been much solid material involved. Based on everything else we have been learning about Europa, the crust probably consists of only a thin layer of ice over liquid water. Convergence of thin ice could occur without leaving much of a topographic record, only subtle markings like those observed. Moreover, the thin ice could be contracted horizontally in this way with very little force. For example, if the ice is thin enough, ocean currents could provide the regional force needed to separate dilational bands and at the same time drive plates together at convergence sites.

The locations of Agenor and the Evil Twin nearly diametrically-opposite one another on the globe may be simply coincidence, but it makes me wonder whether tidal processes have played a role in determining the character and locations of these convergence features in some way. Tidal stresses are symmetrical on opposite sides of the planet. For example, the Wedges region, whose dilation provided the first hint that Agenor might be a convergence feature, does have a counterpart near Agenor's twin, likely due to tides: In the sub-jovian region near the Evil Twin, tightly-curved cracks form circular or boxy blocks ~150 km in scale, similar to the tectonic patterns in the Wedges region. These patterns are consistent with the cycloidal shapes predicted at those locations (Chapter 14). However, near the Evil Twin, there has been less dilation along these cracks than in the anti-jovian Wedges, so in this region there is less obvious evidence of the dilation that balances surface removal by convergence.

Detailed understanding of the surface area budget in the sub-jovian hemisphere around the Evil Twin is hampered by the lack of moderate- to high-resolution imaging. There may be considerable dilation in the region and there are some faint indications of it (e.g., to the north of the Twin), but identification of specific candidate dilational features that correspond to the convergence at Agenor's Twin will likely require improved data. On the other hand, even with better images, unraveling in detail the history of this dynamic, rapidly-changing surface is challenging. There has been evidence for nearly a quarter of a century, since *Voyager*, that new surface area has been created by dilation, and *Galileo* has shown that the new area has been continually created at widely distributed sites of crustal dilation. Now, with identification of specific examples of convergence bands and of the characteristics that identify them, we are beginning to move toward a better understanding of the global surface area budget.

18

The scars of impact

18.1 GAUGES OF AGE AND CRUST THICKNESS

Impacts onto Europa have not played a significant role in shaping the surface that we see today. However, they have provided information critical to our understanding what has gone on there. The numbers of impact features, when compared with our knowledge of the sources of bombardment, constrain the rate at which Europa has been resurfaced. For example, a small number of impact features generally would imply a young surface, but the result depends on the rate of bombardment. If the rate had been slow, even a very old surface might have few signs of impacts. Impacts have also served as probes of the subsurface, leaving scars whose character reflects the nature of what lay below, especially the distance down to liquid water. Although impacts have not been major actors in Europan geology, the features they left behind are measurement tools for the dimensions of time and depth.

In order to use these tools, we need good data, and we have it. For measurements of time, we have excellent counts of the impact features, which were easy to get because the numbers are so small. For the probes of depth, we have excellent images of most of the major impact sites.

Despite all this great information, the precision of the measurements of time and depth is limited, because the impact "gauges" are not well calibrated. In order to translate the statistics of impacts into precise age information about the surface, we would need to know the bombardment rate. Unfortunately, our understanding is still limited regarding the sources and numbers of impacting bodies over time. The good news is that the numbers of craters are so very small on Europa that, even with only educated estimates of bombardment rates, we can be confident that the surface is extremely young compared with the age of the solar system. The other good news is that great progress is being made in the study of small bodies in the solar system, including astronomical investigations of their populations, and dynamical modeling of the orbital evolution that conveys them to targets like Europa. As this progress continues, the data from Europa, already in hand, will give an increasingly accurate assessment of the surface age

Like the measurements of age, the depth-gauging precision of impact features is also limited by our ability to interpret the data. In this case, to make the best use of the detailed images of the morphologies of the features, we need to have a good understanding about what types of scars would be produced by impacts into ice of various thicknesses. Based on extensive experience with investigations of impact craters on the Earth, Moon, and other terrestrial planets, we have a good understanding of the features that are formed by bombardment of solid bodies. However, in order to interpret the record on Europa we will need something more. We will need information about what happens when an ice crust, floating over liquid water, is bombarded. We will need to know what sort of surface feature will result, depending on the thickness of the ice and the size and speed of the projectile. This information can come from experiments and theoretical modeling. If that work is done, we will be able to interpret the observed features as measures of what lies below.

Full use of the data already in hand from Europa, to determine the age of the surface and the thickness of the crust, will require continuation of studies of populations of small solar system bodies and modeling of impact processes on water planets. Vigorous continuation of those efforts will tell us a great deal more about Europa, even without sending more spacecraft there. However, there are intrinsic limitations to these approaches. For age determination, the small-number statistics of impact features will inevitably limit the statistical degree of certainty of any values that are derived. For ice thickness determination, there will always be a subjective, qualitative aspect to the interpretation. Moreover, whatever is learned about the thickness of ice from the character of the impact scars will only apply to the specific place and time of these very few events.

Nevertheless, from what we already know about plausible impacting populations and the physics of impact processes, researchers have begun to apply these useful tools for understanding Europa. Crucial results have been obtained already. However, as we consider this work, it is at least important to bear in mind its limitations, or else, as some people have done, results may be over-interpreted or unduly accepted as definitive.

18.2 NUMBERS OF IMPACT FEATURES: IMPLICATIONS FOR SURFACE AGE

The yardstick for measuring the age of Europa's surface has been developed by a team assembled by Kevin Zahnle, an atmospheric scientist at NASA's Ames Research Center. A crucial part of the development involves understanding the rates at which the orbits of comets, which are believed to be the dominant impactors on the jovian system, can evolve onto collision courses with Europa. Zahnle wisely recruited as partners Luke Dones and Hal Levison (of the Southwest Research Institute), members of an international community of celestial mechanicians who have been revolutionizing our understanding of the dynamics of small bodies in the solar system, through clever and sophisticated computer simulations.

A few years ago, Brett Gladman, another member of that community gave a presentation to a broader audience of planetary scientists where he said that, when he does research, he gets exactly the correct answer. That comment was not very diplomatic, but, as a celestial mechanician myself, I understood what he meant. If a problem is well defined, so that you know exactly where bodies start in the solar system and on what orbits, you can compute how those orbits will evolve, and where the material will go, at least in a statistical sense, with remarkable precision.

If we know exactly where comets are, and how many there are, and how big their solid icy bodies (their *nucleii*) are, and how their numbers vary with their sizes, and how big a crater forms from a given impact, we should be able to know how long it takes to make the numbers of craters on Europa. But, even though Zahnle's *ad hoc* team had the ability to do precise orbital computations, they needed to integrate together information from a wide variety of sources regarding the source population. They considered astronomical observations of the small icy bodies in the outer reaches of the solar system, they considered the statistics of comets that have moved in among the planets, and they accounted for astronomical and spacecraft studies of the sizes of nucleii of comets. They also needed to consider the size distributions of craters on the other Galilean satellites to get an idea of how the numbers of impactors vary with size. And they needed to take into account our best understanding of crater formation processes.

No matter how exact the computations of celestial mechanics may be, there is considerable uncertainty in the other aspects of the problem. Nevertheless, by carefully integrating current best understanding in all these areas, Zahnle et al. estimated in 1998 that the numbers of craters on Europa implied the surface was only 10 million years old. Fortunately, these researchers are quantitatively sophisticated, so they included a factor of five uncertainty as part of their solution. With that uncertainty, their result was that the age of the surface is less than about 50 million years, a value that became available just in time for interpretation of *Galileo* images of the geological record on Europa.

How much this age will change with future research and discoveries is uncertain. Our understanding of the populations of small bodies in the solar system, and how they evolve and migrate, is increasing rapidly. Zahnle et al. themselves revised their age estimate in 2003, because more recent data had shown that the numbers of small comets (those with nucleus diameter <1 km, which make craters smaller than 20 km) is probably smaller than the earlier estimate. In 1998 they used an extrapolation from the sizes of larger comets, but by 2003 there was some observational evidence available. Thus, the age estimate was revised upward to 30–70 Myr, still reasonably consistent with the older 50-Myr upper limit, but given a further sense of the degree of uncertainty.

Zahnle et al. had discounted any significant role for impactors from among the asteroids. Asteroid's (or their collisional fragments') orbits can evolve onto orbits that impact the terrestrial planets, including Earth. This is how we get our meteorites, lose dinosaurs, and trigger other mass biological extinction events. The conventional wisdom is that, compared with comets, asteroids do not hit Europa very often, so they do not affect the crater-based chronology.

Conventional wisdom is sometimes fleeting. On occasion, surprising reservoirs of small bodies in the solar system, whose importance may have been discounted, prove to have significant effects. Recently, a research team from Argentina, led by Adrian Brunini, has suggested that a portion of the asteroid belt known as the Hilda group may bombard Europa nearly as much as the comets considered by Zahnle et al.

The Hildas lie at the outer edge of the main belt of asteroids.[1] Their orbits around the Sun are in resonance with the orbit of Jupiter. The resonance is not exactly like the one among the Galilean satellites, although there are fundamental similarities. In this case, each Hilda asteroid is in an orbit that has 2/3 the period of Jupiter's orbit around the Sun, and its conjunction with Jupiter is locked to the asteroid's pericenter. This lock is similar, for example, to the way Europa's conjunction with Ganymede is locked to its pericenter (Chapter 4). Hildas' orbits are very eccentric, so this alignment keeps the asteroids away from Jupiter and thus stabilizes their orbits.

Brunini et al. pointed out that modest sudden changes in the Hildas' orbits could put them on unstable orbits that could intersect the jovian system. They considered the possibility that collisions among the Hildas could send broken pieces onto such unstable orbits, with some finding their way to the Galilean satellites. By estimating the numbers of small Hildas, the rate of collisions among them, and the sizes and velocities of the debris, they found that the contribution to the bombardment of Europa might be comparable to that from the comets. In that case, the paucity of craters might reflect an even younger age of the surface than Zahnle et al. had inferred.

It is an interesting idea, but it depends on extrapolations of the numbers of asteroids down to sizes too small to have been observed and on assumptions about collisional processes and fragmentation. Also, it is possible that the bombardment of Europa would have come in spurts, corresponding to the individual collisions that sent debris on its way, making this source of craters less of a constraint on Europa's surface age. While it would be premature to accept consideration of the Hildas as lowering the limits on the age, the uncertainties do serve as a reminder that there is more work to be done to refine the estimates of the rate of resurfacing of Europa.

The paucity of craters implies that the surface must be young, but that limit does not necessarily imply that a global resurfacing event wiped the slate clean. More likely, given that the surface appears to have been continually renewed by chaos disrupting previous terrain, by ridges overlying it, and by dilation and convergence, the age given by the crater record represents the timescale over which these gradual renewal processes have been continually replacing the surface.

---

[1] This group is named after asteroid Hilda, the first of their type to be recognized. Unlike features on Europa, the naming of asteroids is completely whimsical and is the prerogative of each one's discoverer. There is even one named after me, more official than my cycloid on Europa. It is a nice honor, placing my name among the gods, lovers, hometowns, and family pets orbiting the Sun between Mars and Jupiter.

18.3 APPEARANCE OF IMPACT FEATURES: IMPLICATIONS FOR ICE THICKNESS

The smallest impact features are perfect little bowl-shaped craters. Many, smaller than a kilometer across, are visible in high-resolution images (e.g., Figures 13.4, 13.5, 16.2–16.5, 16.9, 16.16). Most of these tiny craters are certainly secondaries, created by ejecta sprayed out from larger impacts. In Conamara, they tend to be in the same portion covered by the bright ray from Pwyll Crater (see Chapter 2), and probably mark part of the same spray of material. In other places (e.g., shown in the high-resolution images of Astypalaea in Chapter 13), distinct clustering of these small craters demonstrate that they probably resulted from swarms of small-particle ejecta from other impact sites. If the impactors that created these craters had come from space, their trajectories would not likely have remained in such tight formation.

Larger craters are fewer and farther between, so they tend not to be found in high-resolution images, which in the limited *Galileo* data set cover only a tiny portion of the surface. Larger craters are found scattered around lower resolution images. For example, craters a couple of kilometers across are seen in Figure 16.8, part of the 200-m/pixel Regional Mapping image set.

The simple bowl shapes of these craters are exactly what we would expect for impacts by small projectiles into a solid body. In fact, all craters smaller than about 5 km across display the simple morphologies of impacts into solid material. Craters this size probably probed to less than a couple of kilometers depth. They confirm what we had already inferred from estimates of tidal heating that the ice is at least that thick.

The largest impact structures are Tyre and Callanish, which happen to lie nearly opposite one another on the globe of Europa (Figure 9.1). These features are enormous multi-ringed structures (Figure 18.1, see color section). The inner rings of Tyre and Callanish are over 40 and 50 km across, respectively, and other rings extend far beyond them.

The literature describing Tyre and other impact structures is remarkably conservative, and great care has been taken to build up that case. It probably did no harm, but was it necessary? Tyre's appearance screams impact scar (Figure 18.1). The multi-ring structure represents the damage that was done as the shock wave from the impact propagated out from the center. Multi-ring craters and basins are well known on other planets, so it is no surprise that the larger impact features on Europa have that characteristic. I find it remarkable that the literature on these structures has been so cautious about declaring them to be impact structures, even though it is obvious, while other fundamental and incorrect claims about Europa have been accepted into the canon without any supporting documentation.

As the cautious case that Tyre is an impact structure has been advanced, it has been compared with multi-ring basins on other planets and with the impact scars in the ice of Ganymede, known as *palimpsests*. Such comparisons have little meaning, however, unless you believe that the ice is at least 30 km thick. An impact that made a structure this big would have penetrated at least that deep into solid ice according to our best understanding of impact mechanics. Even the advocates for the

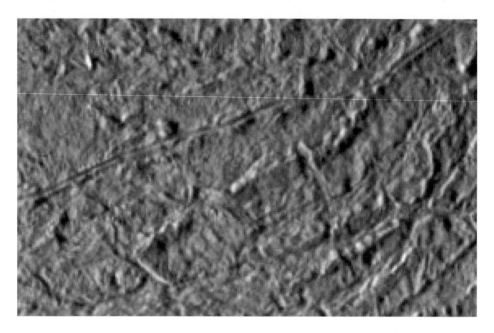

Figure 18.2. Typical chaotic terrain? This area looks like it, but in fact it is an enlargement from Figure 18.1. Chaotic terrain and the interiors of larger impact features are practically indistinguishable because both represent sites of breaches in the crust with temporary exposure of the ocean.

canonically-approved, convecting, thick-ice, isolated ocean model have generally been describing ice thinner than that.

Given that nearly every scientist who has studied Europa admits to, and probably favors, the possibility that the ice is thinner than 30 km, detailed comparisons with the appearance of large impact structures on solid planets are probably irrelevant at best and misleading at worst. A more relevant comparison might be to look for terrain on Europa itself that looks like Tyre. You do not have to look far.

Consider the typical patch of chaotic terrain shown in Figure 18.2. Here we see the usual characteristic features: rafts of older crust; the typical lumpy, bumpy matrix; even a recent crack with incipient ridges crossing the refrozen matrix. The only thing that is not typical about this chaotic terrain is its setting. It is a sample of the surface of Tyre. Aside from the concentric geometry of the larger rafts, Tyre is nearly indistinguishable from other chaotic terrain.[2] Certainly, it is at least as similar to chaotic terrain as it resembles impact structures on any solid planet.

Tyre most likely represents penetration to liquid water, with subsequent tem-

---

[2] Greg Hoppa has conducted many informal tests of the similarity of chaotic terrain and crater interiors on Europa, by showing examples of the latter to various people engaged in Europa image interpretation, and asking them whether it was chaotic terrain. They always got it wrong, with one exception: The only person who could distinguish between images of chaos and of impact penetration was Jeannie Riley, our undergraduate assistant, whom we had trained to do the survey of chaos described in Chapter 16.

porary exposure of the ocean to the surface. Blocks of crust were broken apart and dislodged, then locked into place as the exposed water refroze. From the perspective of the melt-through model of chaos formation, it is no surprise that these terrains look so similar. The only difference is that in one case the crust was open to the ocean by an external impact, and in the other case the crust was opened by heat from below. Details of raft geometry and symmetry differ as expected, but the basic character of the terrain that is produced in the two cases is indistinguishable, because the basic process is the same: temporary exposure of the liquid ocean at the surface.

The similarity of large impact features to chaotic terrain is a very big problem for the canonical thick-ice, isolated ocean model. This similarity, and its disturbing implications for the canonical view, impelled the nomenklatura to drop the official designation of macula for Tyre. It is ironic that the similar darkened appearance of Tyre to other chaotic terrain at low resolution, where the designation macula applies, was denied after the similarity was confirmed by higher resolution imaging of morphology. Nomenclature decisions tended to be strongly influenced by advocates for the canonical thick-ice model.

Even the darkening and coloration are similar. At low resolution, in *Voyager* images and early *Galileo* images, Tyre was indistinguishable from the splotches that represent chaotic terrain. For example, in Figure 2.4 Tyre (at the extreme upper right) and the chaoses Thrace and Thera (at the lower left) appear simply as dark patches and were all given the same designation by International Astronomical Union (IAU) nomenclature gurus: Macula (see Figure 9.1 as well). This time they got it right: Thrace Macula, Thera Macula, and Tyre Macula have more in common than differences, even though one is a major impact feature. The darkening that gave them all the appearance of maculae resulted from this fundamental similarity: They all probably represent penetration to the liquid ocean.

The similarity in darkening extends to coloration. As shown in Figure 18.1, Tyre displays the same orange–brown coloration as most chaotic terrain (e.g., Figure 16.18), as well as those tectonic features, ridges, and bands, where oceanic material has come to the surface. (This darkened material has been associated with hydrated salts, further support for the oceanic source, as discussed in Chapter 21.)

The other very large impact feature, Callanish, is practically the same size as Tyre and has all of the same characteristics (Figure 18.3). The multiple rings of cracks surround disruption of the previous terrain. Aside from the concentric geometry, the rafts between the cracks and the lumpy matrix are indistinguishable from other typical chaotic terrain (Figure 18.4).

Callanish is also darkened in the usual orange–brown, like Tyre and like other chaotic terrain, and in that way similar to other locales where oceanic exposure has likely occurred. The darkening can be detected in Figure 18.3, but is most obvious in images where the illumination emphasizes color and albedo. For example, in Figure 2.2, compare Callanish (the round spot to the left) with the appearance of Conamara Chaos, the other splotches that represent chaotic terrain, and the lines that demark mature ridge systems. All these global-scale markings correlate with terrains and

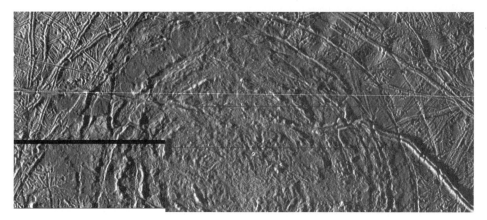

Figure 18.3. Callanish. The area shown is about 50 km from top to bottom.

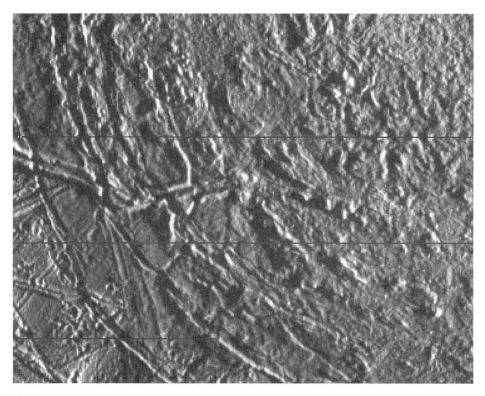

Figure 18.4. Close-up of Callanish, where, aside from the concentric geometry, the rafts between the cracks and the lumpy matrix are indistinguishable from other typical chaotic terrain.

processes (impacts, melt-through, and tidal–tectonic cracks) expected to have involved oceanic exposure at the surface.

The character of the largest and smallest impact features on Europa are consistent with expectations for an oceanic planet covered by a crust of ice. The largest impact structures (Tyre and Callanish), with low topography and terrain similar to chaos, suggest penetration into the liquid layer. At the other end of the size spectrum, craters smaller than about 5 km display the simple shapes expected of impacts into solid material, because they probed no deeper than the icy crust.

Craters of intermediate size (5–30 km diameter) have the most potential for more precisely constraining the thickness of the ice. If their characteristics reveal whether they punctured through to liquid water or not, and if there is a consistent size cut-off for such penetration, they could tell us how thick the ice is. The bad news is that there are not too many in that size range, so a definitive cut-off in properties may be difficult to identify. With statistics of small numbers, differences might have more to do with the local characteristics of each impact site, than with a systematic trend with size. The good news is that there are not too many craters in that size range, so it is easy to get to know them all.

Crater Amergin is about 19 km across (Figure 18.5), and lies at 14°S, 230°W, just at the edge of the region imaged by 200-m/pixel Regional Mapping (E17 RegMap 02). Unlike the large multi-ringed impact features Tyre and Callanish, Amergin has a distinct, single, upraised rim. This crater is clean, upright, and true, a proud example of the Amergin way. But, inside its rim, the terrain is indistinguishable from chaotic terrain elsewhere (Figure 18.6). The morphology is exaggerated because Amergin happens to be near the terminator and the shadows are long. A conveniently-near patch of chaos shows us what such terrain looks like under this same illumination for comparison. Moreover, like Tyre, Callanish, and chaos everywhere Amergin shows as a typical dark spot in images sensitive to color and albedo (e.g., see the spot just adjacent to the A in Amergin in Figure 9.1). Despite the fact that Amergin is far smaller than Tyre or Callanish, its appearance is consistent with penetration to liquid.

Crater Manannán appeared on an image that had been taken with nearly-overhead illumination during orbit G1 at 1.6 km/pixel as a dark spot surrounded by a relatively bright halo that extended outward to a diameter of roughly 150 km. It was then imaged at 218 m/pixel during *Galileo*'s orbit E11 with the sun low enough to reveal its texture and topography (Figure 18.7). The crater is seen to have a diameter of about 21 km, with a lumpy ejecta blanket extending roughly 10 km beyond the rim. The brightening of the terrain further out is probably caused by a spray of fine ejecta, similar to the bright material that composes the prominent ejecta rays of Crater Pwyll (Figure 2.6). Manannán has a less pronounced rim than Amergin, but the boundaries of the crater are well defined. The interior floor, like that of Amergin is indistinguishable from chaotic terrain.

Whereas the Amergin impact occurred in tectonic terrain (in fact, on a dilational band and adjacent ridged terrain), the impactor that formed Manannán must have landed in a pre-existing chaos area. The older terrain surrounding Manannán appears to be chaos, as well as even older tectonic terrain. It is possible that both

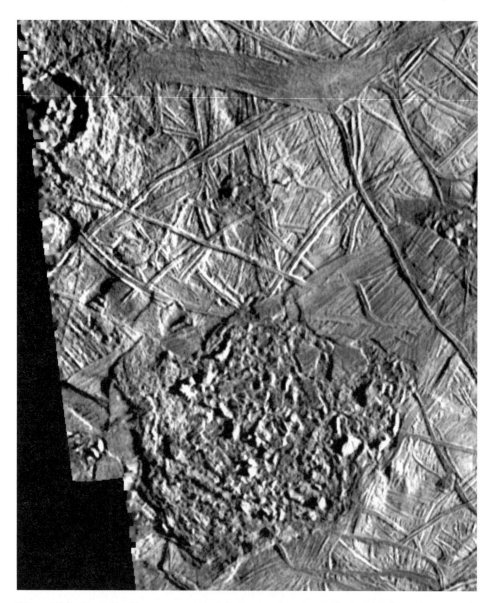

Figure 18.5. Amergin Crater at upper left has an interior indistinguishable from chaotic terrain. The appearance of chaos under the same illumination is shown by the nearby example at bottom center. Amergin's rim is about 19 km across.

Amergin and Manannán represent impacts into unusually-thin ice, because both dilational bands and chaos represent sites where the ocean was exposed and the ice must have remained thin for some time. However, it is unlikely that these sites remained thin for more than a short part of the age of the surface, and the prob-

Sec. 18.3] **Appearance of impact features: implications for ice thickness** 275

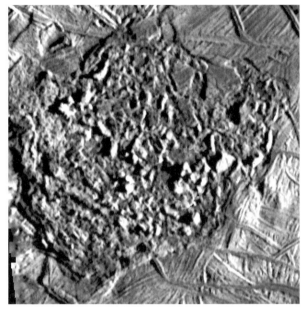

Figure 18.6. Amergin's interior hides in plain sight in the similar terrain of nearby chaos. Here the floor of Amergin has been pasted into the nearby chaos (cf. Figure 18.5).

ability that two 20-km craters (more than 1/3 of the craters near that size) formed in unusually-thin ice seems remote.

With its prominent bright rays, Pwyll is one of the most obvious impact features, even at low resolution (Figures 2.2 and 2.6). The rays suggest that it may be one of the most recent. Otherwise, Pwyll is fairly typical. Again we see the dark appearance of the interior of the crater, extending several kilometers out (Figure 18.8—see color section—as well as Figures 2.2 and 2.6). At 24 km across, it is similar in morphology to the other craters of its size (Figure 18.9). Like the others, its interior resembles chaotic terrain, with interior lumps and bumps and rafts similar to other sites where the ocean has likely been exposed. Like Manannán, the surroundings include chaotic terrain that had disrupted older ridged terrain, and the impact appears to have occurred in that setting.

Pwyll Crater, like Conamara Chaos, was targeted for imaging with reasonable resolution and oblique lighting early in the mission (during the first encounter with Europa in orbit E4 in December 1996), because the global views had shown them to be such prominent and distinctive features. We now know that they were prominent because they are relatively fresh and recent examples of their respective types of feature, but otherwise are fairly typical.

Interpretation of these features was the topic of considerable debate within the Imaging Team. I myself followed the discussion with interest, but detachment. For one thing, no one was interested in my opinion. And, I was only beginning my work on tidal tectonics. I never imagined that following that line of research would lead me to become the standard-bearer (or is it "fall-guy") for thin ice.

Figure 18.7. Crater Manannán, located at 3°N, 240°W is about 21 km in diameter. It lies in a region rich in chaotic terrain (including the nearly 20-km-wide patch at the left in this image), and the interior of Manannán is similar to chaotic terrain. This image was taken during orbit E11 at 218 m/pixel.

Establishment of the party line on Pwyll followed the same process as for Conamara. Conamara had been initially viewed as a site of oceanic exposure, but the thick-ice proponents exerted considerable pressure to ensure that the authorized position of the team favored a solid-state formation process (Chapter 16). In the same time frame, initial interpretations of Pwyll recognized the similarity to chaotic terrain, and the lumps, bumps, and rafts in the crater were hypothesized to be crust material that had floated into position. In fact, there was some specific discussion of how some of the rafts that made up the upraised lumps within the floor of the crater seemed to have drifted inward from the edge, leaving notches in the rim. Again, the politically-powerful and aggressive thick-ice advocates quashed this hypothesis, and the Imaging Team's papers on craters make no mention of any similarity between chaotic terrain and the interiors of impact features.

Rather than consider similarities between impact features and other terrains on Europa, the canonical approach is to fit these features into categories that have been

Sec. 18.3] **Appearance of impact features: implications for ice thickness** 277

Figure 18.9. Pwyll Crater is about 24 km across. This image was taken during orbit E6 at 240 m/pixel. The interior has the appearance of chaotic terrain.

developed to describe craters on solid bodies. Once you constrain the descriptions of Europa's impact features to fit the characteristics of craters observed on solid planets, it should not be surprising that the theorists who model these descriptions conclude that the impacts must have penetrated only into solid material. In other words, if you assume that the ice is thick, then the craters on Europa prove that the ice is thick. The logic is blatantly circular, but craters have been adopted as the key evidence *du jour* for the canonical thick-ice theory.

In accordance with that approach the similarity of Pwyll's interior to chaotic terrain was discounted and it was categorized as a "central peak crater", a well-defined standard class based on a long tradition of crater studies. On solid bodies, small craters are usually simple bowls, while the products of more energetic impacts are "complex" craters, with features like central peaks, flat floors, and terraced rims. These complexities result from rebound or gravitational slumping following the initial impact excavation. Central peaks follow immediately after excavation, as the material under an initial, transient crater flows back inward toward the axis of the impact, where it is concentrated at the center, and rebounds upward. Central peak craters have been examined in great detail in their natural settings on solid planets, and they have been modeled using impact experiments, numerical simulations, and analytical theory. Their formation is very well understood.

Is Pwyll a central peak crater? If we are bound to decide whether it should be classified as a "simple crater" or a "central peak crater", the latter is undeniably a closer fit. But that classification fails to take into account the resemblance of the interior terrain to chaos, and the similarity of the "peaks" in Pwyll to the rafts and ice floes of chaos. It disregards the random placement of these "peaks", which could hardly be described as "central". Even if Pwyll did not look like a site of penetration to liquid, this classification makes the fundamental logical error of ignoring the possibility that complex craters with interior topography could form by any other mechanism than the one familiar from other places. Yet, we are dealing with a satellite that may have a structure completely different from any planetary body we have observed before. To preclude any other possible mechanism is to preordain the conclusion.

A few craters do have their largest lumps concentrated toward the center, but that geometry is perfectly consistent with their similarity to chaotic terrain. It is reasonable that in some cases breakaway crustal rafts have sloshed toward the center during in-fill of the initial transient crater. Examples of craters with lumps near the center include the 20-km-wide Crater Maeve (Figure 14.2), and 18-km-wide Crater Cilix (Figure 18.10, see color section).

The authorized view of the *Galileo* Imaging Team is that the lumps near the center of Cilix, for example, are a "central peak complex".[3] In a cultural vacuum, I would be content to accept that terminology, but this language conveys, and was intended to convey, the notion that the crater formed like other central-peak craters on other planets, by impact into solid material. The topography of Cilix could be at least as accurately described as "crustal rafts displaced toward the center of the crater". That terminology would be more consistent with the appearance of most craters on Europa, and admits the alternative, and more likely, possibility that the character of these craters results from processes very different from what would have happened on a solid planet. The floors of most craters larger than about 10 km in diameter on Europa consist of terrain indistinguishable from chaos.

Even among advocates for the traditional central peak interpretation, there is disagreement about which craters fall into that class. In an influential article, Paul Schenk describes all known craters in the diameter range 3–30 km as being central peak craters, in the sense that they formed completely in solid ice, without penetrating to liquid. Bear in mind that this group of about 20 craters includes Amergin, Manannán, and Pwyll (Figures 18.5, 18.7, and 18.9). Evidently, central peaks are in the eye of the beholder, as is blindness to the similarity to chaos.

Although Schenk saw all 20 of those impact features as central peak craters, Zibi Turtle and Betty Pierazzo (of the University of Arizona) wrote in *Science* that only six of them are. Like Schenk, they included Amergin and Pwyll, but at least they had the good sense to dump Manannán, which would certainly be difficult to defend as a typical solid body central peak crater. Of their six "central peak craters", only four have "central peaks" near the center.

---

[3] The history of descriptions of Cilix has been checkered. The *Voyager* Imaging Team identified it as a crater, but the *Galileo* Imaging Team proclaimed it to be a raised dome or mesa based on stereographically combining *Voyager* and early *Galileo* images. Later images (Figure 18.10) showed it is a crater after all, demonstrating again the uncertainty involved in analytical determinations of topography (see Chapters 16 and 19).

Evidently, only a small fraction of these craters have central bumps, and all the ones bigger than 10 km have floors that look like chaotic terrain. Yet, the canonical position is that all of these craters formed in solid material and none penetrated through the crust.

In order to form central peak craters by the standard mechanism for solid bodies, the ice on Europa would need to be thick enough to prevent interaction with the ocean. Numerical simulations by Turtle and Pierazzo of the formation of $\sim$20-km craters show the early phase of the impact melts downward to 4 km, and subsequent rebound would break the ice through to liquid at $\sim$12 km depth. The ice would have to be thicker than that to avoid penetration to liquid.

These simulations give great insight into the processes of crater formation in ice, and provide quantitative information that is essential to understanding the impact record in the outer solar system, but how well do they constrain the thickness of Europa's ice? If one accepts Schenk's interpretation that all craters from 3 to 30 km are classically-defined central peak craters that could not have involved penetration to liquid, then the implication would be that the ice must be $>$15 km thick. If one prefers Turtle and Pierazzo's identification of only a small fraction of those craters as possible central peak craters, then the theory suggests the ice was thicker than 12 km at the time and place of a couple of impacts.

In the same issue of *Science*, the magazine's staff-writer Dick Kerr reported that Turtle and Pierazzo's work put "a lid on life on Europa" and he disparaged my view of an ice crust thin enough to allow ocean-to-surface connections. He invoked the standard (though false) mantra of the "pits, spots, and domes" (discussed in Chapter 19) to support the idea of thick ice.[4] Had reporter Kerr tried to obtain my view of the story before editorializing for the party line, I would have interpreted the crater modeling differently for him.

The kind of quantitative modeling that Turtle and Pierazzo had done is an essential part of the process of understanding Europa. We need to understand the kinds of processes that might operate over the full range of possible conditions, and their work is a significant step in that direction. However, a chain of logic is only as strong as its weakest link. The numerical modeling of impacts is a strong link. But, the chain of logic that connects a few strange craters to the thickness of ice depends on a belief that typical 20-km craters (e.g., like Amergin) look so much like central peak craters on solid planets that they could only have formed the same way. That belief is a very weak link in the chain.

The key publications that have been used as the justification for the thick-ice, isolated ocean appeared not in regular scientific journals where all the details of a case are made, but rather in the magazines *Science* and *Nature*. The entire observational case for solid-state convection was made in a couple of paragraphs in a letter to *Nature* citing putative properties of small patches of chaos and topographic features, the so-called "pits, spots, and domes" (see Chapters 16 and 19). The case

---

[4] The lower limit to the ice thickness given by Turtle and Pierazzo in that *Science* report was 4 km, not thick enough to rule out the permeable ice layer that our work has implied. Kerr's enthusiatic report of the demise of thin, permeable ice was premature.

for surface corrugations requiring thick ice as the favored mechanism for surface contraction was made in a couple of paragraphs in another letter to *Nature* (see Chapter 17). The crater modeling described above was published as two pages in *Science*. No details can be included in such short papers. Yet these results are accepted as the fundamental supporting evidence for the thick-ice, isolated ocean model, because *Science* and *Nature* are viewed as the voice of scientific authority.

The isolated ocean advocates finally toned down the "pits, spots, and domes" mantra after I and my colleagues actually did a study of them and found that almost nothing that had been said about them was true (Chapter 19). At that point, the thick-ice advocates turned to the craters as their principle argument. Perhaps that is why it was so important for Kerr to set up that new spin on the party line.

The thick-ice advocates welcomed the more recent observational interpretation of the impact record on the Galilean satellites by Paul Schenk. As usual, for key evidence cited in the argument for thick ice, the documentation is sketchy. The case is made in only two pages in a letter to *Nature*, but this format is perfect for the argument for thick ice. It gives the appearance of densely-packed technical detail, with a definitive conclusion that the ice must be thicker than 19 km, seemingly precise to two significant figures. Since few people bother to examine the details, such papers make powerful and persuasive cases for thick ice.

A closer look at the details leads to very different conclusions. Fortunately, Schenk is a responsible and honest scientist, so for those willing to read carefully he did pack in a lot of useful detail for us to consider. In his study, he determined the depths and diameters of craters, as well as classified their morphologies (Figure 18.11). The accuracy and precision of his depth determinations is questionable, but for this discussion I assume that they are acceptable. There are enough other problems with his investigation.

In Schenk's classifications, whether on Europa, Ganymede, or Callisto, at diameters of 3 km there is a transition from simple, bowl-shaped craters to central peak craters, with a corresponding change in the depth-to-diameter slope on a log-log plot. For Ganymede and Callisto, as shown in Figure 18.11, the slope is fairly constant up to second transitional size, above which the depth is fairly constant with diameter, until a third transition (Transition III) at 100 km diameter where it drops quite abruptly for larger impact features. Schenk finds that features larger than Transition III have "anomalous impact morphologies" and include multi-ring structures. He suggests that above Transition III (diameters >100 km) impacts penetrated to a layer of liquid water. Based on his previous impact studies, a 100-km crater corresponds to a transient opening during impact penetrating to 80 km, so he infers that the liquid layer lies at a depth of >80 km below the surface on Ganymede and Callisto. Such speculation has some merit, because the pressure at such great depths might be adequate, combined with the modest heat sources in those satellites, to produce at least a thin liquid layer. Moreover, *Galileo* magnetometer data are consistent with such layers of salty liquid water. According to this story, the transition to penetration to liquid water is marked, in the depth–diameter relationship, by an abrupt change from a flat slope to a downward slope.

For Europa, Schenk believes that the analogous transition occurs at 30-km

Sec. 18.3] Appearance of impact features: implications for ice thickness 281

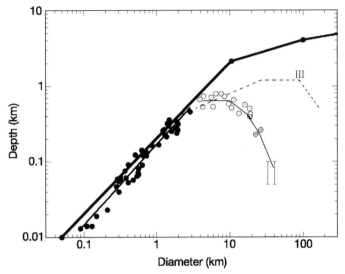

Figure 18.11. Depths, diameters, and classification of impact features are given by the symbols plotted by Paul Schenk: black spots represent simple craters; circles with central spots represent central peak craters; circles with crosses represent "modified central peak craters Mannann'an [sic] and Pwyll" (Figures 18.7 and 18.9), and the two vertical bars represent the diameters and range of possible range of depth of Tyre and Callanish (Figures 18.1–18.4). Classifications are subjective: Moore et al. (2001) did not classify any craters smaller than 5 km diameter as central peak craters, and Turtle and Pierazzo (2001) identified only six of these as central peak craters. The heavy solid line is Schenk's description of the depth–diameter relationship for craters in a solid body, the Moon. I added the dashed line to show the trends of Schenk's results for Ganymede or the nearly-identical ones for Callisto, and a fine solid line that follows the trend of his data points for Europa. These satellites are very different from the Moon The transition to impacts that penetrated to liquid water on Ganymede and Callisto, according to Schenk, is at the bend marked "III". Scheck believes that the Europa curve at 30 km diameter is similar to Transition III, the basis for his argument that the ice on Europa is thick.

diameter. Apparently, to his eye, the depth–diameter relationship for Europa in Figure 18.11 shows at 30 km diameter the same "sharp reduction in crater depths and development of anomalous impact morphologies" as the line for Ganymede and Callisto. This identification of 30 km as the critical-size crater that corresponds to impact penetration to liquid is the basis for the claim that the ice must be at least 19 km thick. Bizarre as that line of thinking may appear, it is now very popular with the thick-ice zealots, especially since their other putative lines of evidence have collapsed.

The problem is that Schenk's own evidence does not support his conclusion. Examination of Figure 18.11 does not reveal the purported similarity between the curve for Europa at 30 km and Transition III for Ganymede and Callisto. There is no abrupt transition in the depth–diameter relationship at 30 km. The trend of

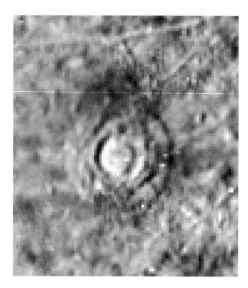

Figure 18.12. The best picture of crater Tegid is from a global-scale sequence taken during orbit E14 at 1.4 km/pixel. The central floor is about 20 km across. Despite the low resolution, a multi-ring structure is visible around this relatively small crater. The similar-sized Taliesin is also a multi-ring crater.

Schenk's data points shows nothing like the "sharp reduction in crater depths" that Schenk found for Ganymede and Callisto at Transition III.

Nor is there evidence to support Schenk's claim that at 30 km Europan impact features follow the abrupt transition to "anomalous impact morphologies" at 100 km as reported for the other satellites. Schenk appears to be using the distinction in appearance between, on one hand, those just smaller than 30 km, specifically the craters Manannán and Pwyll, and, on the other hand, those just larger than 30 km, specifically Tyre and Callanish. Certainly, Pwyll and Manannán are very similar to one another and do not have the multi-ring structures of the larger Tyre and Callanish. However, for Ganymede and Callisto, Transition III was marked (according to Schenk) by a transition to "anomalous impact morphologies", which included a variety of other types of craters as well as multi-ringed features. Schenk himself refers to Pwyll as an "anomalous central peak crater", and if Pwyll is "anomalous" Manannán must be as well.[5] By Schenk's own standards, the transition to crust-penetrating impacts must be to the left of Pwyll and Manannán in Figure 18.11, closer to 20 km diameter.

Moreover, Schenk may not have considered two impact features, Tegid and Taliesin (Figure 18.12), which were only imaged at very low resolution, that are comparable in size to Pwyll and Manannán, but seem to have multi-ring structures.

---

[5] Schenk also calls them "the modified central peak craters Mannannán [sic] and Pwyll". It is not clear whether there is a difference between being anomalous and being modified.

Again, the morphologies that Schenk connects with penetration to liquid water occur at smaller sizes than he admitted.

Schenk's identification of the 30-km diameter on Europa as the equivalent of Transition III for Ganymede and Callisto is unsupportable either on the grounds of the depth–diameter relationship or of morphology.

On the other hand, there is another, much more distinct transition in the Europa data in Figure 18.11 that is a much better analog to Ganymede and Callisto's Transition III. Between about 4 and 10 km, the depths are constant.[6] At 10 km diameter there is an abrupt decrease in the slope for larger craters. If you are looking for a portion of the Europa curve that matches Transition III, it is this bend at 10 km; nothing comparable happens at 30 km. Therefore, Schenk's data suggest that the ice thickness is only a few kilometers.

Moreover, the 10-km diameter corresponds to the transition from typical craters found on solid bodies to those that are indistinguishable from chaotic terrain. If you are looking for a specific change in morphology to something unusual that might be reasonably attributed to penetration to the liquid ocean, that change occurs at 10 km; no such transition occurs at 30 km. Consideration of the character of Europa's craters points to an ice layer only a few kilometers thick.

Schenk's interpretation was exactly the opposite, perfectly fitting the party line that the ice is thicker than 20 km. The conclusion of his letter to *Nature*, which is all that most readers scan, tied this result from the impact record to the conventional evidence for solid-state convection, the well-known "ovoid features". Never mind that a reasonable reading of the impact record suggests that the ice is thin, not thick; never mind that the observational evidence cited for solid-state upwelling had no basis in fact no matter how frequently it was repeated. The time had come to lay to rest the myth of "pits, spots, and domes".

---

[6] Schenk drew a slightly upwardly-sloping line through this region, but a flat slope fits his data points just as well. Also, contradicting his classifications, Moore et al. and Turtle and Pierazzo said, respectively, that no craters smaller than 5 km or 8 km have central peaks.

19

Pits and uplifts

> One need not on that account take the common popular assent as an argument for the truth of what is stated; for if we should examine these very men concerning their reasons for what they believe, and on the other hand listen to the experiences and proofs which induce a few others to believe the contrary, we should find the latter to be persuaded by very sound arguments, and the former by simple appearances and vain or ridiculous impressions.
>
> Galileo, 1615

19.1 UNDENIABLE (IF YOU KNOW WHAT'S GOOD FOR YOU) FACTS

In 1999 an article appeared in *Science* by a young postdoc at Caltech, Eric Gaidos, and a couple of his senior colleagues, Ken Nealson and Joe Kirschvink, that was the first article in the new science of "astrobiology" that got me excited. For most readers, the article must have seemed to put a damper on the raising speculation about life on Europa. Gaidos et al. considered the implications of the canonical *Galileo* result that the ice on Europa was so thick that any ocean must be isolated. They found that, because the ocean was separated from chemical oxidants at the surface, there were severe limitations on the quantity of any kind of biological activity.

Even with speculation about chemistry and energy from possible undersea volcanism, life would be limited at best. Giant squid with "eyes the size of dinner plates",[1] as well as the weird, fanciful, advanced Europan aquatic life that was appearing in popular magazine illustrations, were snuffed under the official thick layer of solid ice.

---

[1] Proposed tongue in cheek by Chris Chyba of the SETI Institute.

I read the paper by Gaidos et al. completely differently. My research group had already gone far enough in study of tidal tectonics that we knew there was an abundance of evidence that the ocean was linked to the surface on a continual basis, through cracks, melt-through, and impact penetration. The implication of the Gaidos et al. paper was not that life was limited in the ocean, but rather that the linkages we were discovering were crucial to the possibility of significant life on Europa. Our work took on a new importance. It also must have presented the potential for a high-profile embarrassment to the advocates of the establishment's thick-ice story.

I met Gaidos for the first time just before a special session on Astrobiology at an American Geophysical Union (AGU) meeting in San Francisco where I had been invited to give a presentation. I told him how much I had enjoyed his article, because it had highlighted the importance of the linkages between the ocean and the surface. He looked at me as if I were insane: How could there be any linkage? Then he repeated verbatim the mantra that he must have heard repeatedly as he studied the known facts about Europa: The "pit, spots, and domes" proved that the ice must be at least 20 km thick.

Once Pappalardo wrote a couple of paragraphs in *Nature* about the so-called "pit, spots, and domes" claiming that they demonstrated convection in solid ice, their existence and implication became an established, fundamental fact. Every authoritative post-*Galileo* review of what we know about Europa lays out the basic facts: 1,560 km diameter; 100–150 km outer layer of H_2O; tidal heating possibly enough to maintain a liquid water layer below the surface; pits, spots, and domes that show solid-state convection in ice thicker than 20 km.

For years, presentations about Europa included among the basic known facts that "the surface is covered with pits, spots, and domes, known as lenticulae, that are rounded, often updomed and cracked across the top, regularly spaced, and typically about 10 km across, demonstrating solid-state convection in ice that is at least 20 km thick".[2] Over 5 years I must have heard this identical mantra dozens of times, to the point that I could silently move my lips along with the speaker's words.

Advocates for the thick-ice model were politically active and well connected, so the "facts" were codified in official reports of NASA and the National Research Council of the National Academy of Science. They became the basis for future mission planning and for decisions on research funding. Science writers were directed to spokesmen who would ensure that the canonical facts about Europa were duly reported in the media (which may explain Dick Kerr's article discussed in Chapter 18).

Naturally, when Gaidos did his background research, to learn what was known about Europa before making his own contributions, it was natural for him to absorb this factual material. So, he knew about the "pits, spots, and domes", and what they implied. It was obvious that I was nuts to consider that the ice was thin enough for

---

[2] Paul Schenk was referring to the roundness featured in the mantra when he called these things "ovoid features" (Chapter 18). Also, the introduction by Pappalardo of this misuse of the term "lenticula" was an effective way to appropriate the fact that real lenticulae are in fact typically about 10 km across by definition (Chapter 2).

ocean–surface linkages. He had no reason to delve into the original sources or to pore over all the *Galileo* images, so he had no way to know that the canonical facts were wrong.

The original source in the literature is Pappalardo et al.'s letter to *Nature*. The letter includes postage stamp cut-outs of putative examples of "pits, spots, and domes". Readers assume that substantial evidence lies behind the claims in such short papers, but unscrupulous operators can abuse that trust. With the evidence of the pictures; the full authority of the *Galileo* Imaging Team implied by the publication format; the prestige of a publication in *Nature*; and reinforcement by the consistent mantra, who could question the facts? The facts were well known, clear, and incorrect.

19.2 THE MYTH OF PITS, SPOTS, AND DOMES

"Pits, spots, and domes" were poorly defined in the *Nature* letter, and remain so; equating them with lenticulae is incorrect; no such classes of features exhibit any of the claimed characteristics (size, shape, spacing, etc.) indicative of subsurface solid-state convection; and their existence was based on images of a small area in a specially-selected location. The definition is so poor and misleading that I prefer to refer to the combined category of "pits, spots, and domes" as PSDs to distinguish them from other usages of the same word, such as actual pits that really do exist on Europa.

The *Nature* letter was based on some of the earliest images with adequate resolution (~200 m/pixel) and illumination for morphological study, from orbit E6, which showed only the region (about 250 km wide by 300 km) around Conamara Chaos (Figure 2.8 shows most of the area in those images). This area represents 1% of the total surface area of Europa, in an unusual locale that was selected for early attention because of the unique appearance of Conamara in earlier low-resolution images. Conamara proved to be an example of very fresh chaotic terrain. Later, during orbits E15 and E17 we got the broad Regional Mapping imagery, which covers ten times as much area (still only a tenth of the surface), under similar lighting and resolution to the Conamara images, but without the bias introduced by selection of a special site. These broader areas are more appropriate for any survey that seeks patterns and generalities, such as our surveys of strike–slip (Chapter 12) and of chaotic terrain (Chapter 16).

Given the constrained nature of the earlier imaging data, any classification scheme and the generalities that are associated with it, should have been regarded as tentative at best. In fact, in the *Nature* letter, PSDs were presented only anecdotally, using a set of six image cut-outs that removed them from their context in the immediate neighborhood around Conamara Chaos (Figure 19.1). In fact, nearly all of the 16 examples of PSDs cited in the literature have been in the immediate neighborhood of Conamara (Figure 19.2).

Of the six original examples in the *Nature* paper, four are actually examples of small patches of chaotic terrain, according to our independent surveys (Chapter 16),

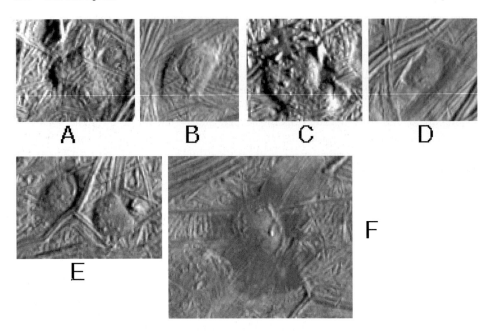

Figure 19.1. The six type examples for PSDs from Pappalardo et al.'s (1998) *Nature* paper. These images were all cut out of the mosaic of images around Conamara taken at 180 m/pixel during *Galileo* orbit E6, and their locations are shown in Figure 19.2, along with most of the other examples of PSDs cited in the literature (see also Figure 2.8). All are shown at the same scale, with the height of frames A–E about 15 km high. A and B are irregularly-shaped uplifts, contrary to the *Nature* paper's statement that all PSDs are "circular to elliptical". The other four are typical examples of chaotic terrain. Contrary to the impression that the *Nature* paper gave by this selection, patches of chaotic terrain are found in all sizes and shapes on Europa, and their is no peak in their size distribution (see the discussion of Figure 16.12).

which were based on consistent criteria which were independent of the size of a feature. When Jeannie Riley mapped these particular features as chaos, she had no idea that they had been used as the type examples for PSDs. The *Nature* letter noted that only one of these patches was "micro-chaos 'material' ". Because chaotic terrain comes in patches of diverse shapes and a wide size range (from as small as could possibly be recognized on *Galileo* images to >1,000 km across), the selection of these four features as typical examples of a class that demonstrates a characteristic size is misleading. One could have selected any number of similar features displaying different sizes and/or shapes. Presumably the 10-km-wide Mini-Mitten chaos in Figure 16.6 would have been counted among the PSDs, but the nearly-identical Mitten chaos in Figure 16.8 puts the lie to the supposedly typical 10 km size. The fact that most PSDs (according to type examples) are chaotic terrain is inconsistent with the statement that this type of feature is typically ~10 km across.

Sec. 19.2] The myth of pits, spots, and domes 289

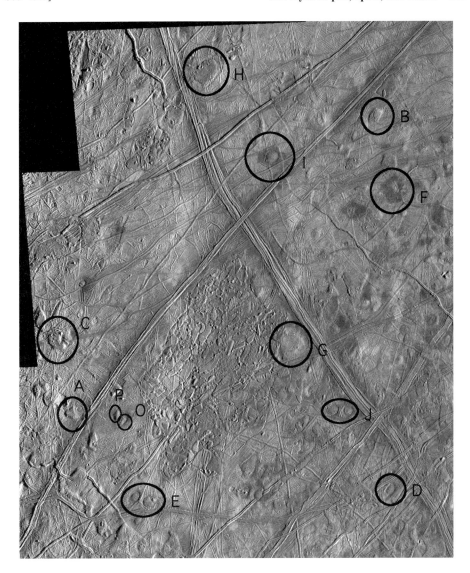

Figure 19.2. Most of the 16 PSD-type examples cited in the literature are in the Conamara region. The region shown is about 250 km across. Examples marked A–F are the same as those shown in Figure 19.1. The small patch of chaos near the top of Figure 2.11 appears here as example I. Compare also Figures 2.6–2.8.

The remaining two example domes in the *Nature* letter (A and B in Figure 19.2) are polygonal or irregularly-shaped uplifted areas. Completely baffling is that, although they were presented as type examples, they do not fit the defining description of PSDs as "circular to elliptical". It is not clear what their appearance has in common with the four other examples.

In fact, these two examples are similar to a type of uplifted feature that is fairly common on Europa. Rather than be subsumed into the ill-defined category of PSDs, uplift features on Europa begged for a systematic survey, which we eventually got around to doing, as described later in this chapter. These uplift features generally appear to be pushed-up plateaus where the surface has been minimally disrupted (if at all) by the rising. Like chaotic terrain, they are seen in a wide range of shapes and sizes, while the *Nature* letter showed only examples that are ~10 km across. As with chaotic terrain, our systematic survey showed that that size is hardly typical.

The only thing that the six selected features in the *Nature* letter have in common is their size. Moreover, these examples were selected on the basis of their size, out of the greater set of similar features that come in a broad range of sizes. Since, via this selection of type examples, these PSD features were defined by their size, the result that they are all about the same size does not have any physical significance. It is simply an artifact of the definition. It certainly should not be taken as evidence for a common formation mechanism. Yet, it has been repeatedly cited (starting with the *Nature* letter) as the factual observational underpinning for the argument for solid-state convection, and hence thick ice and an isolated ocean.

That the PSDs were defined by size is more explicitly stated in papers by Pappalardo, Jim Head, and their students at Brown that were supposed to provide the supporting evidence for the *Nature* letter. In developing statistics, they explicitly excluded features that were not of the selected size and shape. Not surprisingly, they discovered that the features they included in their statistics are all about the same size. This circular procedure provided the only quantitative support for the characteristic size on which the entire edifice of thick ice rested.

If Head's students had not restricted their study to features of a certain size, a very different size distribution would have emerged. Most of the features they measured were (like the type examples in the *Nature* letter) patches of chaotic terrain (i.e., selected examples of a much more general class of terrain that does not have the principal characteristic attributed to PSDs). Jeannie Riley's survey of chaotic terrain showed that there is no preference for ~10-km-sized features and we demonstrated how observational bias (if not properly accounted for) can contribute to a false perception of a peak in the size distribution near 10 km (Chapter 16). Therefore, there is no rationale for a taxonomic separation of ~10-km-size features from other chaotic terrain, nor for including them within the ill-defined PSDs (where they constitute the bulk of the PSDs). That taxonomy has only led to crucial incorrect generalizations, leading to major incorrect implications.

It appears that PSDs were *defined* (albeit vaguely) as those chaos patches that are ~10 km wide and roundish, plus any uplift features that happen to be of similar size. (Whether the latter need to be round or not to fit the definition is ambiguous, because the type examples contradict the stated definition.) The definition arbitrarily excludes the many similar features of other sizes and shapes.

The reported characteristic spacing for PSDs, like their supposed typical size, has also been part of the case for solid-state convection. However, aside from the statements that there is a regular spacing comparable with the characteristic size of PSDs, such a fact has never been demonstrated.[3]

Several other papers by the same players have adopted the PSD taxonomy, and added various descriptions and selections of type examples. In general, the examples in this "PSD literature" have been shown as small cut-outs from larger images, so it has been difficult to keep track of the location and context of each. Adding to possible confusion, the sets of type examples in these papers are each different, although there are overlaps among the sets. Altogether there are 16 of them (including the 6 cited in the *Nature* letter) and most are in the Conamara area (Figure 19.2).

The full set has all the problems of the original set of 6 displayed in the original *Nature* letter, and then some. Of the 16 examples, 10 are patches of chaotic terrain selected for their size and 5 are uplifted areas of various shapes. Weirdly enough, none of the examples of "pits, spots, and domes" is a pit, even though there are plenty of pits on Europa, and they may prove yet to be one of the most important types of feature on Europa.

One of the 16 archetypes appears in Figure 16.5, where we can see it in its context. The neighborhood (Figure 16.5) consists of a jumble of rafts, some of which are barely displaced, with various odd-shaped patches of exposed matrix among them, all near the edge of a fairly large expanse of chaotic terrain in the northern-leading hemisphere. One of these irregular patches of exposed matrix (located about 1/3 of the way up from the bottom of Figure 16.5 and 60% of the way from the left to the right side) is the feature that has been cited as an archetypical PSD. It is in fact, typical of completely standard, lumpy, bumpy chaos matrix material. As with all type examples in the PSD literature, there is no indication as to whether it is supposed to be an example of a pit or a spot or a dome. Of course, it is none of the above. It is not even circular or ovoid or elliptical, the supposedly characteristic traits of PSDs. It is simply a typical jagged little patch of chaotic terrain.

19.2.1 PSDs and lenticulae

Careless use of language, or careful misuse of language, can lead to trouble and false generalizations in a process of scientific classification.

In numerous presentations and publications, Pappalardo, Head, and their coworkers have explicitly equated PSDs with lenticulae. They claim that the word *lenticula* refers to the combination of the "three major classes" (the PSDs), and state that, "Small pits, domes, and dark spots, collectively 'lenticulae', pepper Europa's surface." Recall, however, that the word *lenticulae* (as clearly defined by the IAU) refers only to the small dark albedo spots, typically ~10–20 km wide in the low-resolution images on which the definition was based. The incorrect identification of all of the poorly-defined PSDs with lenticulae allowed the characteristic sizes of lenticulae to convert into a false generalization about the putative PSDs.

---

[3] The title of a paper by the Brown group indicated that it addressed the spatial distribution, but it did not, which is likely to give a false impression.

In fact, as discussed in Chapter 2, lenticulae are one of several classes of Europan albedo features that appear prominently in images taken at low resolution (>1 km/pixel) with relatively vertical illumination, that prove at higher resolution to have some relation to structural features, but that are not actually the structures themselves. Other examples are the "triple-bands" whose dark portions border bright ridge complexes. In a similar way, lenticulae are associated with small patches of chaotic terrain, but are not identical to them. In each case, the triple-bands and the lenticulae, those distinct low-resolution features, appear at higher resolution as faint diffuse markings often extending beyond (or existing only beyond) the margins of the associated structural features, ridges and chaos respectively. (As discussed in Chapter 16, these extended diffuse darkenings could be from a common cause, the warming associated with connection of the ocean to the surface.)

A classification scheme (with its implicit generalizations) that is based on observations made under one set of conditions (resolution, illumination, etc.) should be extended to what is observed under very different conditions only with great caution to be sure that generalizations are not extended inappropriately. So that, while lenticulae tend to have sizes roughly 10–20 km across, it is incorrect to transfer that characteristic to a different set of features observed under different circumstances.

The confusing of PSDs with lenticulae probably devolved from the fact that most type examples of PSDs are actually patches of chaotic terrain. In general, for chaotic terrain the matrix tends to be dark, and darkening often extends a few kilometers beyond the matrix, out onto the surrounding terrain, as a diffuse halo. For example, the small patch of chaos shown as example F in Figure 19.1 has a characteristic dark halo around it. So does the one in Figure 2.11. Noting their positions in Figure 19.2, they can be located in Figure 2.7 as typical lenticulae. Figure 2.7 represents the observing conditions under which lenticulae were defined and for which the term is meaningful. Indeed, lenticulae (using the correct IAU definition) usually prove to be low-resolution albedo manifestations of patches of chaotic terrain.[4]

In fact, almost all chaotic terrain appears as dark splotches similar to lenticulae in all ways except size and shape. For example, Conamara appears similar to the lenticulae in Figure 2.7. We have seen that the size distribution of chaotic terrain runs from large ones over a thousand kilometers across to smaller than a few kilometers across. The reason lenticulae appear with a distinctive, favored size is that most chaos has its darkening extending a few kilometers beyond, so any of the many chaoses smaller than about 10 km across will be marked by a 10–20-km-wide dark spot.

Since *lenticula* is a well-defined (and officially sanctioned) term, it should not be confused with PSDs. Most of the PSD type examples are examples of chaos, a well-defined morphological class of terrain, and most lenticulae are low-resolution

---

[4] A rare exception is the chain of spots along Rhadymanthys Linea, at the lower left in Figure 2.5, which may be genetically closer to triple-bands.

manifestations of chaotic terrain. However, it is illogical to equate PSDs with lenticulae and has been misleading to attribute characteristics of lenticula (especially their typical size) to the ill-defined PSDs.

19.2.2 Are any PSDs pits or domes?

Of the type examples used to describe PSDs that are not chaotic terrain, none are pits. However, non-chaos pits are very common on Europa and represent a potentially-important class of feature that still remains unexplained. These pits have been largely ignored in the literature (except for a brief interval early in the *Galileo* mission when they were confused with craters) and have not been included among the PSDs, despite the use of the expression "pits, spots, and domes". A field of pits is shown in Figure 19.3. They are common, and a description of their character and distribution is a necessary step toward understanding their origins and the implications for the structure and history of Europa. The results of a survey of pits, as well as uplift features, are discussed later in this chapter.

Having noticed that none of the "pits, spots and domes" were pits, I was not completely surprised to find that none of them are domes either. Remember, most of the features that have been identified as PSDs are patches of chaotic terrain. As explained in Chapter 16, chaos can appear qualitatively to be somewhat updomed, and its edges may seem depressed relative to surrounding terrain (Figure 16.13). Some of the examples in Figure 19.1 have this appearance. Quantifying this topography is difficult with available image data: stereo-coverage is limited; and photoclinometry, which assumes that brightness is governed by surface slope, may be confounded by the albedo variations across these features.[5]

The apparent updoming in chaotic terrain (and not just in the PSD-size range) is likely just the result of buoyant rebound to the average surface elevation (Figure 16.13). What appears to be domed upward might be high relative only to the immediate borders of the terrain, and not higher than average for the surrounding terrain. In other words, rather than being a dome relative to the surroundings, the topography generally might be better described as a low moat around the chaos area. That model explains why the small patches of chaotic terrain that dominate the examples of PSDs might have seemed like pits or domes. For PSD-type examples, impressions of topography may be subjective or circumstantial, depending on illumination, albedo patterns, and human perception.

Setting aside those type examples of PSDs in the literature that are chaotic terrain, there are only 5 others (out of the total of 16), and those 5 are variously-shaped uplift areas. In fact, one of them may be an artifact of local albedo variations rather than actual topography, so its uplifted appearance is questionable.

The shapes of these uplift areas do not fit the stated definition of domes in the PSD literature (e.g.) "subcircular to elliptical positive relief features"). On the contrary, an accurate description of the type examples would be "irregular to

---

[5] A continuing problem with research on Europa has been publication of supposed elevation profiles with no information about the procedures involved in obtaining the results nor of the uncertainty in the values.

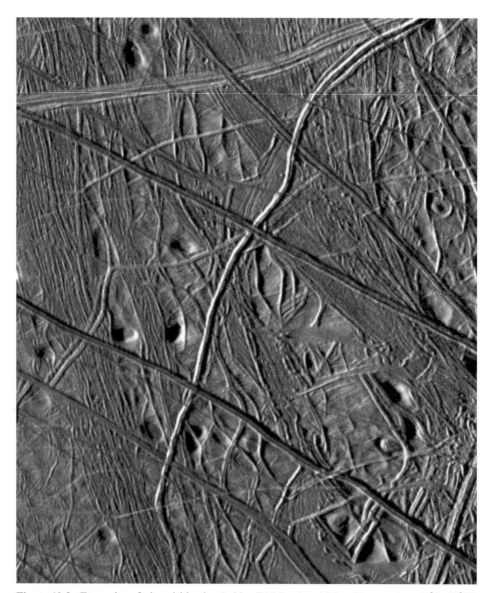

Figure 19.3. Examples of pits within the *Galileo* E17 Regional Map 02 area near 45°S, 90°W. Illumination is from the left (west). The area shown is about 100 km across.

polygonal". In these committee-written papers, the authors responsible for the text must not have talked with the ones preparing the pictures. Given the shapes of the type examples, the term "domes" (with the implication of roundness) seems to be a misnomer. I prefer the term "uplift" features, rather than "domes", and only apply it to those features that are clearly raised above the surrounding terrain (including no examples of chaotic terrain).

Uplift features similar to the five included in the PSD examples are part of a widespread population of such features, so it is important to understand what they are really like. In general, unlike updomed-appearing chaotic terrain, they do not correlate with albedo. They are characterized by hardly (if at all) disrupting the surrounding terrain, which appears to continue right on up and over these topographic features. Occasionally, cracks are noted across the uplift feature (e.g., B in Figure 19.1). The PSD literature makes much of this characteristic as evidence for convective or diapiric solid-state upwelling. On the other hand, just as uplifts are rarely round, hardly any of these features actually have such cracks.

As a widespread phenomenon on Europa, these uplift features seemed to me worthy of attention and of an accurate description. When we carried out a comprehensive survey, we confirmed that they do not fit the general description of PSDs. Inclusion into the PSD category was used to bolster the solid-state convection model, but a more representative description of this population of uplift features is not especially supportive of that hypothesis: They are generally not round. In fact, a roundish "dome" shape is the exception, rather than the rule for these features. They are not confined to diameters $\sim$10 km. In fact, like chaos, they have a wide size distribution. They are rarely cracked on top. In fact, the few that are cracked lie in one region only. Therefore, it has been misleading to have subsumed this class of feature into the PSD category, and to have used incorrect generalizations to support the thick, convecting ice model. How pits and uplifts formed remains an open question, but a first step has been to describe the population accurately.

19.2.3 Farewell to PSDs

The supposed existence of PSDs as a class, and the set of characteristics attributed to them, was widely accepted on the grounds of a massive campaign of repetition, and for years it was the sole putative observational basis for thick ice. Consequently, it is especially important and instructive to summarize what was wrong with this taxonomy.

The PSD classification, which was supposed to include "pits, spots, and domes", is dominated by examples of chaotic terrain, which have characteristics far more varied than those attributed to PSDs, especially in size and shape. The fact that most PSDs are dark spots simply reflects the fact that chaotic terrain is generally dark and often surrounded by a dark halo. Of the PSDs that are not chaotic terrain, most are members of a more general set of irregularly-shaped and variously-sized uplift features, of which few could be described as dome-shaped. The type examples of PSDs do not include any pits other than some examples of chaotic terrain that may or may not actually be depressed. The interesting class of common and widespread pits that are not chaotic terrain does not seem to have been included in the definition of PSDs. These relationships are summarized in the Venn diagram in Figure 19.4.

Accordingly, the PSD terminology and classification should be abandoned unceremoniously. The term "spots" is redundant given the definition of lenticulae; the term "pits" should be reserved for the class of deep, roundish topographic features that are not patches of chaotic terrain (and were not included in the

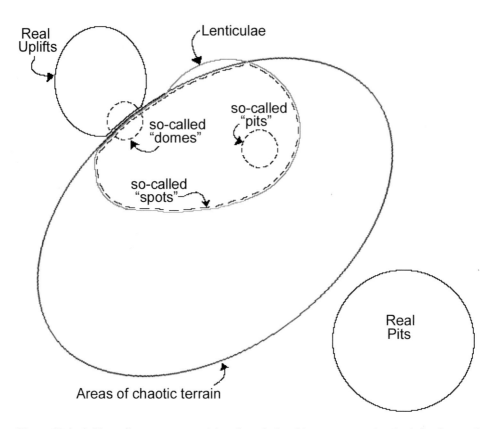

Figure 19.4. A Venn diagram summarizing the relationships among variously-defined sets of features. Chaos is a major type of terrain on Europa, covering nearly half the surface, and patches of chaotic terrain range widely in shape and size. The set of these features is shown by the large oval. Most of the lenticulae (albedo spots according to IAU nomenclature) are the low-resolution, overhead illumination manifestations of small patches of chaotic terrain, so the set of those features lies mostly within the set of chaos features in this diagram. The many chaos features that do not appear as lenticulae in low-resolution albedo are larger; their low-resolution albedo appearance is called "mottled terrain". Only a small portion of the set of lenticulae lies outside the set of chaos features; these include the chain of albedo spots that lies along Rhadymanthys Linea (Figure 2.5). Most of the features vaguely lumped together as "pits, spots, and domes" in the literature (the ill-defined categories that I call PSDs) are also examples of chaotic terrain. PSDs are shown by dotted lines: "spots", as defined by type examples all are patches of chaotic terrain, and as albedo features they are also lenticulae; similarly, "pits" identified in type examples all are chaotic terrain features, which may not even be topographically-depressed; "domes" appear on the basis of type examples to include both supposedly-uplifted chaos features plus other small uplift features. Real pits do exist, and are very interesting, but are not part of the set of features called pits in the PSD literature; they are not chaotic terrain and are not especially dark. Real uplift features are also interesting, but their set does not include chaos patches and does include larger uplifts not included in PSDs. The use of the word *lenticulae* as synonymous with pits, spots, and domes is misleading and incorrect.

PSD classes); and the new category "uplifts" should be used to denote those topographically-high features of various sizes and shapes (usually irregular to polygonal) that are not associated with the formation of chaotic terrain, but rather appear to be raised areas otherwise similar to, and continuous with, whatever terrain surrounds it.

Once we surveyed these features and published these results, an interesting thing happened. The obligatory summaries of facts about Europa downgraded PSDs from established fact to the "controversial pits, spots, and domes", definitely a small step in the right direction. Farewell pits, spots, and domes.

19.3 SURVEY OF PITS AND UPLIFTS

There is a real and distinct class of feature that can be appropriately called pits (e.g., Figure 19.3). Such pits were noted very early in the *Galileo* mission in the low-resolution images taken during orbit G1 of the Europan quadrant centered around longitude 215° in the northern hemisphere (e.g., Figure 2.5). Some early interpretations assumed that these numerous features were small craters, which implied a fairly old surface. However, their numbers, which appeared to peak at a size of about 9 km, were far too great to be consistent with young surface ages inferred from larger craters. Gene Shoemaker, a grand sage of impact crater studies, suggested one way to reconcile the statistics: The larger craters might have disappeared due to viscous relaxation of their topography. A better answer came from the observation that the pits seemed to avoid lineaments, preferentially constrained between these tectonic features, so they probably were not impact craters after all. Subsequent higher resolution imaging has confirmed that interpretation (Figure 19.3): The pits do indeed nestle between ridges and they are not shaped like craters. After that early discussion in the context of cratering statistics, even though many more images become available, the pits were practically ignored in the literature for years.

The pits were not included among the PSDs, except that the word was used as the first name of PSDs. (Remember, among PSDs, the only features that were not chaotic terrain were those that appeared to be topographic highs.) It is possible that these pits are associated in some genetic way with chaotic terrain. For example, collapse associated with the onset of melt-through could create a pit during the process of chaos formation before the melt-through reaches the surface, or during an aborted melt-through (Figure 16.13). However, before we speculate about origins we need to establish observational facts (i.e., what is seen on the surface of Europa), and avoid classification on the basis of theoretical interpretations of genetic processes. The pits differ from chaotic terrain in that there is no evidence of disruption of previous surface. They do not have any of the characteristics we used to identify chaos (e.g., they lack the distinctive lumpy texture). The adjacent terrain, especially fine tectonic structures, seem to continue across the pits uninterrupted except by the change in topography. In a few rare cases (<1%) the bottom of the pit is dark. In some cases the bottom of the pit seems smoother than the surrounding terrain, but this assessment is tentative due to the small area involved and shadowing from the side of the pit.

These pits do exhibit clear and substantial topographic lows, and in that way they differ from all the PSDs. Shadow measurements indicate depths of 200–300 m for pits ~10 km across (smaller pits are not measurable but appear shallower); similar results come from stereographic analysis.

Uplifts tend to be irregular in shape, often polygonal and rarely rounded and also rarely cracked on top. In general, the surfaces of uplifts continue the character of the surrounding surface, as if the crust had been simply punched up from below. A unique case worth highlighting is the enormous uplift just south of the equator near longitude 225°, which is about 40 km long (north-to-south) and 20 km wide (Figure 19.5; see also its location in Figure 12.14, along with other uplifts in the neighborhood). It seems fairly flat on top, is of an irregular shape, and has steep edges dropping down to the surrounding terrain, several hundred meters below. The terrain around this feature includes typical chaotic terrain, as well as tectonic terrain. At the top of the plateau are similar terrain types, seemingly continuous with those around the base. As with most of the uplifts we have mapped, these properties do not seem supportive of a convective upwelling model.

My research group had taken a preliminary look at pits and uplifts, which made us skeptical about the claims regarding "pits, spots, and domes". For example, although we had been shown repeatedly the archetypes of uplifts that were dome-shaped or had cracks, we could find few other examples. Key components of the case for thick ice were clearly bogus. It was aggravating to listen to the "pits, spots, domes" mantra, and frustrating that I did not have in hand a comprehensive survey to refute it.

Then, in late 2000 Martha Leake, a former Arizona student, came to work with us during a sabbatical leave from Valdosta State University in Georgia. Martha's job involves a great deal of teaching, but the sabbatical gave her a chance to apply her expertise in planetary geology. She joined us in carrying out a complete and systematic survey of all the pits and uplifts in the Regional Mapping images, covering the E15 and E17 RegMap 01 and 02, the same areas that had been surveyed by Jeannie Riley for chaotic terrain and by Alyssa Sarid for strike–slip (Figure 12.19).

19.3.1 Pit counts

Figure 19.6 shows the distribution of pits over the areas covered by Regional Mapping images. Recall (from Figure 12.19) that RegMap 01 lies in the trailing hemisphere, and RegMap 02 crosses down the center of the leading hemisphere. In both regions, the parts in the northern hemisphere were taken during orbit E15 and those in the south during E17. Note that there are no pits shown in the leading hemisphere between about 6°N and 25°S due to the 1,300-km-wide chaos feature at that location (Figure 16.10).

A striking global symmetry pattern is apparent in the distribution: The distribution in the northern-leading hemisphere (NL) appears to be very similar to that in the southern-trailing (ST) hemisphere; likewise, a very different distribution is

Sec. 19.3] Survey of pits and uplifts 299

Figure 19.5. Two of the largest uplift features identified on Europa, located just south of the equator in the trailing hemisphere (cf. Figure 12.14). The largest (more northerly) is about 40 km long (north to south). The more southerly one is cracked, very unusual for a Europan uplift feature.

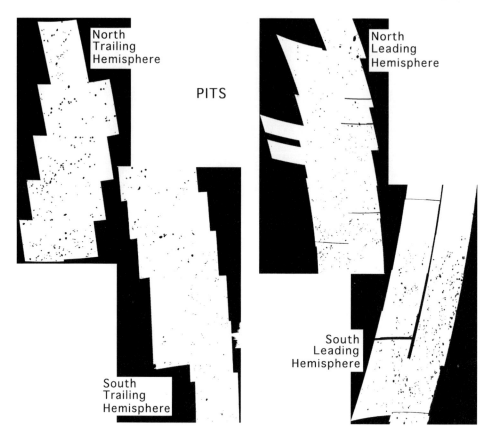

Figure 19.6. The locations, sizes, and shapes of pits in the Regional Mapping areas defined in Figure 12.19.

evident in the southern-leading (SL) hemisphere and northern-trailing (NT) hemisphere. Thus, the symmetry is antipodal (similar distributions are opposite one another on the sphere), but it is oblique to the equator.

In order to quantify the size distributions in these four regions, we made the histograms as shown in Figure 19.7. We plot the number as a function of area, rather than of diameter, because the shapes are variable. A circular pit with a 10 km diameter would have an area of about 80 km^2. The numbers in each case are scaled to the total area mapped in each hemisphere. Numbers per area for the leading hemisphere are depressed due to the giant equatorial chaos area, which occupies approximately 1/3 of the mapped area; so, for meaningful comparison with statistics on the trailing hemisphere, they should be multiplied by about 1.5.

Note that the numbers in each of the four regions (NL, NT, SL, ST) increase with decreasing size down to the limits of recognizability. Pits smaller than about 5 km^2 are unrecognizable due to the resolution limits of the images. Therefore, the

Sec. 19.3] **Survey of pits and uplifts** 301

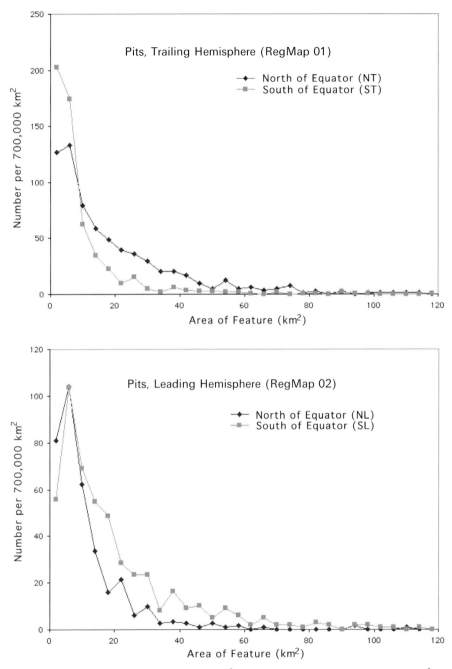

Figure 19.7. Size histograms of pits in 4-km^2-wide bins for features up to 100 km^2. The ordinate shows numbers per area of the mapped surface. Numbers in the leading and trailing hemispheres are comparable when the large equatorial chaos area in the leading hemisphere is taken into account.

peaks at about $6\,km^2$ are not meaningful, but the size distributions are dominated by pits only a few kilometers across or smaller.

The size distributions for the ST region and for the NL region are similar to one another in the following sense: There are few pits larger than about 5 km across ($20\,km^2$), but the numbers increase steeply with decreasing size down to the smallest size that could be recognized in these images. Moreover, the absolute numbers in NL and ST are comparable when the correction for the giant chaos area is taken into account.

Contrast that type of size distribution (found in the NL and ST regions) with the sizes in both the SL and NT regions. In each of the latter, pits >5 km across (area $>20\,km^2$) are several times more numerous than in NL and ST, respectively (but in no cases are there more than a few >10 km across).

This size distribution explains why a squinting view of Figure 19.6 gives the impression that the population in SL and NT is dominated by 10-km-wide ($80\,km^2$) pits, as would a cursory examination of the images of this area. Perhaps, such cursory impressions contributed to the PSD fiasco. However, our counts show that there are plenty of smaller pits even in this population. In fact, the statistics (Figure 19.7) show that, just as for NL and ST, numbers increase substantially with decreasing size, down to the limits imposed by resolution of the images.

The antipodal symmetry of the two types of size distributions is reminiscent of, and likely related to, the symmetry of chaotic terrain in these regions. Our maps of chaotic terrain (Figure 16.10) show that in the SL and NT regions (where the larger pits are common), patches of chaotic terrain are small and cover relatively little of the surface. The surface is predominantly tectonic there. On the other hand, near the equator and in the NL and ST hemispheres, patches of chaotic terrain are large and account for a substantial portion of the surface. This correlation of spatial distributions between chaotic terrain and pits may be trying to tell us something, but I could only speculate what the message is.

The spatial distribution shown in our maps also confirms an earlier-noted appearance of a relationship between pits and tectonics. Alignments of sets of pits are evident in Figure 19.6, especially in the SL and NT regions. Both of these regions seem to have a broad, but time-varying, tectonic fabric, and these alignments may be related to these patterns. Fine-scale lineaments often run uninterrupted by pits, continuing right across their bottoms (e.g., in Figure 19.3). However, pits generally avoid structural tectonic features. Unlike chaos patches, pits tend to lie between ridges. As common as strike–slip displacement is on Europa (Chapter 12), we found only one example of a pit (near 23°N, 218°W) that appears to have been sheared apart (Figure 19.8).

19.3.2 Uplift counts

A map showing sizes and locations of the topographic uplift features mapped by Martha Leake is shown in Figure 19.9. The map identifies features that are simply topographically raised without substantial modification relative to the neighboring terrain, and those that have been cracked, possibly in the course of the uplift process.

Sec. 19.3] **Survey of pits and uplifts** 303

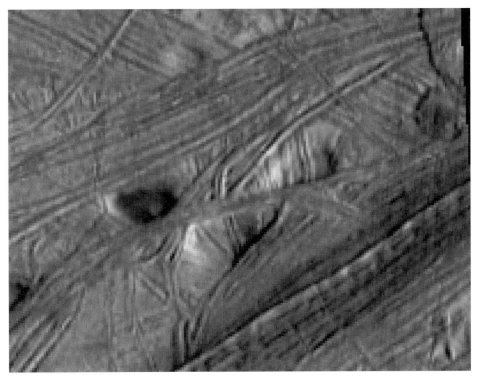

Figure 19.8. A pit that appears to have been sheared apart by a strike–slip fault through its center. In general, pits are found between tectonic lineaments, which are usually identified by ridge pairs.

Histograms of their size distributions appear in Figure 19.10. For uplifts, as in the case of pits, numbers per area mapped are smaller in the leading, compared with the trailing, hemisphere because the leading hemisphere contains the giant equatorial chaos area, which precludes any features of the type we are mapping. Therefore, as with pits, figures in the leading hemisphere should be multiplied by 1.5 for comparison of number density.

As with pits and chaotic terrain, an oblique-to-the-equator antipodal symmetry is evident in the distribution, either by the qualitative appearance in Figure 19.9 or the statistics in Figure 19.10. In the SL and NT regions, all uplifts are seen to be fairly small, while in the NL and ST, a substantial number of larger uplifts are present. In the statistics, the effect shows up strikingly in the numbers of features larger than 100 km^2: in NL there are six uplifts larger than 100 km^2, one of 214 km^2, and one of 438 km^2; and in ST there are 12 uplifts larger than 100 km^2, seven larger than 200 km^2, three larger than 300 km^2, and one of nearly 900 km^2. In contrast, SL and NT have no examples larger than 100 km^2 (in the interests of full disclosure, there are two in NT, but they lie virtually on the equator). This dichotomy (NL and ST vs. SL and NT) is opposite in a sense to that of the pits, for which the larger ones were only present in the SL and NT regions.

Figure 19.9. The locations, sizes, and shapes of uplifts. Uplifts with cracks, nearly all of which lie near the equator on the trailing side, are shown in gray, all others in black. See Figure 12.9 for a key to locations of the five mosaic sections shown here.

Sec. 19.3] Survey of pits and uplifts 305

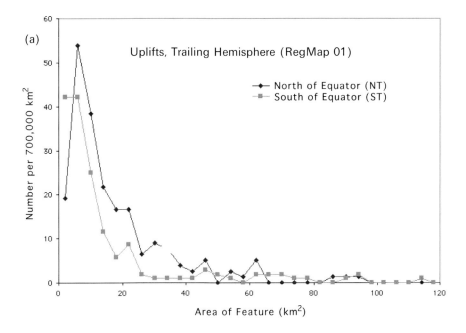

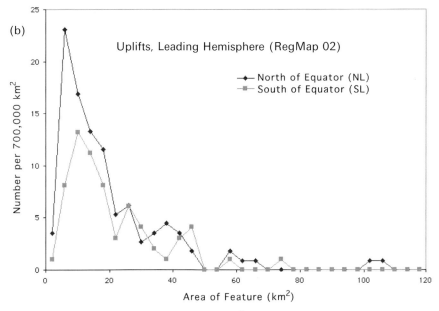

Figure 19.10. Size histograms of uplifts in 4-km²-wide bins for features up to 100 km². (a) Statistics for RegMap 01, with those north of the equator (NT) shown by diamonds and south of the equator (ST) shown by squares. (b) RegMap 02, with those north of the equator (NL) shown by diamonds and south of the equator (SL) shown by squares. To compare numbers between hemispheres multiply leading hemisphere counts by 1.5.

The histograms also show that there are substantially more small uplift features in the NT than in the ST, while in the leading hemisphere there is not a significant difference.

The description of these features as being typically round or elliptical domes, which appears repeatedly in the PSD literature, has been taken as evidence for convective upwelling. Our survey showed the contrary: These features are in fact quite irregular in shape, and often polygonal. Moreover, while the PSD literature invoked a supposed characteristic size of about $80\,km^2$ (10 km across), the actual statistics show no such thing (Figure 19.10). The numbers sharply increase with decreasing size, indefinitely, at least down to the smallest recognizable sizes, just as is the case for pits.

Of the hundreds of uplift features that we have catalogued, about a dozen appear to have surface cracking associated with the uplift (shown in gray in Figure 19.9). Nearly all of these lie just south of the equator in RegMap 02 (lower left in Figure 19.10), at the western end of the Wedges region. Cracked uplifts with a roundish shape are extremely unusual. Considerable misunderstanding of the nature of Europa's surface has followed from the repetitious description of uplifts as round and cracked, because it suggested that upwelling and incipient breakthrough from below was common. In fact, few uplifts are dome-shaped, few are cracked (<3%, and those only in a limited locale), and only one or two examples are both cracked and dome-like.

More typical of cracked uplifts are features where one or two flat plates of crust have tipped slightly, catching extra (or diminished) illumination, which provides the uplifted impression. The fact that most cracked uplifts are of this type and are located in the Wedges region, where there has been considerable crustal displacement, is consistent with the idea that they are simply displaced plates. In any case, the cracked uplifts are quite dissimilar to the other uplifts we have catalogued, and there is no compelling reason to expect a genetic connection.

19.4 FORMATION OF PITS AND UPLIFTS

19.4.1 Survey results vs. the PSD taxonomy

Our study of pits and uplift features over large swaths on Europa's leading and trailing hemispheres shows that the earlier taxonomy of "pits, spots, and domes" (PSDs) was premature and led to generalizations that are not borne out in this more complete study. It is unfortunate that these preliminary generalizations have been repeated and propagated extensively, and widely accepted as fact. The inference that these features are the manifestations of convection cells in Europa's ice crust, and that the crust must be thick, followed from those unsupported and incorrect descriptions.

With a more accurate description of the features on Europa, we can summarize what was wrong with the description in the PSD story. Most PSDs are indistinguishable from areas of chaotic terrain, a much broader class of surface phenomenon that covers a substantial portion of Europa's surface. Patches of chaotic terrain have a wide range of characteristics, including a continuum of sizes from more than 1,000 km across, down to the smallest sizes possibly recognizable given the limits of resolution on available imagery. Most of the PSDs are examples selected from a narrow range in that broad-size distribution The appearance of a characteristic size was an artifact of the selection of only features in that narrow-size range as type examples of PSDs.

Another point of confusion has come from the incorrect application of the word *lenticula* to include all of the PSDs. Lenticulae had been formally defined as spots of a certain size. Thus, the repeated misuse of the word *lenticula* supported the impression that the vaguely-defined classes of features called PSDs were all of this size.

Setting aside the cases that are patches of chaotic terrain, those PSDs that have been presented in the literature as type examples include roundish uplifts ("domes"), a couple with cracks on them. In that way the literature establishing the PSD taxonomy gave the impression that these were archetypes of common kinds of features. This generalization supported the idea that widely-distributed, roundish, uniform-sized subsurface upwellings had bulged the lithosphere upward, seeming to provide substantial supporting evidence for the convecting thick-ice model.

In fact, we have found that the type examples of domes (cracked or not) are practically unique. They are not archetypes of any class. Most uplifts are irregular in shape. Contrary to the PSD literature, typical sizes are not ~10 km across. Some are much larger and many are much smaller. The smaller they are, the more there are, down to the limits of resolution. Earlier studies evidently missed the numerous small ones.

Cases of uplifts with cracks are very rare and mostly confined to one geographical region (near the Wedges area). These rare features do not have properties especially suggestive of upwelling processes. Usually they appear similar to tipped rafts that are fairly common in chaotic terrain. Possibly, those cases formed like chaos rafts: If the ice is thin, a section of ice surrounded by cracks would be floating, and subject to the same tipping mechanisms as rafts in chaos, perhaps tipping due to oceanic currents or differential melting of different parts of the plate. Such a tipped plate would fit the appearance of many of the cracked uplifts.

Pits are quite common on Europa. Ironically, no examples of pits are included among the type examples given in the PSD literature, except for examples of chaotic terrain for which the topography is ambiguous. Our survey showed that few pits are more than 10 km across and most are much smaller. The impression given in the PSD literature that there is a characteristic size of about 10 km is incorrect. Like uplift features, the smaller they are, the more plentiful they are, down to the limits of resolution.

19.4.2 What are these things?

The PSD taxonomy had been invented to support the notion that observed features were created by solid-state upwelling from below via convection cells or similar solid-state upwelling ("diapirism"). Theoretical models of convection indicated that the ice needed to be at least about 20 km thick in order for convection to operate, and the putative sizes of the features seemed to match that scale very well. However, the actual features are mostly much smaller. What does that tell us about the ice crust below the surface?

In 1998, Cornell University graduate student Julie Rathbun, working with Prof. Steve Squyres, investigated by theoretical models the upwelling of low-density blobs within an ice layer. Steve had been a major player in Europa studies in the 1980s as a postdoc at NASA Ames in the years just after the *Voyager* encounter.[6] Based on the canonical reports of *Galileo* results, Rathbun and Squyres were led to believe that diapirism had actually been observed (not merely inferred). Julie found that the size of the "observed diapirs" was consistent with upwelling from a source more than 10 km below the surface.

Fortunately, although Rathbun et al. had fallen for the party line regarding the *Galileo* images, they were smart enough to extend their theoretical calculations to other hypothetical cases as well. Their study of rising plumes of warm ice produced a quantitative relationship between ice thickness and the surface diameter of diapirs, which we can apply to the actual observations.

Most of the pits and uplifts on Europa are less than 5 km across (Figures 19.7 and 19.10). According to the calculations by Rathbun et al. (1998), surface diapirs of that size could not have originated at depths greater than 5 km. Moreover, the numbers of pits and uplifts increase even more at smaller sizes, corresponding to diapir origination depths of a few kilometers at most. The actual sizes of pits and uplifts are more consistent with thin ice than with the model of convecting thick ice.

Although the characteristics of pits, uplifts, and patches of chaotic terrain prove not to be suggestive of a common mechanism of diapirism in solid-state convection, it is plausible that these various types of observed features may have some genetic commonality, especially between pits and chaotic terrain. Such a relationship would have nothing to do with the confused PSD taxonomy, which was dominated by selected examples of chaotic terrain. What they might have in common is an origin in the thermal thinning of ice.

Chaos likely forms when the ocean melts through to the surface. In that model, the early thinning of the ice creates a pit before the surface is breached. Observed pits could then be the result of aborted or incomplete melt-through (Figure 16.13).

Melt-through would be especially resisted along ridges where the ice is thicker (Figure 16.15), which would explain why pits are usually nestled between ridges. Randy Tufts envisioned tectonic terrain as a waffle. Ridges on the surface must have corresponding ridges extending down from the bottom of the ice to buoy them up. Randy pointed out in his thesis that heating from below during incipient melt-through would preferentially thin the already thinner parts between the ridges, consistent with the appearance of pit–ridge relationships. This model explains the relationships between pits and tectonics, including the tendency of pits to avoid

---

[6] At Ames, Squyres worked with the group, including Ray Reynolds and Pat Cassen, that with Stan Peale had discovered the tidal-heating mechanism for the Galilean satellites (Chapters 4 and 7). With Squyres, the group pioneered considerations of the possibility of the ocean, the thickness of the ice, and the implications for life.

interrupting ridges and to form alignments with the tectonic fabric. It is also consistent with a genetic relationship between pits and chaos, in that they both represent the result of the thermal thinning of the crust.

Why does the slight excess of 5-to-10-km (20–80-km^2) pits in the north-trailing and south-leading (NT and SL) hemispheres (Figure 19.7) correlate with the regional locations of small chaos areas (Figure 16.12)? Such a location might be where there was considerable crustal thinning, but not quite enough heat to melt through the ice to expose large contiguous areas. Also of interest is that the regions with more extensive chaos (ST and NL) have relatively few small pits in the 10–60 km^2 range (Figure 19.7).

The answer may have something to do with regional differences in ice thickness, although exactly how is unclear. If ice thickness is the key, it may be related to the strange antipodal symmetry (NL and ST vs. SL and NT). For example, if a thickening of the crust near the poles occurred due to less tidal heating there, it may have caused polar real estate to migrate toward the equator (Chapter 12). If polar "land" only migrated as far as the mid-latitudes it would explain the current antipodal symmetry pattern. In fact, our investigation of strike–slip features suggested that polar real estate migrated into the NL and ST hemispheres relatively recently (Chapter 12). Whatever the cause, the observed distributions display strong patterns suggestive of some significant underlying cause.

Whether uplifts are related to chaos or pits is unclear. Some seemingly-domed features are actually examples of chaotic terrain, but we excluded such features from the uplift category under discussion here. Some uplifts from among the "cracked uplift" category appear to be formed from crustal plates that have been tipped, similar to rafts in chaotic terrain. Relations of other uplift features to chaos are less clear. It also remains possible that the various types of pits and uplifts have nothing in common with one another or with chaotic terrain.

Determination of how uplift features formed is an important outstanding issue. Perhaps diapirs formed due to local heating, but such processes would have been very close to the surface according to Rathbun et al.'s calculations. Such a process would form a continuum with our models of pits and chaos formation, in the sense that all would involve local heating. However, the polygonal shapes of many uplifts need to be explained. Perhaps some represent old rafts from an earlier epoch of widespread chaos formation, subsequently reworked until they became continuous with the surrounding terrain. Or perhaps the lithosphere was punched up by blocks of ice that had somehow been stranded at depth and subsequently popped up buoyantly from below. Richard Ghail, a planetary geophysicist at Imperial College, London, came up with the notion over lunch one day that blocks of clathrate[7] may have randomly broken loose from the bottom of the ocean, floated quickly upward, and punched up the bottom of the ice crust. The appearance of uplifts

---

[7] Clathrate is ice laced with molecules of carbon dioxide, methane, or other light hydrocarbons. The gas molecules are locked into cages made up by the crystalline structure of the ice. Clathrates are plausible on Europa, because the substances involved were likely part of the internal compositional mix, and they would be expected to rise toward the surface as part of the compositional differentiation of the satellite by density, including perhaps volcanism.

certainly looks exactly as if something punched up the thin crust, so I hope Richard or someone will follow up on that idea. These ideas are speculative, presented only to stimulate further investigation.

The formation of pits, chaos, and uplifts likely depends on an appropriate thickness of ice. Thermal formation of both pits and chaos requires ice to be fairly thin, consistent with the various other lines of evidence, especially from tidal tectonics. Depending on the thickness of the ice, size–frequency distributions of these features could be greatly affected. The striking antipodal symmetry, in which distributions of pits, of uplifts, and of chaos are similar in the northern-leading (NL) hemisphere to the southern-trailing (ST) hemisphere and similar in the northern-trailing (NT) hemisphere to the southern-leading (SL) hemisphere, may be related to issues of ice thickness. It may have been due to wander of polar real estate part-way toward the equator, something for which we have found other strong evidence.

The properties of PSDs that have been widely cited as primary evidence for convective upwelling in thick ice (e.g., that uplifts are generally dome-shaped and often cracked; that pits and domes are regularly spaced; that there is a typical diameter of $\sim$10 km) were premature and not supported by subsequent data.

The origins of pits and uplifts on Europa remain uncertain, as do their implications for the bigger story of the structure and history of the satellite. The fiasco of "pits, spots, and domes" resulted from prematurely concluding what they meant before we knew what they were like. An accurate description and characterization of the population of these features should have been the first step toward their interpretation.

Part Four

Life on Earth and Europa

20

The bandwagon

> They were no less ridiculous than those who in great number opposed my first celestial discoveries, persuading themselves (as is usual in noisy altercations of idle words) that by texts, authorities, syllogisms, and their foolishnesses they could force the course of nature to conform to their dreams. Malignity, envy, and ignorance are unconquerable beasts, and I see by daily experience that my contradictors, though overthrown by a hundred confrontations and past experiments, and made certain that new opinions introduced by me and at first denied by them are true, do not cease to oppose themselves to other things day by day propounded by me, still hoping some day to have me on the hip and that some one small error of mine will cancel all the other true teachings I have introduced. Now, you must let the vulgar shout, and just continue in conversation with the Muses, enemies of the tumultuous rabble.
>
> Galileo, 1630

The bandwagon for the convective, thick-ice, isolated ocean model had tremendous political momentum, making it advantageous to be onboard. The peer review system controls the fate of proposals for research funding and of manuscripts submitted to scientific journals, so being in the mainstream has practical benefits. It also ensures comfortable and friendly interactions at scientific conferences, whether during casual conversation or formal presentations. Those who bucked the party line could count on intimidation, attacks, and scorn.

Being off the bandwagon was unhealthy for a career, especially for younger scientists. Greg Hoppa, who had made many of the key discoveries about Europa, who probably knew more than anyone about the contents of the *Galileo* data set, and was one of the finest minds among the new generation of planetary scientists, had no hope of a professional future in his area of expertise that would support his growing family, so he left the field. Other young people who toed the party line were moving up in their careers, getting faculty jobs, serving on high-profile policy committees, and being honored for their contributions.

Young researchers faced a moral challenge. Doing good science was not necessarily the road to success, especially if their work yielded results that might be consistent with thin, permeable ice. On the other hand, it was evident that, if they could skew their research to support the party line, they could expect to be rewarded.

At the same time, the thick-ice dogma was not faring very well against the facts. The standard mantra of pits, spots, and domes had finally come under scrutiny (Chapter 19). The fallback to evidence from craters was hardly definitive (Chapter 18).

With the foundation of the party line needing shoring up and in this unwholesome social context, several papers appeared in the literature that purported to support the thick-ice model, but which, closer inspection shows, used surprisingly distorted logic to reach those conclusions. Most remarkably, in several of these papers, analyses that provided good corroborating support for thin, permeable ice were hidden between introduction sections that denigrated evidence for thin ice and conclusion sections that only paid homage to the thick-ice, isolated ocean model. I review these papers here because they are part of the current case for thick ice, and more importantly they demonstrate how a seemingly detailed, quantitative analysis can be twisted to conform to a party line.

20.1 STRIKE–SLIP IN THICK ICE

Francis Nimmo and Eric Gaidos developed a computer simulation of strike–slip shear in a thick-ice model. Strike–slip has been a thorn in the side of the thick-ice people because our model of tidal walking, which fits the observed distribution of strike–slip displacement so nicely, requires thin ice (Chapter 12). So, Nimmo and Gaidos scored very big points when they published their results. They concluded that strike–slip could be driven in thick ice. They also came up with a bonus: There would be enough friction for substantial heating under the length of a fault, providing a *post factum* rationale for Head and Pappalardo's story of linear diapirism as a way to build ridges. Warm ice would rise evenly from below, for the full length of a crack, addressing at least one of the several problems with that model (Chapter 10), and a step toward reconciling the existence of ridges with the model of thick ice. Moreover, Nimmo and Gaidos's paper displayed all the usual trappings of a detailed quantitative study.

Detailed scrutiny shows the conclusions to be flawed, however. First, consider some characteristics of the thick-ice model. There are a couple of kilometers of cold ice near the surface, with the temperature rapidly increasing with depth. This is the lithosphere, where the material is elastic and brittle. Exactly at what temperature the transition from elastic to viscous occurs depends on the rate of deformation. Also, the results of modeling depend on various assumptions about the temperature profile, which determines the viscosity of the ice as a function of depth through the ice, to the water interface 20 km down. There is a huge amount of uncertainty about the temperature profile that would be appropriate for a thick-ice model, and even more regarding the viscosity of the material under those conditions, so I could be nit-picking about these details. I cannot resist mentioning that they assume a "Newtonian" viscosity (i.e., that strain is proportional to stress, which is grossly incorrect for ice). But, we can ignore these issues, because the problems with this work are so much more fundamental.

In their model, the brittle, elastic layer (the lithosphere) is cracked vertically, and the region is placed under tidal shear stress, with reference to the calculations of tidal stress and strain by me and my students (Chapter 6). In response, slip occurs along the fault in the lithosphere, and viscous shear takes place in the warm ice below. There is so much dissipation involved in distorting the ice that a great deal of heat is generated, raising the temperature by 66°C. (These results are delightfully precise, conveying a great deal of credibility.) That heat is exactly enough, they find, to buoy up the lithosphere by 100 m, comparable with the height of the larger double-ridges.

As in terrestrial global economics, sometimes it is helpful to follow the energy. If there is so much heating going on, something must be pushing very hard to displace and distort the ice against all that friction. But Nimmo and Gaidos did not model things in those terms. They simply imposed on their computer simulation a predetermined speed for the shear displacement of the lithosphere of about 10 cm/day for the cases that they cite in their conclusions. That is mighty swift, and it is difficult to imagine what could drive it against the resistance of all that viscous ice.

They do explain where they get that value for the speed, however. It comes from our calculations of tidal stress and strain in an elastic lithosphere (Chapter 6). Remember, we had found that an elastic shell around Europa would distort by amounts ~1 m per 100 km on a diurnal timescale ($\sim 10^5$ sec). Shear would occur at that rate only at some locations. In our considerations of tidally-driven strike–slip, we did envision shear to be as fast as 1 m/day, but only at those locations, and only during the phase of the diurnal tidal-walking cycle when the crack is open and shearing. The average rate is about 10 cm/day. In our model, the displacements can only be as fast as the strain when there is no shear resistance. Nimmo and Gaidos imposed this rapid shear rate on a model where there is tremendous resistance from 18 km of solid ice, rendering their model completely unrealistic and self-contradictory. If their thick-ice model were really driven by tides, the actual rate of shear displacement would be orders of magnitude slower, and heating would be negligible.

At its heart, the conclusions of Nimmo and Gaidos are an example of circular reasoning. In their computer simulation of thick ice, they force the displacement to go as fast as we had found it might go for thin ice. On that basis, they conclude that rapid strike–slip displacement can occur with thick ice.

Paul Geissler pointed out to me another way to interpret the results of Nimmo and Gaidos, to the extent that they have any meaning at all. Most of the heating occurs at the base of the brittle layer, so one might imagine applying the same result to our thin-ice model, in which there is no convection, so the viscous layer is comparable with, or thinner than, the brittle, elastic layer. In that case, this heating mechanism proposed by Nimmo and Gaidos would help explain how the bottom of a crack could be linked to the ocean. If tidal stress cannot get it all the way down, melting along the bottom of the crack might complete the connection.

Given that my research group had pioneered theoretical work on strike–slip displacement on Europa, one might have expected Nimmo and Gaidos to discuss our work with us, before trying to apply our results in a publication. We could have helped them avoid the inappropriate use of some of our numbers, which was central to their story, and made sure that they understood the mechanism of tidal walking.

20.2 OVERBURDEN FLEXURE

My student Randy Tufts had pioneered a technique for probing the structure of the lithosphere on Europa by studying the flexure due to loads on the surface. He investigated, as part of his PhD dissertation, ridges whose weight appeared to have warped down the surface on either side of it. Analysis of the dimensions of the downwarp, especially how far away from the burden the lithosphere bends down, can be used to infer the thickness of the lithosphere. Randy found for cases near Conamara that the elastic supporting layer was about 300 m thick.

Recently, Francis Nimmo got together with Bob Pappalardo, driver of the thick-ice bandwagon, and with Bernt Giese (of the German space agency, DLR), who had been inferring topography from stereo *Galileo* images, to consider what they interpreted as downwarping of the lithosphere in another location (Nimmo et al., 2003). Using a general approach similar to Tufts', they concluded that the ice on Europa must be at least 25 km thick. Actually, they found that the ice could be thin, but with an Orwellian twist, they wrote as their conclusion that the ice must be thick.

Here is the story. Giese's map of the elevations of the surface, based on stereo-information from two sets of images, showed a slightly-elevated region just southwest of the impact crater Cilix (Figure 20.1, see color section). Topography can be derived from stereo-views, because images taken from different directions allow us to triangulate on distances. Our brains do this automatically when they combine the information from both eyes to create a three-dimensional image of the world.

What makes construction of a topographic map, like Figure 20.1, difficult is that the separate images were taken at different resolutions and under very different lighting conditions, so the elevations must be fairly uncertain. According to Nimmo et al., the "vertical resolution is 30–60 m". Which is it, 30 m or 60 m? Usually, the uncertainty is expressed as a single value (e.g., meaning that the elevation might be off by as much as 60 m). The meaning of uncertainty given as a range of values is not clear at all. Nimmo et al. give no description of how they determined the degree of uncertainty. Given the type of images used to generate this map, and the way the vertical precision is described, there is reason for suspicion that the uncertainty may actually be much greater than 60 m. But, setting aside that issue, there are deeper problems with this paper.

Nimmo et al. noticed a 40-km-wide raised area centered about 30 km SSW of Cilix Crater. This topography is very subtle. There are a large number of substantial ridges crossing this area, intersecting and overlapping one another, so that when their height is discounted, this "plateau" is about 150 m higher than the surroundings. Such a slight rise over such a large area is a very subtle topographic feature, and the height is not much greater (if at all) than the uncertainty in the elevation.

The plateau is not recognizable in any image, so it is impossible to determine its geological character and there is no basis for any assumption about its origin, if there is any significance to it at all. Nevertheless, Nimmo et al. make a specific geological interpretation that this plateau was emplaced in such a way that it loaded down the lithosphere. That is quite a far-fetched interpretation, given that there is no information to support it. It raises the question of why other raised topographic features that we can actually see (e.g., Figure 19.5) show no signs that they weighed down the surrounding lithosphere.

Attempting to demonstrate that this plateau weighed down the lithosphere, they constructed topographic profiles across the region along the black diagonal lines in Figure 20.1. The interesting part is to the northwest of the plateau, where Nimmo et al. thought they found evidence that the lithosphere was bent down. They combined the six profiles in the following way. First, they shifted each one so that the northern edge of plateau was at the same point on each one. Then, they averaged the profiles by summing them and dividing by 6. They found that the averaged profile shows a drop from the surrounding terrain (20 km to the NW) down toward the base of the plateau. This drop was interpreted as a sagging due to the overburden of the plateau and formed the basis of the estimate for ice thickness.

Let us take a closer look at the data that went into that average topographic profile. Looking at the area scanned by the six profiles to the northwest of the plateau, we see that it is dominated by ridges, except for a triangular-shaped area between the plateau and the ridges. Now, if you take a set of profiles across the drop-off from the high levels of the ridges, and shift the profiles so that the drop-off occurs at a slightly different point in each profile, then the average profile will look like a gradual sag. The appearance of a structural flexure could well be in part an artifact of the averaging process. In any case, the drop-off in elevation, from 20 km NW of the plateau to the edge of it, is dominated by a transition down from several substantial tectonic ridges. Moreover, the amount of this drop-off in the average profile of Nimmo et al. is only about 100 m, barely more than the uncertainty in the elevations.

Thus, it is very uncertain whether the topography interpreted by Nimmo et al. is real and, if it is, whether it represents a downwarping under a load. But let us give them the benefit of the doubt and consider what they found when they fit the supposed topography to a physical model of a burdened icy crust. They computed that the ice must be within the range between 6 and 35 km thick. In other words, their result was equally consistent with either the canonical thick-ice model or with our model, in which the ice is thin enough for the ocean to be linked to the surface.

Remarkably, the stated conclusions of the paper are completely different from the result of that analysis. The abstract of the paper, which is the only part ever read by many readers, summarizes the conclusion: "Combined with independent estimates we infer a probable shell thickness of ≈ 25 km." The paper concludes by linking models of convective thick ice, Schenk's crater paper, and their own work, and giving the final sentence: "These combined results together suggest a uniform present-day shell thickness for Europa close to 25 km." Never mind that the results of their paper were that the ice could be as thin as 6 km, the authors needed to stay on the thick-ice bandwagon.

20.3 MELT-THROUGH BASHING

I first heard of a young scientist named Jason Goodman after he did theoretical work that provided strong support for our melt-through model for chaos formation, and then presented it as evidence that our model was wrong.

Actually, I later noticed his name in the acknowledgements section of a paper that had come out of Jim Head's Brown University group in early 2000. That paper, with the objective-seeming title "Evaluation of models for the formation of chaotic terrain on Europa", was an advocacy piece for the thick-ice model. Its lead author was graduate student Geoffrey Collins.

Collins must have found himself in a tough spot. The work reported in his paper showed that the evident mobility and tipping of rafts in Conamara required extensive melt just below the surface. This result was clearly at odds with Pappalardo and Head's belief that solid-state convection formed chaotic terrain. In fact, it confirmed my arguments about the appearance of chaotic terrain and provided good support for our model of melt-through. However, to defend the dogma of the isolated ocean and still have melt under the chaos, they came up with the idea of isolated local blobs of melt within the thick ice.

The paper included a reprise of the canonical argument that melt-through was impossible because it would require many calories of heat. They did not show that the required amount of heat was prohibitively large, only that it seemed like a big number. In fact, as discussed in Chapter 16, only modest concentrations of plausible amounts of tidal heat can do the job.

Collins et al. raised another objection to the melt-through model. They claimed that even with localized heat at the ocean floor, the warm water would spread as it rose, so that there would be insufficient heat concentration at the base of the ice to melt through a chaos site. As with Collins et al.'s other claims, this one had not held up to quantitative scrutiny. As discussed in Chapter 16, Thomson and Delaney demonstrated the plausibility of a plume of warm water rising from the bottom of the ocean up to the top, and remaining confined enough that it could melt the ice.

In preparing their polemic on chaos, Collins et al. connected with Jason Goodman, then a graduate student in oceanography at M.I.T. Later, as a postdoctoral associate at the University of Chicago, Goodman seemed to some colleagues in oceanography to be determined to disprove our model. He considered different assumptions about the unknown conditions in the ocean from those of Thomson and Delaney, and he performed laboratory experiments to help understand the rising of warm buoyant plumes. He found that plumes, even after rising all the way up from the deep ocean floor, could remain as tightly confined as 25 km in diameter as they reached the bottom of the ice. That result provides good support for the melt-through model.

While the heart of Goodman et al.'s paper on the subject supports our melt-through model, he sandwiched it between an introduction and a final discussion that twists the story to give the opposite impression. The introduction purports to review the case for an ocean on Europa without ever citing the first (and, for many, most convincing) observation-based evidence, which was our explanation of the formation of cycloids. Next, the paper shows the analysis that seems to support our melt-through model, because the rising plumes of warm water stay tightly confined. But, in the discussion of this result, Goodman et al. contort that result to conclude that our model is wrong. They note that my group's survey had shown that most patches of chaotic terrain are smaller than 10 km across. Therefore, because Goodman's calculations yield a plume of warm water that is greater than 25 km wide, they conclude that it could not possibly be the cause of chaos. That argument extends his results to a domain where they are simply not appropriate. His calculations do not address what would happen at the surface of an ice layer if a plume of warm water pressed against its base. If the plume of warm water thinned the ice to nearly breaching the surface, it seems likely that at a fine scale, based on local anomalies of ice characteristics, topography, or ocean eddies, very small break-throughs could open to the surface. Goodman's models show that melt-through is plausible; they hardly prove that small-scale breakthrough is impossible.

Moreover, Goodman et al. explicitly chose to disregard a suite of interesting crust breakthrough phenomena that follow from an unusual characteristic of liquid water: It becomes denser as it approaches freezing point, which may enhance the melt-through process according to work by Jay Melosh et al. Instead, Goodman et al. restrict their work to an assumed range of salinity that avoids that effect. Given that we have no idea what the ocean is like, arguing that small-scale oceanic exposure is impossible on the basis of a narrow range of particular assumptions is unjustified.

Then, to further disguise the fact that Goodman's calculations actually support the idea of melt-through, their discussion goes onto a tangent with a spurious critique of Dave O'Brien's simulations of the melt-through process. Remember, O'Brien's work had addressed and disproved Pappalardo and Collins' unsupported claims that tidal heat could not provide enough energy to melt through the ice.

Their attack on O'Brien et al. takes off from the obvious fact that in an idealized situation you can never have melt-through because no matter how much heat you have, some thin layer of ice will be maintained due to surface radiation. If you double the heating rate at the bottom of the ice, then ice becomes half as thick (see Chapter 7). Goodman et al. claim on this basis that that melt-through is impossible because there is always a thin skim of ice, and the ocean can never be exposed. That argument is rather desperate and silly. Once the ice is thin enough, local anomalies, dynamics of plate and ice block motion, and all sorts of fine-scale effects that have not been considered could readily open the ice and expose liquid on a temporary basis. For O'Brien et al.'s simulation, we had chosen a spatial resolution of 100 m, because we had inadequate information about Europa at a finer scale, and we knew from familiarity with topography that unaccountable random effects must exist on that fine scale. Moreover, the point of O'Brien et al.'s work had been to address the false claims of Pappalardo and of Collins that there was insufficient tidal heat to melt away kilometers of ice. Goodman's claims that O'Brien had made a fundamental error were nonsense.

If Jason Goodman had presented the results of his oceanic plume study objectively, he would have offered a useful contribution to understanding of the possible physical processes on Europa.

20.4 CONVECTION MODELS

A number of theoretical geophysicists have been engaged in developing models of convection in the ice of Europa ever since the initial *Voyager* era realization that there might be large amounts of tidal heat flowing from the interior. Consideration of whether there might be an ocean depended on whether heat would flow out slowly enough to maintain a liquid layer. The consensus of these models is that convection is unlikely if the ice crust is less than about 15 or 20 km.

Beyond that broad result, other conclusive results are elusive. They are strongly dependent on a large number of assumptions about unknown conditions. For example, the nature of the convection, if there is any, depends sensitively on the details of the rheological (flow) properties of the material. Theorists need to guess at plausible ice grain sizes, activation energies, stress-to-strain-rate relationships for the material, heat fluxes, relative dissipation sources, etc., none of which are known, but any of which can affect the results substantially.

Detailed study and intercomparison of the various theoretical models in the literature is of considerable interest from the hypothetical perspective. However, given the evidence that the ice on Europa is probably less than about 10 km thick, and the fact that these theoretical convection models are so strongly dependent on unknown parameters, the outcomes of the modeling have not been very useful for understanding Europa.

In general, the theoretical modelers have not demonstrated the kinds of blatant biases discussed in the earlier examples in this chapter. Rather, because they are deeply involved in their theoretical specialty, they do not have the time or expertise to independently examine and interpret the images of Europa. They must rely on others for interpretations of the observations, and with faith in the system they naturally accept the authoritative reports. Theorists become the first victims of a perversion of the system.

What the theorists do is start from the observed and certified fact that convection is manifested in the surface of Europa, with a well-defined spatial scale of about 20 km. Then, they adjust the various unknown parameters that describe the ice layer in order to construct a model that fits the observation. In this way, they can infer, or at least constrain, the character of the interior.

Theoretical modeling is difficult. Ideally, theorists should become familiar with the data that form the motivations and constraints for their work. More often, they

Sec. 20.4]   **Convection models** 321

must trust authoritative sources regarding the observations. The last thing that a researcher wants to hear after buying in to the establishment description of Europa and doing all that theoretical work is that the basic observational premise was false. Perhaps a reluctance to accept a correction to the underpinnings of their work is understandable.

The Spanish researchers Javier Ruiz and Rosa Tejero (Universidad Complutense de Madrid) obtained fairly typical theoretical results. They found that convection on a scale suggested by the mythical PSDs would require very specific properties for the ice, including a particular value (49 kJ/mol) for a parameter Q that describes the non-linear viscosity of the ice, ice grain size of 1 mm, and negligible heat flow from the interior of Europa. These requirements are quite narrow and unrealistic, so Ruiz and Tejero, in a preprint of their paper, wrote: "Thus, it is not possible to reconcile the heat flow results with a convective origin for the lenticulae."[1] Furthermore, they added: "the interpretation of the lenticulae as features related to convection is not the only explanation", and they cited our melt-through hypothesis.

As published in *Icarus*, however, the conclusion changed course by 180°, despite the fact that the results of the theoretical work were exactly as in the preprint. The sentences quoted above were replaced by: "Thus, it is possible to reconcile the heat flow results with the lenticulae spacing predicted from a convective origin for these features ..." With this new spin, Ruiz and Tejero were comfortably on the bandwagon.

Some convection modelers do try to address the actual characteristics of pits and uplifts. Adam Showman[2] (University of Arizona) and Lijie Han (Lawrence Berkeley Lab.) found that production of topographic features is possible only if the brittle layer is thin. Even then, production of uplifts is negligible. The model can generate pits, but they are usually larger than 10 km wide. The far more common small pits (Figure 19.7) are not produced by convection. Convection does not produce the population of pits and uplifts observed on Europa.

There is no evidence that the ice on Europa is convecting, nor is there any other evidence that it is thicker than 10 km. It is time to get off the bandwagon.

---

[1] There, *lenticulae* was used in accord with the incorrect usage introduced by Pappalardo et al. to mean the PSDs.

[2] Adam has a hard time avoiding the facts about pits and uplifts. Shortly before he joined the faculty in my department, I had taped huge prints of Martha Leake's maps of pits and uplifts on the wall in the corridor near my office (2-m-tall color versions of Figures 19.6 and 19.9). Adam happened to move into the office exactly opposite the maps. He cannot walk out his door without getting hit in the eye by brightly-colored pits and uplifts. Other than that, I have no particular influence over him.

21

The biosphere

21.1 DREAMS OF LIFE

In 1979 the simultaneous revelation of the appearance of Europa's surface and the realization that it is significantly heated by tides, transformed the icy satellite into a leading contender to be a home for extraterrestrial life. Ray Reynolds, part of the original team that had recognized the likely amount of tidal heating, led much of the research in the 1980s on the implications of that heat, and what we saw on Europa, for the possibility of life. Ray and his co-workers at NASA's Ames Research Center recognized the likelihood of a global ocean and, using early estimates of tidal heating and of the response of the ice, calculated that the ice crust must be less than 10 km thick. They interpreted the linear features seen clearly in the *Voyager* images (e.g., Figure 14.1, and on a global scale similar to Figure 2.2) as cracks linking the ocean to the surface, and they wrote in detail about how such a physical setting could support life.

That work was exciting stuff. Randy Tufts by that time had spent many years in social work as a community organizer, and his stewardship of Kartchner Caverns was becoming less intense as control was handed over to the state. At age 40, he was looking for a direction for the next phase of his life, when he read about Europa's ocean and recognized its implications for life. He had an undergraduate degree in geology, he knew that *Galileo* was on its way to Jupiter, and he recognized that the new images of the surface would tell us whether Europa really was a contender for life. He returned to the University of Arizona, as a graduate student, and joined my research group as the *Galileo* data began to arrive, prepared to explore the possibility of life on Europa.

The possibility of life was exciting, but the consensus of the planetary science community was more conservative. In general, the existence of the ocean was still considered an unresolved issue, let alone the idea that the ice might be thin enough to provide a habitable setting. Ray Reynolds and his collaborators were highly respected, but their work on Europan life was generally considered to be intelligent speculation.

The sanguine vision of Reynolds et al. seemed to be crushed when the *Galileo* Imaging Team reported what it saw in terms of an ice shell so thick that any ocean must be isolated. The official story did not have much evidence behind it, but it had the voice of authority. So, researchers who wanted to consider the possibility of life on Europa were driven to the bottom of the sea, trying to find plausible ways that life might be supported directly by the tidal heat emerging from the rocky interior. They looked to the analogy of undersea volcanic vents on Earth, where ecosystems get energy from the internal heat of our planet. These vents on Earth had already attracted considerable attention, especially because a plausible case had been made that terrestrial life might have originated in such places. The evolutionary tree of life seems to have branched from primitive organisms that prospered in such extremely hot places and were anaerobic, living without needing oxygen. At sub-oceanic vents, the chemistry, the energy source, the protection from solar ultraviolet radiation and from impacting bodies may have offered a safe place for the formation of life in the early solar system.

Theorists' efforts to describe life in the ocean met with varying degrees of success. They were stuck with two intractable problems. First, with the ocean isolated from the surface, life without oxygen was limited at best, according to Gaidos et al. (Chapter 19). What is more, we simply have no evidence about conditions 150 km below the surface of Europa, down at the bottom of the sea.

21.2 THIN ICE ON A WATER WORLD

We do have information from the *Galileo* spacecraft about the conditions and processes in the crust above the ocean, and it turns out the crust is much more like what Reynolds et al. had envisioned, thin enough for the ocean to be linked to the surface. As a member of my research group, Randy Tufts found himself playing a key role in developing this picture, along with myself, Greg Hoppa, Paul Geissler, and my other students and postdocs. Several lines of evidence, based on what Europa really looks like, point to conditions in which life is made possible by a thin, permeable ice crust.

Lineaments on Europa fit the predictions of tidal stress calculations. These agreements range from broad, global-scale patterns to specific detailed shapes, such as the ubiquitous, distinctive cycloids, unique to Europa, that run for many hundreds of kilometers across the surface. This correlation tells us that these lines are cracks, the products of tensile stress produced by tidal distortion of the global ice shell. The magnitude of the required stress means that there must be a thick liquid layer, many tens of kilometers thick, under the ice. Otherwise the surface shell would not be distorted very much. Moreover, the ocean must be global, to produce the stress patterns that have been observed.

The sequence of crack formation evidenced in the geological record, combined with our understanding of how and where cracks of various shapes and orientations should form, supports the theoretical prediction that Europa rotates non-synchronously. Non-synchronous rotation results in an important part of the tidal stress that produces cracks, in addition to diurnal tides. Our best estimate for the rotation period relative to the direction of Jupiter is 50,000 years. This is an astronomical number, in the sense that it has an uncertainty of a factor of five or more. In any case, Europa rotates very slowly (relative to Jupiter) compared with the equivalent 1-day period of the Earth, but very fast compared with the infinite period of the Moon.

Strike–slip displacement is very common on Europa, as on Earth, but rare elsewhere. Observations fit the predictions of our tidally-driven "walking" model, which depends on cracks extending downward all the way to the liquid layer so that adjacent plates are free to slide. Our strike–slip survey also indicates that Europa undergoes occasional slippage of its entire shell, as a whole, relative to the direction of the spin axis.

Ridges probably formed by diurnal pumping of slush through cracks open to the ocean. Once a crack forms, it will be worked by tides and remain active as long as the tides continue to open and close it on a daily basis, so that the crack does not freeze shut. Some cracks are driven open by ocean currents or some other process that causes large-scale plate movements, gradually exposing ocean water at the surface, as the crack continues to be worked by diurnal tides.

Chaotic terrain probably formed by melt-through from below, briefly exposing liquid water until it refroze, creating a matrix lumpy with bits of the old crust, lying between larger, often displaced, rafts that still display the previous surface. Chaos formation has been very common throughout the age of the currently-visible surface, so that recognizable chaos covers a substantial fraction of it. If such melt-through occurs that commonly, then the thickness of the ice must have varied significantly with time and place, even where liquid exposure did not occur.

The history of Europa's surface has been a competing and continual interplay of comparable amounts of resurfacing by chaos formation and by tectonics, each destroying or covering what was on the surface before. Each of these types of processes involve direct interaction of the ocean with the surface.

Even modest-scale impacts can puncture through the ice and briefly expose the ocean, leaving terrain that is similar to chaos, although more circularly symmetric. But such features are rare and impacts have played a minor role in shaping the surface and exposing the ocean, compared with cracking and melt-through.

The paucity of craters tells us that resurfacing of Europa occurred recently. Little of the remaining surface is more than 50 million years old. Because resurfacing has been gradual and continual, much of the surface may be much younger. In addition, *Galileo* magnetometer data suggest not only that Europa had an ocean, which we knew from the cycloids, but that the ocean is there now. Given that so much activity has continued into the most recent 1% of the age of the solar system, it would be surprising to learn that it ceased shortly before terrestrial spacecraft arrived there. It seems most plausible that the activity linking the surface and ocean is continuing.

21.3 SUBSTANCES ABOVE AND BELOW

The permeable ice layer between the ocean and the surface allows an exchange between the substances that are likely to fall on, or be created in, the top of the ice and the substances likely to have found their way into the ocean water from the deep interior. This access may overcome the difficulties that oceanic life would face if it were isolated by a thick, impenetrable ice crust. Moreover, life generally thrives at the interfaces between different physical conditions. The richness and diversity of tide pool life is a good example, as is the extra density of weeds along the side of a road. Such settings allow the system to exploit the benefits of diverse conditions, and the disequilibrium between them.

The available substances at the surface and in the ocean are quite different from one another, because they come from different places. Even the thin crust keeps them from mixing freely. At the surface, molecules come apart or recombine under the bombardment of all the energetic particles and radiation within Jupiter's magnetosphere. Oxidants are continually produced by disequilibrium processes, such as photolysis by solar ultraviolet radiation, and especially radiolysis by charged particles. Significant reservoirs of oxygen have been detected from the spectrum of reflected light in the form of H_2O_2, H_2SO_4, and CO_2 on Europa's surface. Moreover, molecular oxygen and ozone are inferred from the existence of Europa's oxygen atmosphere, sparse as it is. Their likely presence is confirmed by the detection of these compounds on other icy satellites, such as Ganymede. When Gaidos et al. argued against life on Europa, it was not because they doubted that there was plenty of oxygen, but rather that they had been told that the ocean was isolated from it.

Impact of cometary material should also provide a source of organic materials and other fuels at the surface. Such substances have not been directly observed on Europa, but they can be reasonably inferred because of the expected continual bombardment by tiny cometary particles as well as by the larger comets that produce observable craters. Moreover, such substances have been detected on the other icy satellites. In addition, significant quantities of sulfur and other materials may be continually ejected and transported from Io to Europa.

The ocean contains endogenic substances (from the interior), such as salt, sulfur compounds, and organics, as well as surface materials that may be transported through the ice. Evidence for oceanic composition is found at those surface locations where oceanic material most likely has been exposed, especially along the major ridge systems and around chaotic terrain. In color representations of Europa, this material appears as orange–brown. Unfortunately, the images taken by *Galileo* were all in the visible-to-near-IR wavelengths. Even with separate images taken through different filters, this portion of the spectrum is not diagnostic of composition.

However, *Galileo*'s near-infrared mapping spectrometer (NIMS) did provide spectra in a wavelength range that can be used to infer composition, and that instrument had very high resolution in wavelength, as well as sufficient spatial resolution to distinguish the composition at different locations. The areas that appeared orangish-brown in images have spectra in the near infrared that suggest hydrated, sulfur-based salts in frozen brines. Alternative interpretations had included various granularities of water ice, or sulfuric acid and related compounds, but the salt water result seems most widely accepted. (Interpreting spectrograms can be subjective, so we often must wait for a consensus to emerge from the specialists, always bearing in mind that sometimes politics can determine a party line.) The orangish-brown appearance at visible wavelengths at the same locations may be consistent with organics, sulfur compounds, or other unknowns, in addition to the salts detected by NIMS. The correlation with sites of oceanic exposure suggests that the ocean possibly contains a wide range of biologically-important substances.

The ice at any location may contain layers of oceanic substances deposited at the top during ridge formation. But then, those materials, as well as any substances that formed at the surface or that arrived there from space, must work their way deeper as subsequent ridge building places new strips of material on top. As the ice thickens by deposition at the top, it maintains its average thickness by melting at the bottom. This process can bury surface materials, gradually working them downward, and eventually feeding (or recycling) them into the ocean. Occasionally, a melt-through event speeds up the delivery of chemicals into the ocean, and then resets the ice as a single layer of refrozen ocean.

For oxidants and organic fuels, rapid burial is especially important to prevent their destruction by radiation at the surface. Impact "gardening" (digging and mixing of the material near the surface) may help with the initial burial. However, ridge formation may provide the dominant mechanism for continually covering and protecting oxidants and other surface materials, and for gradually conveying all materials in the crust downward into the ocean as strip upon strip of new material is deposited on top, along the edges of cracks.

Chemical disequilibrium among materials at various levels in a crack is maintained by production at the top and the oceanic reservoir at the bottom, while the tidal ebb and flow of water continually transports and mixes these substances vertically. This transport runs through a variation in ambient temperature from $0°C$ at the base of the crust to about $170°$ colder at the surface, a gradient of roughly $0.1°$ per meter of depth.

21.4 LIFE IN THE CRUST

What we were learning about Europa restored the prospects for life in the ocean, because, consistent with the vision of Reynolds et al., the crust was proving to be permeable. We now had evidence that multiple mechanisms linked the ocean to the surface, that they all had acted continually for as long as we could tell, and that all this happened recently, probably continuing today. Delicious-looking[1] orangish-brown stuff lined the sites where connections had opened. Moreover, with what

---

[1] Each to his own taste, but, from the perspective of a hypothetical organism in what might otherwise seem a hostile environment, the substances plausibly in the mix could be attractive.

we now know, it seems likely that the crust, as well as the ocean, is in a physical condition that could conceivably support life.

The key physical processes act on a range of timescales that make this idea plausible. On a daily basis, warm tidal water is pumped up and down through the active cracks, thanks to diurnal tides. Over tens of thousands of years, rotation carries each crack to a different location where tidal stress variation is different. Every few million years, exposure of open water occurs at any given location, because chaotic terrain covers much of the satellite and there has been multiple resurfacing.

While Gaidos et al. had described fundamental problems with supporting life on Europa, their analysis was based on the prevailing belief that the ocean was isolated. Our evidence was converging on a very different picture, where the ocean is linked to the surface in a set of specific ways.

Consider the profile of a crack, opening and closing with the daily tide (Figure 21.1, color section). No organisms could survive near the surface, where bombardment by energetic charged particles in the jovian magnetosphere would disrupt organic molecules within ~ 1 cm of the surface. This bombardment is not very good for spacecraft electronics either. The only reason *Galileo* survived as long as it did was that it spent most of its time in orbit around Jupiter much further out than Europa.

A few centimeters down, organisms would be adequately shielded by the ice from deadly radiation damage. Nevertheless, sunlight adequate for photosynthesis could penetrate that far and more, down a few meters below the surface. Thus, as long as some part of the ecosystem of the crack occupies the appropriate depth, deep enough to be safe from damaging radiation but shallow enough to catch the light, it may be able to exploit photosynthesis. Such organisms would benefit from anchoring themselves at an appropriate depth where they might photosynthesize (symbolized by the plants in Figure 21.1), although they would also need to survive the part of the day when the tide drains away and temperatures drop.

Other non-photosynthesizing organisms (symbolized by bugs in Figure 21.1) might anchor themselves at other depths, and exploit the passing daily flow. Their hold would be precarious, as the liquid water could melt their anchorage away. Alternatively, some might be plated over by newly frozen water, and frozen into the wall. The individuals that are not anchored, or that lose their anchorage, would go with the diurnal flow. Organisms adapted to holding onto the walls might try to reattach their anchors.

Other organisms (symbolized by jellyfish in Figure 21.1) might be adapted to exploiting movement along with the mixing flow. A substantial fraction of any floating organisms would be squeezed out of the crack each day as it is closed by tides. Some might be entrained in the slush that builds ridges along the rims of the crack (Figure 10.3) leading to probable death at the cold and irradiated surface. Luckier individuals would be squeezed into the ocean during the diurnal crack closures, and then flow up again with the next tide. Organisms and organic debris that are squeezed into the ocean could help feed any oceanic ecosystem. Of course, the linkage between the crust ecology and life in the ocean would go both ways, as organisms and detritus are sucked up out of the ocean into the crack during the diurnal opening phase. Oxidants, organics, and other substances from the crust, as well as endogenic substances from below the ocean, would be mixed and transported through the crack. The daily cycles of tidal variation and day vs. night would nurture and govern the processes of life.

Life might be comfortable and routine in and below an active crack. Conditions would be uncomfortable by human standards, but other terrestrial organisms that can extract oxygen from water and function well in temperatures at the freezing point are common. A given crack is likely to remain active for thousands of years, because rotation is nearly synchronous and the crack remains in the same tidal strain regime. The stable daily cycle in a crack on Europa might allow organisms to multiply and the ecosystem to prosper.

Over the longer timescale of tens of thousands of years, non-synchronous rotation will carry a given site to a substantially-different tidal strain regime. At some point, the tidal working of any particular crack is likely to cease. The crack would seal closed, freezing any immobile organisms within it, while some portion of its organism population might be locked out of the crack, trapped in the ocean below.

For the population of a deactivated crack to survive, organisms must have adequate mobility to find their way to a still active (generally more recently created) crack. Alternatively, or in addition, the portion of the population that becomes frozen into the ice must be able to survive until subsequently released.

The frequency of chaotic terrain formation tells us that, at any given location, an organism would need to wait only a few million years, on average, before a melt event releases it. Survival in a frozen state for the requisite few million years seems plausible, given evidence for similar survival in Antarctic ice. Frozen organisms liberated by melting would float free in the ocean and perhaps find their way into a habitable niche.

Some frozen trapped creatures might be freed in another way. During the period of non-synchronous rotation (less than a million years), tidal stress would create fresh cracks through the region, which would likely cross the paths of the older refrozen cracks (in some cases reactivating older cracks), liberating organisms into a niche similar to where they had lived before. That advantage is counterbalanced by the fact that release only comes at the intersections of new cracks with old ones, releasing fewer captives.

In any case, the need to survive change may provide a driver for adaptation and mobility. The timescales for tidal change may help support natural selection. Those individual organisms that are able to survive the hard times and return to an active crack would then have thousands of years to reproduce in a relatively-stable niche. Thus, the best adapted individuals would have an opportunity to multiply. Conditions in Europa's crust may be comfortable enough for life to prosper, but challenging enough to drive adaptation and evolution

The physical conditions in Europa's crust create environments that may be suitable for life. Moreover, they provide a way for life to exist and prosper in the ocean as well, by providing access to necessary oxidants and linkage between oceanic and crustal ecosystems. Oceanic life would be part of the same ecosystem as organisms in the crust. Components of the ecosystem might adapt to exploit sub-oceanic conditions, such as possible sites of volcanism, but they would not need to depend on such uncertain resources.

If there is an inhabited biosphere on Europa, it most likely extends from within the ocean up to the surface. While we can only speculate on conditions within the ocean, we have observational evidence for conditions in the crust, and the evidence points toward a potentially-habitable setting.

Even if a setting is habitable, there would be no life unless it originated in some way. Our understanding of the possible processes of the origins of life on Earth has been improving rapidly, but for Europa we are a long way from being able to make a definitive case that conditions were conducive to spontaneous origins. One intriguing idea that may be especially relevant to a cold planet is that RNA, the simplest reproducible genetic material, may have required low temperatures to form in the first place. According to that argument, higher temperatures would break down the complicated chemical structures before they grow into stable forms. If that story is correct, it may make the origin of life on Europa more plausible. In addition to water, one commodity Europa has in abundance is the cold.

Another intriguing idea about the origin of life on Earth is that the development of the first biomolecules capable of replication may have been aided by, or even required, tidal cycling, which alternately soaked and dried the materials. Europa's crust also provides tidal cycling, though further consideration is needed to determine whether they might enhance the initiation of life.

Moreover, if the origins of life require sites analogous to sub-oceanic volcanic vents on Earth, there is a good chance that Europa provided those as well. Almost certainly, there has been plenty of heat generated in the interior. Volcanism at the sea floor is plausible, although we have no direct evidence for it.

Another source for the first life on Europa might have been transport from elsewhere. If some sort of "pan-spermia" seeded the early Earth, it likely would have seeded Europa as well. If life initiated independently on Earth or some other terrestrial planet closer to Jupiter, there is some small probability that colonists might have been transported to Europa, riding on impact ejecta. Some bacteria have been found to survive surprisingly well in space. If asteroids prove to be a significant part of the population of bodies colliding with Europa (Chapter 18), it would increase the plausibility (perhaps from completely insane to amusingly nutty) of transport from the inner solar system.

Ultimately, the way organisms are most likely to be transported to Europa from Earth will be by hitch-hiking on spacecraft. NASA was admirably cautious in terminating the *Galileo* mission by crashing the spacecraft into Jupiter in the autumn of 2003, rather than let it drift around in orbit with the chance of colliding with a satellite. The habitable zone of Europa, its biosphere, extends upward from the ocean to within a few centimeters of the surface. Europa would be vulnerable to infection if living organisms plopped down on it.

21.5 PLANETARY PROTECTION

21.5.1 The possibility of contamination

When isolated ecological communities come into contact, the result can be opportunity or catastrophe, depending on your point of view. Some organisms find a new world to colonize; others are decimated or destroyed. The changes are dramatic and last for ever.

When people are involved, access to new resources has been a benefit (at least to some of the people), while introduction of pests and disease have been obvious problems. Not only humans, but all organisms, are likely to be affected.

Because extraterrestrial life may exist, planetary exploration could bring trouble if people are not careful enough. This danger was well recognized decades ago, when astronauts ventured to the Moon. When the first crews returned, they were quarantined to prevent "back-contamination", the hazard that some infectious organism might have hitch-hiked back with them. The safety procedures were largely symbolic. Who knew the incubation period for hypothetical extraterrestrials? Whether the returning hardware and samples needed sterilization was also largely a matter of speculation.

Subsequent planetary exploration has not involved astronauts, nor have samples or hardware been returned, so back-contamination has not been an issue yet. However, forward-contamination (i.e., the infection of alien organisms or ecosystems by terrestrial organisms hitch-hiking on a spacecraft) is a distinct possibility. Consideration of forward-contamination has focused on Mars since the 1960s. Now, with Europa laid open like a planetary Petri dish, the possibility of forward-contamination there seems very real.

By definition, forward-contamination does not affect the Earth, so why should it be of concern to us? To a large extent this question is one of ethics: Is it morally right to endanger life elsewhere? There are practical dimensions as well. One is the far-out possibility that we might inadvertently threaten and thus antagonize potentially-proactive enemies. That hazard seems remote, especially given that people have damaged so much life on Earth without so far having provoked conscious retaliation.

Another practical concern of forward-contamination is the more plausible prospect that an exploration campaign would contaminate a planet before completing the objective of characterizing native life there. If we destroyed or modified extraterrestrial life before we could find out about it, we would fail to achieve one of the most exciting and motivating goals of planetary exploration.

Whether by forward- or back-contamination, the issue has been recognized at the international level. By treaty, space-faring nations are required to appear to take precautions. In accordance with that agreement, the U.S. government retains a Planetary Protection Officer, a very nice man named John Rummel, who certainly has the best job title on our planet.

For independent assessments of policy matters related to scientific issues, the U.S. government often turns to the National Research Council. NRC is an arm of the National Academy of Science, a federally-chartered honor society, which selects new members on the basis of whom the current members know and think deserve it. It is the most prestigious scientific honor in the U.S. When NASA contracts with the NRC for advice regarding planetary exploration, it gets the prestige of the National Academy, but it does not get its members. Instead, NRC puts together *ad hoc* committees, selected by the power-brokers of the planetary science community, which produce the reports that NASA pays for. Everyone wins. NASA gets the policy guidelines it wants, with nominally gold-plated credentials, the scientists who serve on the committees for free get to influence policy and get the prestige of having been tapped by an arm of the Academy for their supposed expertise. At the same time, the Academy does a public service, while bringing in some cash.

In 2000, NASA had recognized the excitement of the discoveries by *Galileo* about Europa, and a campaign of missions specifically targeting the satellite was being considered. The NRC was hired to assemble a Task Group on the Forward-Contamination of Europa, charged with setting the standards for protecting Europa from germs that might ride along on future spacecraft. Naturally, the party line of thick ice was well represented. The NRC panel knew for a fact that the ice was so thick that the ocean, the only place where there might be life, was isolated from the surface.

When NRC issued its report to NASA, I could hardly have been less interested. At that point, I expected, given the players, that any authoritative panel would ignore the evidence that the ice was permeable, so the report was just one more insult. Moreover, I knew that any actual landing on Europa was decades away, so that the politics driving the party line of 2000 might not be as effective by then. In the meantime, any concern I might have had about the short-term well-being of the imaginary creatures of Europa was addressed by NASA's actions, if not by its advisors. *Galileo* was going to be eliminated as a threat.

Randy Tufts saw things differently. He had joined the business of planetary science precisely because of his interest in life on Europa. Before that, his life had been devoted to political activism and to preservation of the natural world, especially of Kartchner Caverns. Nothing about Randy could let that NRC report get by. As my former student and now postdoctoral associate, he pestered me to take a careful look at the report with him.

21.5.2 Standards and risk

The NRC made a specific recommendation regarding the maximum acceptable level of risk for contaminating Europa that NASA should adopt in planning and carrying out its exploration program: 0.01%. For the basis of this number, the report cited a 1964 resolution of the Committee on Space Research (COSPAR) of the International Council of Scientific Unions. At Randy's insistence, we tracked the number back to its source.

In the early 1960s Carl Sagan and Sidney Coleman derived a quantitative requirement for the sterilization of spacecraft to be used in an anticipated campaign for the exploration of Mars. The basis of the calculation was not an ethical standard with an objective of protecting Martian life, but rather a practical requirement that these missions would characterize life there for us before significantly contaminating the planet. Sagan and Coleman chose a target probability of success of 99.9%. In other words, they considered a 0.1% probability of messing up Martian life before it had been characterized by humans to be acceptable. Ethical issues aside for the moment, these numbers were completely arbitrary.

Then, assuming conditions relevant to Mars and making educated (early 1960s) guesses about the programmatic strategy that might be used in a campaign to detect life there, they showed how to compute the spacecraft sterilization requirement. The result was an evaluation of the maximum acceptable probability, 0.01%, that a single viable organism be aboard any vehicle intended for planetary landing.

That evaluation was adopted by COSPAR. Although the result was given that impressive international imprimatur, it rested on a rather shaky foundation. It derived from a notion that planetary protection is to protect the interests of scientists, rather than alien organisms. It selected the level of acceptable risk arbitrarily. It relied on pre-Space Age understanding of the planet Mars and on crude assumptions of how well a terrestrial organism might survive the trip and colonize the new setting. Finally, the prescription was based on a 1964 guess about the specifics of a future multi-mission campaign for exploring Mars.

Whether the COSPAR policy was ever appropriate for Mars is therefore highly questionable. Randy and I were totally baffled as to why the NRC would have applied it to Europa, a completely different planet, in 2000.

While the specifics of Sagan and Coleman's analysis are irrelevant, it provides a template for setting planetary protection criteria. It demonstrates that the basis for any evaluation must be a moral or philosophical principle. The one introduced by Sagan and Coleman could be applied to Europa with appropriate calculations. But that principle is self-serving, in that it does not address the well-being of life on that planet, except that it should survive long enough to satisfy human curiosity.

At the opposite extreme would be another principle: that non-interference with life on other planets would take absolute priority. Randy and I called this concept the "Prime Directive" ideal, borrowing the term from *Star Trek*, where it originated contemporaneously with Sagan and Coleman's work and the COSPAR resolution. Actually, in *Star Trek* the "Prime Directive" usually only applied to protecting alien societies, and even then it was readily discarded as needed to advance the plot line. Yet, the phrase seems appropriate here because it conveys a certain absolutism. The problem with this principle is that, if rigorously applied, it would likely bring exploration of some of the most interesting moons and planets to a halt.

Randy and I considered whether another moral principle might provide a more rational basis for developing a planetary protection standard, that would be more objective than the self-serving principle of Sagan and Coleman, and less constraining than the absolute isolationism of the "Prime Directive". We came up with one that is objective, but not absolute. It is based on the idea that there is already a process of natural cross-contamination among planets, something so far mostly quantified in the context of terrestrial planets, which are thought to exchange chunks of crust from time to time after an asteroid hits and sends ejecta off into space at escape velocity. Living cells could conceivably survive such a journey. After all, many kinds of delicate organic molecules (including, perhaps, the very molecules that allowed life to develop here in the first place) are regularly carried to Earth within meteorites.

As long as the probability of people infecting other planets with terrestrial microbes is substantially smaller than the probability that such contamination happens naturally, exploration activities would be doing no harm. We called this concept the "Natural Contamination Standard".

For Europa this Natural Contamination Standard may seem nearly as strict and confining as the Prime Directive, because natural transport of viable organisms from Earth may be so difficult that it provides an impossibly-stringent criterion. On the other hand, it may be equally difficult for organisms from southern California or Florida, where most planetary spacecraft are built and launched, to survive a trip and prosper in the cold, icy environments of Europa. It may be a factor as mundane as that one that makes meeting a stringent standard possible. We will not know the operational requirements that devolve from any underlying standard until after careful scientific study.

The Natural Contamination Standard has considerable merit, but there may be other good candidates as well. The point is that before anyone can compute spacecraft sterilization requirements or any other operational requirements in a meaningful way, some fundamental principle, based on ethical and philosophical considerations, will be required. Only then will scientists be able to begin to translate the principle into quantitative requirements for mission design and operations.

21.5.3 Getting it right

NRC's report did not address the fundamental issues that must underlie any risk assessment, nor did it present a quantitative assessment. It certainly did not take seriously the evidence that Europa's icy crust is permeable to the ocean below. Nevertheless, at the heart of the report is a recommended specific, quantitative standard for planetary protection: "The probability of contaminating a Europan ocean with a viable terrestrial organism at any time in the future should be less than 10^{-4} [i.e., 0.01%] per mission." It cited the COSPAR resolution as the source of that number, which in turn had come from Sagan and Coleman. However, that number had not been intended to be an overall criterion for protecting a planetary ecosystem, but rather the sterilization standard for each vehicle. NRC applied the number to a completely differently-defined probability, with no rationale. That error made it all the more ironic when their report left to NASA the responsibility for computing the sterilization requirements.

More fundamentally, wherever the number came from, it had nothing to do with any considerations of Europa.

Randy and I discovered these problems in a draft of the report that had been distributed for comment. When we dutifully offered our comments they were not welcomed. The NRC staff-person in charge of the project was positively livid. The staff made it clear that further consideration was impossible, the committee had been disbanded, and NASA's payment had all been spent. There was no money left to fix the report. I cannot help but wonder why they asked for comments on the draft. The report was subsequently published by the National Academy Press.

Randy and I published a piece on this issue in *Eos*, the news and policy journal of the American Geophysical Union. The editors included with it a response by Larry Esposito, the planetary astronomer who chaired the NRC committee. Larry is a decent person, who obviously had taken on a tough job. Regarding the 0.01% figure he wrote, "the task group members reached a consensus on this value based on their collective experience and judgment ... The best justification is that it is the result of thoughtful deliberations of the task group members." Several members of the committee privately told me the same thing, that the recommended value did not really come from the COSPAR resolution that they cited, but rather it was simply a compromise among the subjective judgments of the members of the group. The reference to the COSPAR resolution was added afterward to lend an appearance of objectivity.

Although irrelevant, the 0.01% standard continues to have a life of its own. NASA has contracted with JPL to do advanced planning for future missions to Europa.[2] As part of that planning, JPL is directed (and paid) to develop strategies for meeting the standards set by the NRC report. Engineers are assigned to determine what must be done to be sure the 0.01% limit is not exceeded, even though the number has no quantitative basis and what it is supposed to specify is ill-defined.

Before NASA proceeds too much further with planning its Europa campaign, the scientific community needs to reopen the development of guidelines for preventing forward-contamination. The process can proceed on a rational and quantitative basis as outlined above, once an underlying ethical or philosophical basis is selected. That basis might be Sagan and Coleman's preserve-it-until-we-are-done-with-it, or the Prime Directive, or Tufts and Greenberg's Natural Contamination Standard, or some other principle that nucleates a consensus. Once such a principle is in place, quantitative standards for mission design, construction, and operations can be developed by straightforward scientific analysis.

Another important reason for continuing discussion is that knowledge of Europa has grown since the NRC report, which was issued under the influence of the party line developed during the initial quick-look at *Galileo* data. The evidence that Europa's icy crust has cracks and openings that connect with the ocean is now more widely understood and appreciated, and it has profound implications for the possibility of contamination. The Europan biosphere, if it does contain life, may well extend to within centimeters of the surface, in which case it is far more vulnerable than if it is confined to the ocean under more than 20 km of ice. Any assessment of the probability of forward-contamination, and of standards for mitigating it, should include the developing understanding of these conditions.

The likelihood that the ice is thin and permeable, with potentially-habitable niches in the crust as well as the ocean, makes Europa an even more inviting

---

[2] The mission would be carried out under NASA's *Jupiter Icy Moons Orbiter* program, discussed in Chapter 22.

target for exploration. Active, dynamic resurfacing could be observable and organisms might be available for sampling at or near the surface. On the other hand, for the same reasons, Europa would be even more vulnerable to forward-contamination than most planetary scientists might have thought a few years ago. NASA should not proceed with *in situ* exploration of the jovian system until it has grappled with these issues in a more serious and objective way than it has in the past. It seems likely that there will be plenty of time for it.

22

The exploration to come

22.1 PLANS FOR FUTURE SPACE MISSIONS

During the *Galileo* mission NASA decided that Europa was one of its highest priority objectives in planetary exploration. *Galileo* Project Scientist Torrence Johnson consistently cited our explanation of cycloids as the most convincing evidence that there really is, or has been recently, an ocean under the surface. That ocean made Europa the sexiest planet in the solar system. Of course, the thick-ice party line made it less attractive than it should have been, but there were enough plausible fantasies and speculation about life in the ocean that Europa had sufficient allure.

Mars, of course, has long had its own advocates. The case for life there is equally provocative. Mars also has the advantage of being easier to reach, because it is closer and because a spacecraft does not have to slow down as much when it gets there.

What is not widely known is that NASA has already run a successful mission specifically intended for the exploration of Europa, known as the GEM. Here is the story.

As the *Galileo* mission was approaching the end of the operational period that had been funded by Congress through 1997, the project management was getting nervous that further funding would be cut off and an expensive, functioning spacecraft would be turned off. The concern might seem silly. Why would Congress pull the plug on a $2 billion robot that was orbiting Jupiter and still taking pictures and gathering data? For a tiny percentage of the original investment, the project could double its return. Yet, such wasteful decisions had been made before. The classic example was turning off in the 1970s the functioning package of scientific instruments that had been installed on the Moon by the Apollo program and that, like *Galileo*, had been returning exciting and important data.

The project management at JPL and the planetary program people at NASA headquarters in Washington wanted to sell the idea of a *Galileo* Extended Mission to run for an additional couple of years, at least, for as long as *Galileo*'s fuel lasted and its electronics resisted deterioration in Jupiter's radiation field. In the process of making their case, the perception was that the word "Extended" would not sell well with Congress, which is not favorably disposed toward requests for more money for projects they thought had already been fully funded. So, in a brilliant move in a silly game, they renamed the extension the *Galileo Europa* Mission. The name capitalized on the new sex appeal of Europa, the extension was magically transformed into a new mission, and the cute acronym GEM was saved. Congress bought it and we got two more years of operations for GEM. The taxpayers got more than their money's worth.[1]

At the same time as the project management was playing games with names, NASA was developing an ambitious plan for future missions to explore Europa, building on what had been learned from *Galileo*. Of course, what they were told had been learned was that Europa may have an ocean, but it is probably isolated under a thick layer of convecting ice.

The exploration strategy that emerged was based on a set of three spacecraft missions. The *Europa Orbiter* would get to the satellite in the first decade of the new millennium. It would orbit around Europa (in contrast to *Galileo* which orbited Jupiter and occasionally encountered a satellite), for a month, before its electronics got fried by radiation. During that time it would survey the surface, measure gravity, and attempt to confirm the existence of the ocean and characterize the crust.

Then in the following decade, a lander would be placed on the surface of Europa to make *in situ* measurements in preparation for the third mission. The third, culminating mission in the 2020s would attack what was perceived as the crucial challenge, drilling down through the thick ice to explore the ocean below. In fact, the entire three-mission strategy was built around the assumption that thick ice was a major factor in the effort to reach the objective: the ocean with its possible life.

In preparation for dealing with the perceived crucial challenge of getting down through the ice, NASA began to fund various experimental schemes for drilling through thick ice and communicating back to the surface. Prototype robots (called "cryobots") that would drill downward or that would melt their way down through the ice were designed and tested. Communications schemes involving trailing a wire to the surface or acoustic transmissions were developed. Capabilities for navigating and observing within the ocean began to be planned for these penetrators (Figure 22.1). Their successes and failures were followed closely by the media, by NASA management, and by the scientific community. The focus on these technologies with a 30-yr lead time reflected the significance of the thick ice of Europa as a barrier to the beckoning ocean.

An under-ice lake near the south pole in Antarctica became the definitive terrestrial analog. Lake Vostok is mighty interesting itself, independent of any supposed similarity to Europa. It is one of the largest lakes in the world, over

---

[1] After GEM, *Galileo* got an additional extension, also formally defined as a distinct mission, called the *Galileo* Millennium Mission, because it carried the project forward from late 1999 into the new millennium, finally ending with the spacecraft's deliberate crash into Jupiter in the autumn of 2003.

200 km long, with a volume of liquid water similar to Lake Superior's. Lying below 4 km of solid ice, its liquid water was cut off from the Earth's atmosphere, and the rest of the biosphere, about 15 million years ago.[2] It may well contain preserved samples of the Earth's ancient atmosphere and even of life forms from those times. To those believing that Europa's ocean is isolated from the surface, Vostok seemed like a perfect analog. Its exploration would require innovative instrumentation and penetration technologies, and strategies for exploration without contamination. Development of technology for Europa began to be tied to exploration of Lake Vostok.

In fact, however, the analogy was probably irrelevant. What made Vostok so interesting is that it has been isolated from the surface of the Earth, but we have found that Europa's ocean has probably been intimately linked to its surface.

Because NASA depended on advisors who were wedded to the dogma of thick ice, they were probably addressing the wrong problem. Instead of developing mission strategies and technologies for drilling through thick ice in the distant future, they should have been considering whether they could find places where the ocean is brought up to the surface, or has been at the surface recently. Rather than worry about drilling 20 km down, they could simply scoop up samples of the ocean at the surface.

That possibility would suggest a different strategy. With rapid resurfacing, any landing site would provide oceanic samples; the trick will be to find the freshest ones. The *Europa Orbiter* should have had as a primary objective the location of the most likely sites for recent or current oceanic exposure. Then the lander mission could sample and examine the ocean right on the surface. The objective of reaching and exploring Europa's biosphere would be attained at least a decade earlier and 20 km of ice closer.

The issue became moot with the cancellation of the *Europa Orbiter* in 2002. After 3 years of planning, and preparation of instrument proposals by several competing teams, the mission cost estimates by JPL had gone so high that the project was deemed unaffordable.

In order to restore a return to Europa to the plan for planetary exploration missions, NASA came up with a clever strategy, and/or a cynical one, depending on how you look at it. The space agency, in collaboration with the U.S. Department of Energy, wanted to develop a nuclear electric propulsion system. The motives are a matter of debate. Having this capability would certainly facilitate outer planet exploration, removing the fuel constraints of conventional propulsion systems, like *Galileo*'s, as well as producing huge amounts of electricity to power powerful instruments for scientific exploration.

On the other hand, nuclear-powered propulsion is of considerable interest to the military. While NASA insists that the Department of Defense has "no active interest" in the work, dubbed Project Prometheus, the Air Force is working on

---

[2] Back then, Europa's surface geography would have been quite different from the way it looks today, although many features and expanses of terrain might be unchanged in that time, and the types of features and terrains were probably similar to what *Galileo* and *Voyager* imaged.

related technology and interacting with NASA on the development. When asked by the media, NASA certainly has not ruled out transferring the technology to the military.

NASA's Project Prometheus needed a presentable purpose, and at the same time the mission to Europa was too expensive for NASA's Solar System Exploration division. A marriage was probably inevitable, and the offspring is the *Jupiter Icy Moons Orbiter* (JIMO), now the flagship mission of Project Prometheus and its new, nominal driving purpose. The deal is not fooling everyone. As a reporter for a British newspaper noted, "Nuclear reactors in space would be critical to any robust space-based weapons system. To what degree", he asked me, "am I being paranoid to see the JIMO as a cloak for development of this technology?" I would not know. However, planetary scientists who advise NASA generally accept the deal as the best they could get under the circumstances. A powerful technology will be developed for a planetary mission, where otherwise there might not have been any mission at all for the foreseeable future.

The negative side is that any return to Europa will have to wait until after this ambitious system is designed, built, tested, and perfected. Moreover, electric propulsion, where charged particles are ejected from the spacecraft instead of exploding gases, takes a long time to accelerate and travel times may be long. The earliest return to Europa may not be for 20 years or more. Another wild card is the environmental issue. Environmentalists, at least a specialized fringe, protested the launching of the nuclear-powered electric generators on *Galileo* and the *Cassini* spacecraft to Saturn. Those generators were tiny compared with what would be launched by Project Prometheus.

Project Prometheus may be our best hope for getting back to Europa with the U.S. space agency, but when it will happen is uncertain. Like Prometheus of myth, you never know whose liver the eagle will eat.

22.2 LOOK IN THE ICE

Rather than obsessing on reaching the ocean, planners for Europa exploration should take advantage of the accessibility of oceanic materials at the surface and throughout the icy crust. Everything on Europa's surface and within its ice was brought up from the ocean at some time. If we were to land at a random place, we could hardly go wrong, because anything we sample came up from the ocean relatively recently. Even better, if we bear in mind the geological processes that bring oceanic materials to the surface, process them through the crust, and then cycle them back into the ocean, we can develop strategies for more meaningfully sampling the depths, still without going far below the surface.

Chaos, which has resurfaced nearly half of Europa during the past 50 million years, replaces a section of crust with frozen ocean. The freshest chaos probably includes a representative sample of fairly recent oceanic materials, including any biological detritus or frozen organisms, if they exist. If changes in the ocean have occurred over millions of years, they would be revealed by comparison of the contents of the crust in patches of chaotic terrain of various relative ages, which can be determined by the degree of degradation of the terrain.

Ridges are probably made of oceanic slush squeezed out of cracks and spreading a few hundred meters across the top of the adjacent crust. The process of freezing in a crack may purify the water somewhat, leaving behind some of the salts or other impurities. Ridges do tend to be bright relative to chaotic terrain or to terrains that evidently have had the impurities concentrated by warming and sublimation at the surface (e.g., ridge complexes are brighter than the adjacent dark margins that form "triple-bands"). Also, densely-ridged terrain has covered the surface with relatively bright ice. Thus, ridge material may not represent the composition of dissolved oceanic components as well as chaos does. Nevertheless, the physical extrusion process would likely include a good representation of biological materials in the ridge material. The chances of a spacecraft landing on a ridge are comparable with those of landing on chaos. If it does, it will be able to sample that stuff.

New surface, made of fresh oceanic material, is also created as the gaps open up during the dilation of cracks, creating dilational bands. Within a given band, samples could be taken that represent a time sequence of emplacement. Organisms trapped near the edge would represent the first creatures to take advantage of the young, active crack, while those in the middle would represent the bugs that were present just before the opening sealed closed. In this way, a time sequence of ecological change could be preserved near the surface.

There might well be variation in the biological record from place to place, reflecting the time or place that the material was brought up. Even within a given ridge, there might be the possibility of detecting ecological change. Near the top, organisms would have come from the time when the ridge was about to freeze shut. A few tens of meters below, they would represent life in a newly-active crack.

Rather than worrying about the daunting, perhaps impossible, task of drilling down tens of kilometers through solid ice, we should be preparing to sample just below the surface and perhaps 100 m farther down.

The same permeability of the ice crust that offers the possibility of life on Europa makes that life accessible to us. The ocean is linked to the surface and frequently exposed. Its materials are placed on and in the crust. As ridges build on top, the lower layers of ice are forced down until they melt back into the ocean, releasing their contents. Occasionally, areas of crust melt through, releasing substances and organisms back into the ocean, followed shortly by refreezing of part of the ocean as a new patch in the crust.

Unlike Mars, where scientists must select candidate landing sites based on the probability that water was once near the surface, on Europa any landing site will do. All the material on the surface was at one time in the not-too-distant past in the ocean. Moreover, every square kilometer on the surface currently has liquid water only a few kilometers down beneath it. Some landing sites would be better than others, with more recent oceanic material, or even with active exposure, but any place would be fine. If there is life on Europa, it may be difficult to overlook it.

Even the special hazards of Europa's surface could be readily avoided. The rugged terrain, especially in interesting places like chaos, would make a soft landing difficult. (For that reason, on Mars NASA has always looked for flat, boring landing sites.) Then, any conventional electronics that survive the landing would be fried by intense radiation at the surface within a few days. But Greg Hoppa has pointed out that an easy way to get around these problems is to use a penetrator, a hardened projectile designed to ram into the ice upon impact. You could land wherever you want, and the scientific instrumentation, sampling tools, analytical devices, and avionics would be buried, safe from the nasty radiation. In fact, down past the top few centimeters, there might be a chance of finding organisms or oceanic substances undamaged by the radiation. "Use a periscope to put up your antenna and to look around the surface," Greg suggests.

22.3 MOTHBALLED DATA

As the likelihood that Europa's ice is thin and permeable has become increasingly apparent, politically-astute planetary scientists have begun to modify their positions. It is no longer categorically repeated that observational evidence shows the ice is thick and convecting. Instead, the line is that the ice might be thick or thin, but we will not know until spacecraft return to Europa. That line is very clever and effective. It recognizes the strong evidence that the ice is permeable, while avoiding acknowledging that the thick-ice paradigm never had an observational basis. Most importantly, it sets up a scientific framework and rationale for advocating future missions.

However, the emphasis on the need for future spacecraft missions disregards the huge amount of information that we already have in hand from *Galileo*. In fact, more generally, there has always been a tendency for most of the attention to spacecraft data, especially images, to be paid during a short period after they are returned to Earth. Quick interpretations are made based on a small portion of the data. After that, the crowd moves on to the latest and greatest images from the next big mission.

We have seen the pitfalls of accepting quick-look interpretations. But, it is also important to understand that when a spacecraft returns tens or hundreds of thousands of images, most of that data is stored away and never studied in detail or at all. Occasionally, a graduate student or other persistent scholar might revisit old data as part of a research project, but there is relatively little funding or motivation for digging through that material. And the older the data get, the more difficult it is to retrieve and understand. Instead, the big money and attention are lavished on the fresh new images from the latest mission. It is no wonder that the politically-powerful figures in planetary science connect themselves to missions, or that their superficial pronouncements become scientific dogma.

The fact is we might not need to wait for future missions to determine the true character of Europa's crust. Already, a careful assessment of the images from *Galileo* has shown a completely different picture than the convective, thick ice promoted by the Imaging Team based on an initial qualitative look at a few early pictures. Many images and features remain to be studied in detail, and quantitative studies could go far beyond what we have done so far.

Emphasizing the need for future missions, likely decades in the future, to resolve these issues may be a disservice. Increased support and encouragement for detailed engagement with the data already in hand will certainly tell us a great deal more than

Sec. 22.4] Weird features: The exceptions that hold the keys 343

we already know, and may well resolve the scientific issues as they have been framed by the political process.[3]

There will certainly be plenty of time to do it right. It may be decades before the necessary money or the technology, from Prometheus or whatever, will be available. In the meantime, a careful program of scientific analysis of the expensive and hard-earned data already in hand will allow us to make best use of the resources for future exploration. If we confirm that we might put a lander at just the right place next to an active crack so that within a few hours fresh sea water will slosh to the surface, or that frozen sea life are spread over the surface by the variety of ways that oceanic water reaches the surface, then we would design very different mission strategies than if we believe that everything interesting is sealed off more than 20 km down.

One way to get a feeling for how much remains in the *Galileo* images that we do not yet understand, and the potential for learning fundamental things about Europa even before follow-on missions, is to consider a variety of features that remain puzzling, mysterious, or not yet fully exploited. Here are a few of my personal favorites.

22.4 WEIRD FEATURES: THE EXCEPTIONS THAT HOLD THE KEYS

22.4.1 The many-legged spider of Manannán

A set of very-high-resolution images aligned across Manannán Crater show its interior to look exactly like chaotic terrain imaged at comparable resolution (Figure 22.2). That much is perfectly consistent with other images of Manannán and of other craters of its size, and consistent with puncture through to liquid water as discussed in Chapter 18.

But something else appears in that image that is unlike anything seen elsewhere on Europa. A set of radial dark markings radiates outward from a point near the center of Manannán. It does not seem to be associated with any pronounced topography, and it is difficult to tell whether the radial lines represent fine indentations in the surface or dark material. This spideroid is surrounded by cracks that form circles about 2 km in diameter around it.

What is this strange feature? It is anomalous, but that is no reason to neglect it. It deserves more attention. However it formed, it is an important clue and constraint regarding the character of Europa's crust.

22.4.2 Disruption in the Sickle

Within that otherwise classic example of dilation, the Sickle (Figure 11.3a) lies an interesting combination of disrupted terrain shown in Figure 22.3. Near the northern edge of the band the disruption is a patch of chaotic terrain about 7 km wide and

---

[3] Of course, there is plenty of value in new data. For example, at the end of the "*Galileo* nominal mission", just before the start of the GEM, I bet Greg Hoppa, Randy Tufts, and Paul Geissler that nothing new would be discovered during the GEM. I did pay up with a pitcher of beer when we saw the many-legged spider of Manannán, described in Section 22.4.

Figure 22.2. The interior of Manannán was imaged at very high resolution (20 m/pixel) during orbit E14, and proved to be indistinguishable from chaotic terrain even at this resolution. This image shows an 8-km-wide central portion of the 21-km-wide crater (Figure 18.7). The lumps and bumps are illuminated from the east (right), and have an appearance similar to chaotic terrain (such as that shown at similar resolution in Figure 16.3). A bizarre 24-legged spider pattern is seen just above the center in this picture. It is surrounded by concentric cracks. Note the similarity of the terrain on the floor of Manannán to chaotic terrain imaged at similar resolution (Figure 16.3). Both probably record opening of the ice and temporary exposure of the ocean.

about 15 km from north to south. The chaos has distinct cliff-like borders, typical of cases where rafts have separated from the side. Just south of the chaos, spanning the middle of the band, is a feature that has the same size, irregular shape, and orientation as the patch of chaotic terrain.

This latter feature was presented as an archetypical example of a PSD by Pappalardo. As always, it was not clear whether it was supposed to be a pit or a spot or a dome, but in fact it is none of these. Rather, it seems to be a site where the surface broke into crustal plates, or rafts, but they are not displaced and there is not much exposed matrix, if any, between them. Some of the plates seem to be tilted, as chaos rafts often are, giving the feature a slightly-raised appearance. In our survey of

Sec. 22.4] Weird features: The exceptions that hold the keys 345

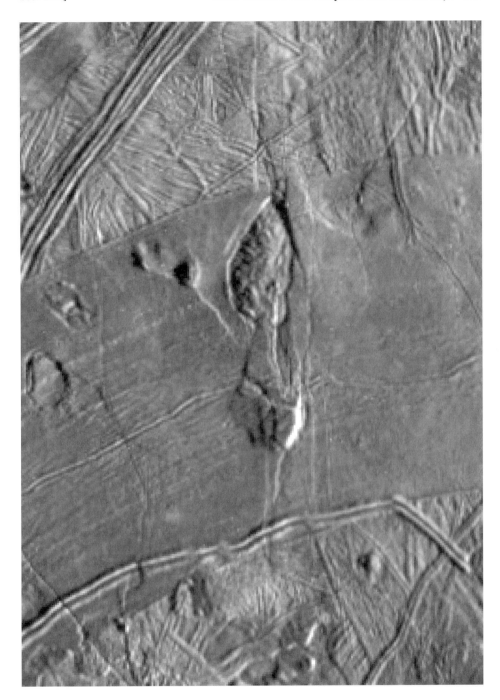

Figure 22.3. A portion of the Sickle dilation band (Figure 11.3a). Is there a genetic relationship between the types of disrupted terrain and the dilation that created this band?

pits and uplifts, we classified it as an uplift with cracks, based on that raised appearance, but the description of a few tilted rafts fits just as well.

The context of these features may be helpful. Cracking is not confined to the lower feature. Actually, cracks run through and near both the chaos area and the tilted plate feature, crossing from north of the band to south of it. And 25 km to the west, near the edge of Figure 22.3, another plate has been cut out of the surface and tilted slightly.

The disruption within the Sickle band can be consistently interpreted as examples of chaotic terrain. The northern component being fairly typical of chaos with only small rafts and lumps, while the southern part is an extreme of many rafts that have hardly moved.

Nevertheless, intriguing questions remain. Why are these features so different from one another? Did one, or the other, or both form by melt-through while the crust in the dilational band was still thin? Why do they both have the same size and shape? Is it coincidence that they displaced from one another in the north–south direction, the same as the direction of dilation in the band?

A genetic relationship seems plausible. Perhaps the juxtaposition of two similar-shaped features resulted from an interaction of the process of chaos formation with the dilation of the band in which it is sited. However it formed, understanding this feature could be a real key to understanding geological processes on Europa.

22.4.3 Short, curved double ridges within Astypalaea

The large parallelogram at the northern end of Astypalaea contains more than the regular furrows that characterize most of the pull-apart zones on the strike–slip fault (Chapter 13). Within its boundaries lie several sets of short, squiggly double-ridges, some with parallel S-shaped forms (Figure 22.4).

Randy Tufts believed that the S-shaped features in his beloved fault were caused by the strike–slip shear that dominated the displacement of the terrain on either side of Astypalaea. However, these squiggly double-ridges lie within the largest pull-apart zone along the fault, where they would not be subjected to shear. I wonder whether these areas represent blocks of older terrain that were carried into the dilating pull-apart zone, where their pre-existing cracks were worked and distorted during the dilation, while diurnal working built the double-ridges that line the cracks.

Coming to understand the process that created these distinctive features within a setting where the dominant dynamic is well understood will offer important constraints on the character of the crust.

22.4.4 Isolated tilted rafts

Tilted rafts are fairly common in chaotic terrain. It seems reasonable that in the dynamic process of oceanic exposure, blocks of ice would tip somewhat. But why are isolated rafts of ice tilted in the middle of tectonic terrain (see Figure 22.5)? These features look like rafts that were cut from the tectonic terrain that surrounds them, and tipped in place, exposing a bit of chaos matrix formed from the liquid below. We

Sec. 22.4] Weird features: The exceptions that hold the keys 347

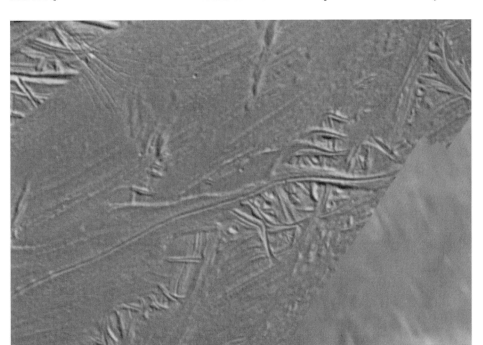

Figure 22.4. Terrain in the large parallelogram pull-apart is dominated by fine furrows typical of dilational bands, except here, where strange, short, curved, and S-shaped double-ridges are found. This image is an enlarged portion from the top of Figure 13.1. What is this stuff? Note that the distinctly fuzzier area at the lower right was imaged at much poorer resolution. The distinctly different terrain at the upper left predates the displacement (cf. Figure 13.1).

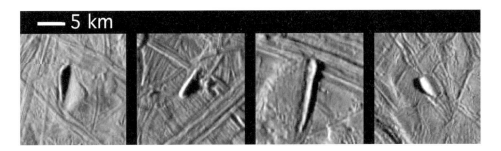

Figure 22.5. Isolated fins appear to be large, tilted rafts in small patches of chaos.

22.4.5 Horsetail of Agenor

Agenor Linea, which is probably a convergence feature, as discussed in Chapter 17, has a tail of fine arcuate double-ridges curving southward from its east end (Figure 22.6). It has been suggested that these cracks indicate that strike–slip displacement occurred along Agenor, creating tension that ran obliquely from the end of Agenor, which caused these cracks in the pattern of a horsetail curving down to the southeast.

That story has problems. First, the cracks show no sign of dilation, so if there was some strike–slip along Agenor, it was not displaced very far, assuming these cracks are actually associated with Agenor. They may not be associated with Agenor, however. They include cusps, which show they are cycloids, so the tension that created them was probably due directly to diurnal tides, not a response to the neighboring shear displacement. Moreover, a very similar set of cycloids and arcuate cracks and ridges seems unrelated structurally to Agenor.

That so many similar cycloidal cracks are located in the same place, all with similar modest, simple double-ridges, is very interesting. This cluster of cycloids may provide information about the stress history in this region, but we do not know yet what it is trying to tell us.

22.4.6 Multiple-cusp cycloids

In other places, cycloids have multiple cusps like those shown in Figure 22.7. Our working hypothesis is that, after a cycloidal ridge is formed, similar tidal stresses at some later time reactivate the old cycloids, rather than create new ones nearby. If that is the case, then this cycloidal crack was reactivated several times, each time following the trajectory of the original crack, except for a slight divergence at the cusp. Whatever the cause, this appearance is very common in cycloids. The example in Figure 22.7 is only one of many.

There are many examples of what might be crack reactivation in the tectonic record, aside from these cycloidal cases. They all have the potential to tell us a great deal about how well cracks anneal after the initial working and ridge building. The results could place important constraints on the character of the ice crust and its thermal and tectonic processing. The reactivation of cracks may also have biological implications. Organisms frozen into an inactive crack might not need to wait a million years to be released by melting if the crack simply reactivates the next time the diurnal strain is appropriately oriented.

22.4.7 Old-style bands

Geological mappers have claimed that there have been changes over time in the style of geological activity, especially in a putative tendency for chaotic terrain formation

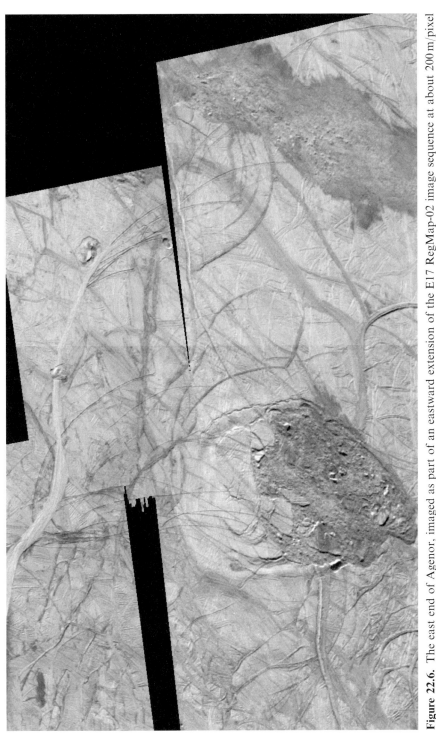

Figure 22.6. The east end of Agenor, imaged as part of an eastward extension of the E17 RegMap-02 image sequence at about 200 m/pixel shows a horsetail of double-ridges. A couple of these ridges have cusps that show they are cycloids. The others probably are as well. Similar features appear further to the west in this mosaic. For the location see the lower right corner of Figure 17.1. The dark chaoses at the south are Thrace and Thera (shown in Figure 16.18).

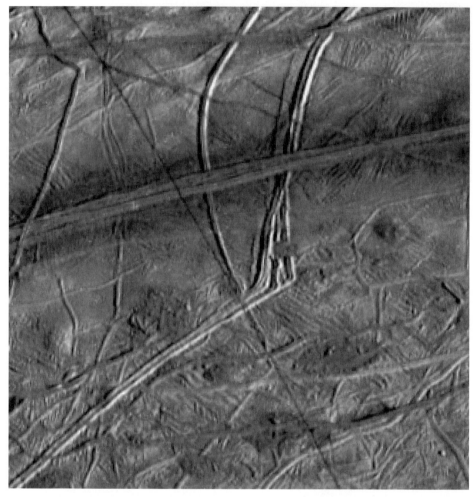

Figure 22.7. A cycloid that has been reactivated, but each new crack followed a slightly different path at the cusp. This example is in the northern hemisphere in the E15 RegMap 01. Crossing from left to right is the major "triple-band" Cadmus Linea. This figure is an enlargement of part of Figure 12.10b.

to have become more common recently. We have seen that such effects could equally well be explained by observational selection effects. Ockham's razor then favors the simpler model that things have to go on without change during the period recorded in Europa's surface geology.

However, there is one type of feature that always seems to be at the bottom of any cross-cutting tectonic sequence, and thus may indicate some type of long-term change. The oldest tectonic bands (e.g., Figure 22.8) usually have a fine-furrow structure that is somewhat different from more recent dilational bands. Rather

Sec. 22.4]   **Weird features: The exceptions that hold the keys** 351

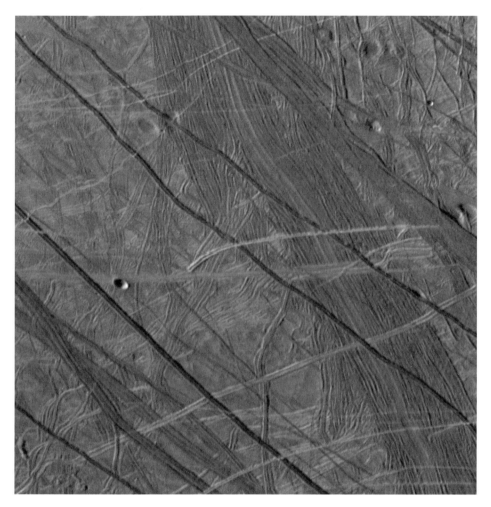

Figure 22.8. Cracks, ridges, and bands in the southern-leading hemisphere (E17 RegMap 02). This 150-km-wide area is enlarged from a portion of Figure 12.9. Lineaments here are found in sets, each with a particular azimuth. The oldest set (see Figure 12.9) has a characteristic appearance, represented here by the 50-km-wide band that runs from the top center down to the lower right. This ancient band runs almost directly north–south, but the meridian line is oblique because of the perspective of the globe from the *Galileo* camera.

than have furrows that are fairly straight and parallel, the pattern from above is more wavy and braided.

Do these ancient bands indicate a systematic change in the tectonic process and style over the past 20–50 million years? Or are we simply seeing the effects of aging of dilational bands that originally looked like the modern ones? I tend to lean toward the latter, that these are simply old dilational bands for two reasons. First, Ockham's razor impels me toward the simpler explanation, stressing the similarity to more recent dilational bands. And, second, where we have been able to reconstruct these old bands, they do seem to be dilational (e.g., Figure 11.5). These questions remain open, however, and they will be answered as we dig back more deeply and carefully in the sequential record of tectonic processing that we already have in hand.

We will also learn much more as we tie that record back into the global perspective. For example, the sets of lineaments in the southern-leading hemisphere (Figures 22.8 and 12.9), each with a distinct azimuth in the region, are actually both associated with a distinct globe-encircling set of cracks. They could be aligned, each with the next older one, by a series of global reorientations. In other words, these global lineaments probably record a succession of polar wander events, in which the entire shell of ice surrounding Europa's ocean slipped into a new orientation relative to the poles and to the direction of Jupiter.[4] This record is in pictures here on Earth, ready to be deciphered.

A return to Europa by spacecraft would be wonderful, but many fundamental issues may be resolved if we fully exploit the data already in hand, rather than mothball it and move on to the next trendy topic.

22.5 SELF-CORRECTING SCIENCE

Future missions to Europa will be needed to build on what we learn from the data that we have already obtained with great effort and expense. Before that we will need to make best use of what is already available. We will certainly have plenty of time to carefully consider what we have obtained for the *Voyager* and *Galileo* missions before proceeding with further exploration; but, unless a conscious effort is made to change the usual policy of neglect, that may not happen. Whether or not there is any future mission, it will be a disgrace not to exploit what we already have in hand.

Future spacecraft may be needed to confirm that the ice crust is permeable, but with adequate, unbiased analysis there is a good chance that the issue will be resolved long before then. Already, quantitative analysis of what has really been seen on Europa has consistently shown that the several processes for resurfacing of Europa—ridge building, dilation, and melting—are best explained by direct linkage of the ocean with the surface, and emplacement of oceanic material on the surface.

The structuring of the research agenda as a need to resolve two competing models, thick ice vs. thin ice, never should have happened. The ice may ultimately prove to be thick, but right now there is no evidence for it. The facts point to a permeable crust, while the political process has gelled around an otherwise unsupported belief in thick ice isolating the ocean deep below. The origins of these models

---
[4] The geometry is best visualized with a globe of Europa, which can be constructed from an appropriately-projected map produced by the U.S. Geological Survey. On such a globe, one can readily track the extensions of the sets of lineaments in Figure 22.8 (and Figure 12.9) and envision the sequence of global reorientations.

are rooted in completely different types of human enterprises, the political vs. the scientific. It is not surprising which one gained broadest acceptance coming out of the gate.

My former student Bill Bottke, who works on the dynamics of asteroids and comets and who has followed the Europa story from some distance, keeps reminding me that science is self-correcting. In other words, over the long term the human scientific enterprise progresses toward the truth, even if politics creates short-term divergence. Eventually, spacecraft missions will provide enough information that the true character of Europa will be undeniable. At that time, the most politically-astute will get the credit for being right all along, no matter what they said earlier. But, Bill is right: Science would eventually lead to a correct understanding of reality.

But what if humanity never gets its act together for a return to Jupiter? Space missions are not expensive relative to the U.S. budget, but in tight times they can be seen as an expensive frill. Continued exploitation of the images and other data from *Galileo* becomes all the more important, as long as it does not continue to be smothered by powerful political operators.

Science is self-correcting in the long term, but not fast enough for some of the individuals who made the greatest contributions to understanding of the true nature of Europa. By the spring of 2000, Paul Geissler had moved his focus to Io and other planets. That May, Randy, Greg, and I visited the Arctic with a Discovery Channel TV production crew for a documentary about Europa, where we finally had a first-hand look at chaotic terrain, cracks, and ridges in thin ice. The media recognized our work, but Greg and Randy were astute. Having bucked the party line, they understood that getting the science right was not going to be rewarding for them.

Realizing that his prospects in academia were poor, and needing to support his growing family, that autumn Greg Hoppa took a job in the aerospace industry, unrelated to planetary science. The field lost one of its most capable and creative young scientists.

Randy too was contemplating an alternative career. When we got back from Alaska he complained about aches in his back and legs. A product of the desert, he had repeatedly rolled his snowmobile during our expedition on the Arctic ice (several times on top of me or Greg), but the pain was more serious. He was diagnosed with a bone marrow disorder. Randy Tufts died on April 1, 2002.

Science is self-correcting, but science is a human construct. Self-correction is too slow to correct damaged careers; self-correction is too slow to correct the flow of too much money to teams that do notoriously bad science; self-correction is too slow to prevent hustlers from recognizing opportunities to exploit the system; self-correction is too slow to prevent talented young researchers from concluding that planetary science does not reward careful high-quality work, or keep them from moving on to more rewarding pursuits.

If we are complacent and wait too long for science to self-correct, there is a real danger that permanent harm will come to the social and intellectual institutions that carry on the tradition of the 16th-century Galileo, making it increasingly difficult to resist canonical control. Then, only in the very long run, if exploration continues, will humans find out what Europa is like. In the meantime, Europa abides.

The broader scientific community may still be waiting to understand Europa; but, for my small research group, the last few years of the 20th century were the most exciting and enjoyable that we could ever imagine. Disappointed as we were by the antagonistic reception and attempted marginalization of our work, we will never regret doing what we did. The *Galileo* space mission, an amazing technological feat, was essential to our success: *Galileo* obtained fantastic images of Europa; its technical failures reduced the number of images to a size that even a small, independent group could handle; and our political marginalization ironically gave us a strong advantage: Our tiny group was free to follow the evidence where it led us, without compromise or accommodation to the social and political constraints that governed so many others. Certainly, future research and new evidence will test and pick at what we have developed, refining our understanding, and perhaps reversing some of our results. But we had the chance to synthesize the information from an amazing set of images within a strong, quantitative theoretical context, and to understand the workings of a planet that we came to love. The real reward in science is the personal satisfaction of discovery, and we hit the jackpot. For us, we had discovered a new world.

> I believe that [good philosophers] fly, and that they fly alone, like eagles, and not in flocks like starlings. It is true that because eagles are rare birds they are little seen and less heard, while birds that fly like starlings fill the sky with shrieks and cries, and wherever they settle befoul the earth beneath them ... To put aside hints and speak plainly, dealing with science as a method of demonstration and reasoning capable of human pursuit, I hold that the more this partakes of perfection the smaller the number of propositions it will promise to teach, and fewer yet it will conclusively prove. Consequently the more perfect it is the less attractive it will be, and the fewer its followers.
>
> Galileo, 1632

References

REFERENCES TO CHAPTER 1

Support for the depiction of Europa in Chapter 1 is laid out in the chapters that follow. The picture of life in a tidal niche is developed in Chapter 21. The evidence that such an environment might exist is developed in Part Three, based on the theory of tides developed in Part Two. But first, Part One lays the groundwork, describing the general appearance of the satellite (Chapter 2), and the scientific process as played out in the real world of human intellect, ambition, arrogance, and politics (Chapter 3).

A short version of our Europa story appears in Greenberg (2002), and an overview in a more scholarly format appears in Greenberg and Geissler (2002), Greenberg et al. (2002), or Greenberg et al. (2001). A pioneering (and prescient) post-*Voyager* work on life on Europa was Reynolds et al. (1983). My group first discussed the implications for life based on evidence for permeable ice and put together the case in Greenberg et al. (2000).

This book is about Europa, not about astrobiology in general. If life is discovered on Europa, other speculation will be eclipsed, but in the meantime Grinspoon (2003) is a great overview of the issues involved. Another perspective comes from Ward and Brownlee (2000) who argue that life may be rare in the universe, making us more special. But their standards for habitability may be overly constrained. Europa would not be habitable by their standards (too far from the Sun), but they forgot about the conditions around giant planets that can yield a moon like Europa. Ironically, they use an image of Conamara Chaos on Europa as their frontispiece.

Whether there is life there or not, Europa is exciting because so much has happened so quickly and so recently. The main constraint on the age comes from the paucity of craters (Chapter 18), for which Zahnle et al. (2003) is the current definitive reference.

356 References to Chapter 2

Greenberg, R. (2002) Tides and the biosphere of Europa. *American Scientist*, **90**(1), 48–55.
Greenberg, R. and Geissler, P. (2002) Europa's dynamic icy crust: An invited review. *Meteoritics and Planetary Science*, **37**, 1685–1711.
Greenberg, R., Geissler, P., Tufts, B.R., and Hoppa, G.V. (2000) Habitability of Europa's crust. *Journal of Geophysical Research-Planets*, **105**(E7), 17551–17561.
Greenberg, R., Tufts, B.R., Geissler, P., and Hoppa, G.V. (2001) Europa's crust and ocean: How tides create a potentially habitable physical setting. In: *Astrobiology* (a volume of the series Lecture Notes in Physics), pp. 111–124, Springer Verlag, New York.
Greenberg, R., Geissler, P., Hoppa, G.V., and Tufts, B.R. (2002) Tidal tectonic processes and their implications for the character of Europa's icy crust (invited review). *Review of Geophysics*, **40**(2, 1), 1–33.
Grinspoon, D. (2003) *Lonely Planets*. HarperCollins, New York.
Reynolds, R.T., Squyres, S.W., Colburn, D.S., and McKay, C.P. (1983) On the habitability of Europa. *Icarus*, **56**, 246–254.
Ward, P.D. and Brownlee, D. (2000) *Rare Earth*. Copernicus Books, New York.
Zahnle, K.L., Schenk, P., Levison, H., and Dones, L. (2003) Cratering rates in the outer solar system. *Icarus*, **163**, 263–289.

REFERENCES TO CHAPTER 2

Telescopic indications that the surface is water ice were pioneered by Kuiper (1957) and confirmed by Pilcher et al. (1972), and the gravity analysis that demonstrated the thick outer layer with the density of water was by Anderson et al. (1998). Theoretical modeling of the formation of the Galilean satellites in a gaseous nebula around the very young Jupiter, which showed why plenty of water should be expected, was by Consolmagno and Lewis (1976). For interesting consideration of what lies below the water see Ransford et al. (1981) and Ghail (1997, 1998).

Galileo's drawings are from his manuscript for *Sidereus Nuncius*, reprinted in 1892.

For a description of the *Galileo* mission see Harland (2000) or Fischer (2001). Harland also includes a summary of nomenclature of features. The definitive description of the surface of Europa from *Voyager* images is Lucchitta and Soderblom (1982), one of many chapters in Morrison (1982) that cover the post-*Voyager* state of knowledge of the Galilean satellites.

Misuse of various terms that were originally defined in the context of low-resolution images is very common. The egregious and deliberate abuse of *lenticulae* is discussed in Chapter 19. Continued use of the low-resolution descriptive term *bright plains*, even after their character as densely-ridged terrain was revealed, has been misleading, especially when the expression morphs into background plains (e.g., Prockter et al., 1999 and Figueredo and Greeley, 2000) to bolster unsupported chronologies of changing resurfacing processes (Greeley et al., 2000).

Ideas about the composition of the dark material on the surface are discussed in Chapter 21, and in Geissler et al. (1998). The latter also discusses how surface colors are affected by ice grain size as well as composition.

Anderson, J.D. et al. (1998) Europa's differentiated internal structure. *Science*, **281**, 2019–2022.

Consolmagno, G.L. and Lewis, J.S. (1976) Structural and thermal models of icy Galilean satellites. In: T. Gehrels (ed.), *Jupiter* (pp. 1035–1051). University of Arizona Press, Tucson, AZ.

Figueredo, P.H. and Greeley, R. (2000) Geologic mapping of the northern leading hemisphere of Europa. *Journal of Geophysical Research-Planets*, **105**, 22629–22646.

Fischer, D. (2001) *Mission Jupiter*. Copernicus, New York.

Geissler, P. et al. (1998) Evolution of lineaments on Europa: Clues from Galileo multispectral imaging observations. *Icarus*, **135**, 107–126.

Galilei, G. ([1610], 1892) *Sidereus Nuncius*. In: *Le opere di Galileo Galilei* (Edizione Nazionale, Vol. III, Part 1). Tipografia Barbera, Firenze, Italy [in Italian].

Ghail, R.C. (1997) The composition and internal structure of Europa. *Lunar and Planetary Science Conference*, **28**, #407.

Ghail, R.C. (1998) Ice crust thickness and internal composition of Europa. *Lunar and Planetary Science Conference*, **29**, #1766.

Greeley, R. et al. (2000) Geologic mapping of Europa. *Journal of Geophysical Research-Planets*, **105**, 22559–22578.

Harland, D. (2000) *Jupiter Odyssey*. Springer-Praxis, Chichester, UK.

Kuiper, G.P. (1957) Infrared observations of planets and satellites. *Astronomical Journal*, **62**, 245.

Lucchitta, B.K. and Soderblom, L.A. (1982) The geology of Europa. In: D. Morrison (ed.), *The Satellites of Jupiter* (pp. 521–555) University of Arizona Press, Tucson, AZ.

Morrison, D. (ed.) (1982) *The Satellites of Jupiter*. University of Arizona Press, Tucson, AZ.

Pilcher, C.B., Ridgeway, S.T. and McCord, T.B. (1972) Galilean satellites: Identification of water frost. *Science*, **178**, 1087–1089.

Prockter, L.M. et al. (1999) Europa: Stratigraphy and geological history of the anti-Jovian region. *Journal of Geophysical Research-Planets*, **104**, 16531–16540.

Ransford, G.A., Finnerty, A.A., and Collerson, K.D. (1981) Is Europa's surface cracking due to thermal evolution? *Nature*, **289**, 21–24.

REFERENCES TO CHAPTER 3

The exquisitely well-timed paper that revolutionized our expectations regarding the Galilean satellites days before we got to see them was Peale et al. (1979), and this discovery is described in more detail in Chapter 4. Upper limits on internal heating rates in Europa, and implications for conducting ice thickness are discussed in O'Brien et al. (2002). The specific requirements on grain size in order to have convection were noted by McKinnon (1999), one reason that convection is theoretically unlikely (Spohn and Schubert 2001). Post-*Voyager* consideration of the possibility that the ice crust is permeable, linking the ocean to the surface, was championed by Reynolds et al. (1983) and Squyres et al. (1983). However, when evidence was presented that plates of surface ice are highly mobile by Schenk and McKinnon (1989), publication was blocked for years. More conventional perspectives on the history of the *Galileo* project than presented in this book appear in the books by Harland (2000) and by Fischer (2001). The story of the discovery of volcanism on Io is described in more detail in Lopes and Gregg (2004). The history of the Copernican revolution and Galileo's work is described by Koestler (1959), Reston (1994), and Sobel (1999). The opening Galileo quotation is from his *Dialogue concerning the two chief world systems*, quoted in Drake (1970).

Drake, S. (1970) *Galileo Studies* (p. 71). University of Michigan Press, Ann Arbor, MI.
Fischer, D. (2001) *Mission Jupiter*. Copernicus, New York.
Harland, D. (2000) *Jupiter Odyssey*. Springer-Praxis, Chichester, UK.
Koestler, A. (1959) *The Sleepwalkers*. Macmillan, New York.
Lopes, R. and Gregg, T.K. (2004) *Planetary Volcanism*. Springer-Praxis, Chichester, UK.
McKinnon, W.B. (1999) Convective instability in Europa's floating ice shell. *Geophysical Research Letters*, **26**, 951.
Peale, S.J., Cassen, P., and Reynolds, R.T. (1979) Melting of Io by tidal dissipation. *Science*, **203**, 892–894.
O'Brien, D.P., Geissler, P., and Greenberg, R. (2002) A melt-through model for chaos formation on Europa. *Icarus*, **156**, 152–161.
Reston, J., Jr. (1994) *Galileo: A Life*. HarperCollins, New York.
Reynolds, R.T., Squyres, S.W., Colburn, D.S., and McKay, C.P. (1983) On the habitability of Europa. *Icarus*, **56**, 246–254.
Spohn, T. and Schubert, G. (2001) Internal oceans of the Galilean satellites of Jupiter (Abstract). *Jupiter Conference, Boulder, CO*.
Squyres, S.W., Reynolds, R.T., Cassen, P., and Peale, S.J. (1983) Liquid water and active resurfacing on Europa. *Nature*, **301**, 225–226.
Schenk, P. and McKinnon, W.B. (1989) Fault offsets and lateral crustal movement on Europa: Evidence for a mobile ice shell. *Icarus*, **79**, 75–100.
Sobel, D. (1999) *Galileo's Daughter*. Walker, New York.

REFERENCES TO CHAPTER 4

Many of the effects of tides mentioned in this chapter are described in more detail in Chapter 5 (effects on rotation), Chapter 6 (effects on stress in the crust), Chapter 7 (tidal heating) and Chapter 8 (effects on long-term orbital evolution). The classical expression for tidal deformation is developed in many geophysics textbooks (e.g., Hubbard, 1984), but the original source is the work of Love (e.g., 1911, 1927). A good introduction to tides and their effects is Burns (1977). Tables that give the wrong values for orbital eccentricities of the Galilean satellites are common. The ones mentioned in the text are in Morrison et al. (1977) and Morrison (1982), despite my simultaneous reviews describing the correct eccentricity (e.g., Greenberg, 1977 and 1982). I discussed the effect of eccentricity damping in Greenberg (1978). The constraints on the interior density layering are from (Anderson et al., 1998). The height of the tide for various assumptions about interior structure is shown in Moore and Schubert (2000) and in Yoder and Sjogren (1996), although details of the calculation method are lacking. The Laplace resonance is described in Laplace (1805) and more recently in Greenberg (1982). It is described in the context of other orbital resonances among satellites in Greenberg (1977). The paper that revolutionized thinking about the heating of the Galilean satellites was Peale et al. (1979).

Anderson, J.D. et al. (1998) Europa's differentiated internal structure. *Science*, **281**, 2019–2022.
Burns, J.A. (1977) Orbital evolution. In: J.A. Burns (ed.), *Planetary Satellites* (pp. 113–156). University of Arizona Press, Tucson, AZ.
Greenberg, R. (1977) Orbit-orbit resonances among natural satellites. In: J.A. Burns (ed.), *Planetary Satellites* (p. 157). University of Arizona Press, Tucson, AZ.
Greenberg, R. (1978) Orbital resonance in a dissipative medium. *Icarus*, **33**, 62–73.
Greenberg, R. (1982) Orbital evolution of the Galilean satellites. In: D. Morrison (ed.), *The Satellites of Jupiter* (pp. 65–92). University of Arizona Press, Tucson, AZ.
Hubbard, W.B. (1984) *Planetary Interiors*. Van Nostrand Reinhold, New York.
Laplace, P.S. ([1805], 1966) *Mécanique Céleste 4*. Courcier, Paris. Translated by N. Bowditch and reprinted by Chelsea, New York.
Love, A.E.H. ([1911], 1967) *Some Problems of Geodynamics*. Dover, New York (reprint of Adams Prize Essay, University of Cambridge, Cambridge, UK).
Love, A.E.H. ([1927], 1944) *A Treatise on the Mathematical Theory of Elasticity*. Dover, New York (reprint of Cambridge University Press edition).
Morrison, D. (ed.) (1982) Introduction to the satellites of Jupiter. *The Satellites of Jupiter* (pp. 3–43). University of Arizona Press, Tucson, AZ.
Morrison, D., Cruikshank, D.P., and Burns, J.A. (1977) Introducing the satellites. In: J.A. Burns (ed.), *Planetary Satellites* (pp. 3–17). University of Arizona Press, Tucson, AZ.
Moore, W.B. and Schubert, G. (2000) The tidal response of Europa. *Icarus*, **147**, 317–319.
Peale, S.J., Cassen, P., and Reynolds, R.T. (1979) Melting of Io by tidal dissipation. *Science*, **203**, 892–894.
Yoder, C.F., and Sjogren, W.L. (1996) Tides on Europa (Abstract). *Europa Ocean Conference* (p. 89). San Juan Capistrano Institute, San Juan, CA.

REFERENCES TO CHAPTER 5

A description of the general effects of tides on rotation is in Peale (1977), and a clear concise mathematical description of the breakdown of tides into the components introduced by orbital eccentricity is in Jeffreys (1961). The idea that the Galilean satellites, especially Io and Europa, might not rotate synchronously was introduced by Greenberg and Weidenschilling (1984). The distinctions between the MacDonald and the Darwin tidal models are discussed in that paper. The original sources are MacDonald (1964) and Darwin (1880).

The misconception regarding the need for the shell to be decoupled from the interior for non-synchronous rotation has been practically universal and is repeated frequently by authorities. For example, it appears in the *Galileo* Imaging Team's official report to COSPAR—Belton et al. (1998). This error, among others, was also propagated by Pappalardo et al. (1999). Fischer's (2001) history of the *Galileo* project repeats this misconception about rotation. It continues to pervade the thinking of scientists, repeated recently by Kattenhorn (2002) and Zahnle et al. (2003), for example. This misconception about non-synchronous rotation also appears in a definitive report by NASA's *Europa Orbiter* Science Definition Team (1999) entitled the *State of Knowledge of Europa*. While that report was fairly accurate about the state of knowledge, including transmitting such misconceptions, it was not very accurate about the state of Europa.

Belton, M.J.S. and the *Galileo* Imaging Team (1998) *Results of the Galileo Solid State Imaging (SSI) Experiment*. Committee on Space Research of the International Council for Science Conference, Nagoya, Japan.

Darwin, G.H. (1880) On the secular change in the elements of the orbit of a satellite revolving about a tidally distorted planet. *Philosophical Transactions of the Royal Society, London*, **171**, 713–891.

Europa Orbiter Science Definition Team (1999) *State of Knowledge of Europa*. NASA, Washington, DC.

Fischer, D. (2001) *Mission Jupiter*. Copernicus, New York.

Greenberg, R., and Weidenschilling, S.J. (1984) How fast do Galilean satellites spin? *Icarus*, **58**, 186–196.

Jeffreys, H. (1961) The effect of tidal friction on eccentricity and inclination. *Monthly Notices of the Royal Astronomical Society*, **122**, 339–343.

Kattenhorn, S.A. (2002) Nonsynchronous rotation evidence and fracture history in the Bright Plains Region, Europa. *Icarus*, **157**, 490–506.

MacDonald, G.J.F. (1964) Tidal friction. *Review of Geophysics*, **2**, 467–541.

Pappalardo, R.T. et al. (1999) Does Europa have a subsurface ocean? *Journal of Geophysical Research*, **104**, 24105–24056.

Peale, S.J. (1977) Rotational histories of the natural satellites. In: J.A. Burns (ed.), *Planetary Satellites* (pp. 87–112). University of Arizona Press, Tucson, AZ.

Zahnle, K.L., Schenk, P., Levison, H., and Dones, L. (2003) Cratering rates in the outer solar system. *Icarus*, **163**, 263–289.

REFERENCES TO CHAPTER 6

The fundamental equations governing tidal stress come from Vening Meinesz (1947) and appear also in a clear and accessible form in Melosh (1980). Tidal stresses on Europa's crust were described by Helfenstein and Parmentier (1983, 1985). Our plots of the complete stress field were introduced in Greenberg et al. (1998a, b). Many of the details appear in Greg Hoppa's (1998) thesis. A recent review is Greenberg et al. (2003). Elastic parameters for ice are from Gammon et al. (1983). The tendency for failure in tension, rather than shear, is discussed in Tufts' (1998) thesis, with reference to Suppe (1985). An example of a geological interpretation that depends on the implausible notion that elastic stress could build up over many tens of degrees of rotation is the corrugation model critiqued in Chapter 17.

Gammon, P.H., Klefte, H., and Klouter, M.J. (1983) *Journal of Physics and Chemistry*, **87**, 4025.

Greenberg, R. et al. (1998a) Tectonic processes on Europa: Tidal stresses, mechanical response, and visible features. *Icarus*, **135**, 64–78.

Greenberg, R., Geissler, P., Hoppa, G., Tufts, B.R., and Durda, D. (1998b) Tidal stress on Europa: Celestial mechanics meets geology. In: D. Lazzaro et al. (eds), *Solar System Formation and Evolution* (Conference Series, Vol. 149, pp. 149–165). Astronomical Society of the Pacific, San Francisco.

Greenberg, R., Hoppa, G., Bart, G., and Hurford, T. (2003) Tidal stress patterns on Europa's crust. *Celestial Mechanics and Dynamical Astronomy*, **87**, 171–188.

Helfenstein, P. and Parmentier, E.M. (1983) Patterns of fracture and tidal stresses on Europa. *Icarus*, **53**, 415–430.

Helfenstein, P. and Parmentier, E.M. (1985) Patterns of fracture and tidal stresses due to nonsynchronous rotation: Implications for fracturing on Europa. *Icarus*, **61**, 175–184.

Hoppa, G. (1998) Europa: Effects of rotation and tides on tectonic processes. PhD thesis, University of Arizona, Tucson, AZ.

Melosh, H.J. (1980) Tectonics patterns on a tidally distorted planet. *Icarus*, **43**, 334–337.

Suppe, J. (1985) *Principles of Structural Geology*. Prentice-Hall, Englewood Cliffs, NJ.

Tufts, B.R. (1998) Lithospheric displacement features on Europa and their interpretation (288 pp.). PhD thesis, University of Arizona, Tucson, AZ.

Vening Meinesz, F.A. (1947) Shear patterns of the Earth's crust. *Transactions of the American Geophysical Union*, **28**, 1–61.

REFERENCES TO CHAPTER 7

Post-*Voyager* papers that addressed tidal heating rates and implications for the existence of an ocean and thickness of the ice include Cassen et al. (1979, 1980, 1982), Squyres et al. (1983), Ross and Schubert (1987), and Ojakangas and Stevenson (1989). More recent considerations of upper limits to the heat flow were described by Geissler et al. (2001) and O'Brien et al. (2000, 2002). Measurements of the heat flux from Europa's surface were reported in Spencer et al. (1999). Estimates of the ice thickness required for convection have been made by McKinnon (1999), Reynolds and Cassen (1979), Wang and Stevenson (2000), and Spohn and Schubert (2001).

Cassen, P., Reynolds, R.T., and Peale, S.J. (1979) Is there liquid water on Europa? *Geophysical Research Letters*, **6**, 731–734.

Cassen, P., Peale, S.J., and Reynolds, R.T. (1980) Tidal dissipation in Europa: A correction. *Geophysical Research Letters*, **7**, 987–988.

Cassen, P.M., Peale, S.J., and Reynolds, R.T. (1982) Structure and thermal evolution of the Galilean satellites. In: D. Morrison (ed.), *Satellites of Jupiter* (pp. 93–128). University of Arizona Press, Tucson, AZ.

Geissler, P.E., O'Brien, D.P., and Greenberg, R. (2001) Silicate volcanism on Europa. *Lunar and Planetary Science Conference*, **32**, #2068.

McKinnon, W.B. (1999) Convective instability in Europa's floating ice shell. *Geophysical Research Letters*, **26**, 951.

O'Brien, D.P., Geissler, P., and Greenberg, R. (2000) Tidal heat in Europa: Ice thickness and the plausibility of melt-through. *Bulletin of the American Astronomical Society*, **32**, 1066.

O'Brien, D.P., Geissler, P., and Greenberg, R. (2002) A melt-through model for chaos formation on Europa. *Icarus*, **156**, 152–161.

Ojakangas, G.W. and Stevenson, D.J. (1989) Thermal state of an ice shell on Europa. *Icarus*, **81**, 220–241.

Reynolds, R.T. and Cassen, P.M. (1979) *Geophysical Research Letters*, **6**, 121–124.

Ross, M.N. and Schubert, G. (1987) Tidal heating in an internal ocean model of Europa. *Nature*, **325**, 133–144.

Spencer, J.R., Tamppari, L.K., Martin, T.Z., and Travis, L.D. (1999) Temperatures on Europa from Galileo PPR: Nighttime thermal anomalies. *Science*, **284**, 1514–1516.

Spohn, T. and Schubert, G. (2001) Internal oceans of the Galilean satellites of Jupiter. *Jupiter Conference, Boulder, CO*.
Squyres, S.W., Reynolds, R.T., Cassen, P., and Peale, S.J. (1983) Liquid water and active resurfacing on Europa. *Nature*, **301**, 225–226.
Wang, H. and Stevenson, D.J. (2000) Convection and internal melting of Europa's ice shell. *Lunar and Planetary Science Conference*, **31**, #1293.

REFERENCES TO CHAPTER 8

Descriptions of the Laplace resonance are described in the references cited for Chapter 4. Possible scenarios for the long-term evolution of the resonance have been considered by Yoder (1979), Greenberg (1981, 1982, 1987, 1989), Ojakangas and Stevenson (1986), and Peale and Lee (2002). Observational evidence regarding long-term change in orbits is described by Greenberg (1989) and Aksnes and Franklin (2000). The formation of satellites in a circum-planetary nebula was described by Pollack et al. (1991) and more recently considered by Mosqueira and Estrada (2003). The idea of satellite formation near the end stages of the nebula is by Canup and Ward (2002). The Imaging Team's initial paper on chaos was Carr et al. (1998).

Aksnes, K., and Franklin, F.A. (2000) Io's secular acceleration derived from mutual satellite events. *Bulletin of the American Astronomical Society*, **32**, 1117.
Canup, R.M. and Ward, W.R. (2002) Formation of the Galilean satellites: Conditions of accretion. *Astronomical Journal*, **124**, 3404–3423.
Carr, M.H. et al. (1998) Evidence for a subsurface ocean on Europa. *Nature*, **391**, 363–365.
Greenberg, R. (1981) Tidal evolution of the Galilean satellites: A linearized theory. *Icarus*, **46**, 415–423.
Greenberg, R. (1982) Orbital evolution of the Galilean satellites. In: D. Morrison (ed.), *The Satellites of Jupiter* (pp. 65–92). University of Arizona Press, Tucson, AZ.
Greenberg, R. (1987) Galilean satellites: Evolutionary paths in deep resonance. *Icarus*, **70**, 334–347.
Greenberg, R. (1989) Time-varying orbits and tidal heating of the Galilean satellites. In: M.J.S. Belton, R.A. West, and J. Rahe (eds), *Time-Variable Phenomena in the Jovian System* (Special Publication #494). NASA, Washington, DC.
Mosqueira, I. and Estrada, P.R. (2004) Formation of the regular satellites of giant planets in an extended gaseous nebula, *Icarus*, **163**, 198–255.
Ojakangas, G.W. and Stevenson, D.J. (1986) Episodic volcanism of tidally heated satellites with application to Io. *Icarus*, **66**, 341–358.
Peale, S.J. and Lee, M.H. (2002) A primordial origin of the Laplace relation among the Galilean satellites. *Science*, **298**, 593–597.
Pollack, J.B., Lunine, J.I., and Tittemore, W.C. (1991) Origin of the Uranian satellites. In: J.T. Bergstralh, E.D. Miner, and M.S. Matthews (eds), *Uranus* (pp. 469–512). University of Arizona Press, Tucson, AZ.
Showman, A.P. and Malhotra, R. (1997) Tidal evolution into the Laplace resonance and the resurfacing of Ganymede. *Icarus*, **127**, 93–111.
Yoder, C.F. (1979) How tidal heating in Io drives the Galilean orbital resonance locks. *Nature*, **279**, 767–769.

REFERENCES TO CHAPTER 9

Refer back to Chapter 6 for a general discussion of stress theory. Interpretations of the lineament patterns on Europa were first described by Helfenstein and Parmentier (1983), and followed up by McEwen (1986). The tendency of ice to fail in tension, rather than in shear, is discussed in Tufts (1998). The use of the variation of crack azimuths with time to infer non-synchronous rotation was introduced by Geissler et al. (1998a, b), and refined in Greenberg et al. (1998). The issue of time sequences of lineament azimuths and their meaning is explored further in Chapter 15. The *Voyager–Galileo* comparison was described by Hoppa et al. (1999), referencing the geological map by Lucchitta and Soderblom (1982). Astronomical evidence for the near-synchronous rotation period is from Morrison and Morrison (1977), and considerations of crater distributions relevant to rotation were discussed by Shoemaker and Wolfe (1982).

Geissler, P. et al. (1998a) Evidence for non-synchronous rotation of Europa. *Nature*, **391**, 368–370.
Geissler, P. et al. (1998b). Evolution of lineaments on Europa: Clues from Galileo multi-spectral imaging observations. *Icarus*, **135**, 107–126.
Greenberg, R. et al. (1998) Tectonic processes on Europa: Tidal stresses, mechanical response, and visible features. *Icarus*, **135**, 64–78.
Helfenstein, P. and Parmentier, E.M. (1983) Patterns of fracture and tidal stresses on Europa. *Icarus*, **53**, 415–430.
Hoppa, G.V. et al. (1999) Rotation of Europa: Constraints from terminator and limb positions. *Icarus*, **137**, 341–347.
Lucchitta, B.K. and Soderblom, L.A. (1982) The geology of Europa. In: D. Morrison (ed.), *The Satellites of Jupiter* (pp. 521–555). University of Arizona Press, Tucson, AZ.
McEwen, A.S. (1986) Tidal reorientation and the fracturing of Jupiter's moon Europa. *Nature*, **321**, 49–51.
Morrison, D. and Morrison, N.D. (1977) Photometry of the Galilean satellites. In: J.A. Burns (ed.), *Planetary Satellites* (pp. 363–378). University of Arizona Press, Tucson, AZ.
Shoemaker, E.M. and Wolfe, R.A. (1982) Cratering timescales for the Galilean satellites. In: D. Morrison (ed.), *The Satellites of Jupiter* (pp. 277–339). University of Arizona Press, Tucson, AZ.
Tufts, B.R. (1998) Lithospheric displacement features on Europa and their interpretation (288 pp.). PhD thesis, University of Arizona, Tucson, AZ.

REFERENCES TO CHAPTER 10

The Arctic analog for ridge building on Europa was suggested by Pappalardo and Coon (1996). The model of ridge formation described in this chapter is from Greenberg et al. (1998). The estimate that newly-exposed liquid water would freeze to $\frac{1}{2}$ m thick in a few hours was made by Reynolds et al. (1983). The crack-slamming model was proposed by Turtle et al. (1998a, b). The argument for a solid-state process of long uniform, linear diapirs, with the misidentification of the so-called triple-ridge in Figure 10.6, was made by Head et al. (1999). The misidentification of the set of double-ridges in Figure 10.7 was by the same group (Spaun et al., 1998), with the entire *Galileo* Imaging Team listed as co-authors. Randy Tufts' work on the downwarping of crust under a ridge appears in his thesis (Tufts, 1998). Greeley's idea about dark material being sprayed out of fissures was described in Greeley et al. (1998). Fagents' idea about darkening by warming of the surface from below is in Fagents et al. (2000), and is discussed further in Chapter 16. The various ideas about how cracks might penetrate deeply to the ocean are found in Leith and McKinnon (1996), Ojakangas and Stevenson (1989), and Crawford and Stevenson (1988).

Crawford, G.D. and Stevenson, D.J. (1988) Gas-driven water volcanism and the resurfacing of Europa. *Icarus*, **73**, 66.
Fagents, S.A. et al. (2000) Cryomagmatic mechanisms for the formation of Rhadamanthys Linea, triple band margins, and other low-albedo features on Europa. *Icarus*, **144**, 54–88.
Greeley, R. et al. (1998) Europa: Initial geological observations. *Icarus*, **135**, 4–24.
Greenberg, R. et al. (1998) Tectonic processes on Europa: Tidal stresses, mechanical response, and visible features. *Icarus*, **135**, 64–78.
Head, J.W., Pappalardo, R.T and Sullivan, R.J. (1999) Europa: Morphologic characteristics of ridges and triple bands from Galileo data (E4 and E6) and assessment of a linear diapirism model. *Journal of Geophysical Research*, **104**, 24223–24236.
Leith, A.C. and McKinnon, W.B. (1996) Is there evidence for polar wander on Europa? *Icarus*, **120**, 387–398.
Ojakangas, G.W. and Stevenson, D.J. (1989) Polar wander of an ice shell on Europa. *Icarus*, **81**, 242–270.
Pappalardo, R.T. and Coon, M.D. (1996) A sea-ice analog for the surface of Europa. *Lunar and Planetary Science*, **27**, 997–998.
Reynolds, R.T., Squyres, S.W., Colburn, D.S., and McKay, C.P. (1983) On the habitability of Europa. *Icarus*, **56**, 246–254.
Spaun, N.A., Head, J.W., Pappalardo, R.T., and the *Galileo* Imaging Team (1998) Geologic history, surface morphology and deformation sequence in an area near Conamara Chaos. *Lunar and Planetary Science*, **29**, #1899.
Tufts, B.R. (1998) Lithospheric displacement features on Europa and their interpretation (288 pp.). PhD thesis, University of Arizona, Tucson, AZ.
Turtle, E.P., Melosh, H.J., and Phillips, C.B. (1998a) Europan ridges: Tectonic response to dike intrusion. *Eos, Transactions AGU*, **79**, S203.
Turtle, E.P., Melosh, H.J., and Phillips, C.B. (1998b) Tectonic modeling of the formation of Europan ridges. *Eos, Transactions AGU*, **79**, F541.

REFERENCES TO CHAPTER 11

Dilation was included in the displacement discovered by Schenk and McKinnon (1989), which met unjustified opposition to its publication. Numerous dilation bands were identified and documented in Tufts' thesis (1998). A systematic classification and interpretation was presented in Tufts et al. (2000).

Schenk, P. and McKinnon, W.B. (1989) Fault offsets and lateral crustal movement on Europa: Evidence for a mobile ice shell. *Icarus*, **79**, 75–100.

Tufts, B.R. (1998) Lithospheric displacement features on Europa and their interpretation (288 pp.). PhD thesis, University of Arizona, Tucson, AZ.

Tufts, B.R., Greenberg, R., Hoppa, G.V., and Geissler, P. (2000) Lithospheric dilation on Europa. *Icarus*, **146**, 75–97.

REFERENCES TO CHAPTER 12

The first strike–slip description was in Schenk and McKinnon (1989). The discovery that Astypalaea is a strike–slip fault was in Tufts et al. (1999). The tidal-walking theory was developed by Hoppa et al. (1999). Surveys of strike–slip were described in Hoppa et al. (2000) and Sarid et al. (2002). The ease of polar wandering was emphasized by Goldreich and Toomre (1969). A way to drive polar wandering by variations in ice thickness was proposed by Ojakangas and Stevenson (1989). An effort to find evidence in *Voyager* data for polar wander was made by Leith and McKinnon (1996).

Goldreich, P. and Toomre, A. (1969) Some remarks on polar wandering. *Journal of Geophysical Research*, **74**, 2555–2567.

Hoppa, G.V., Tufts, B.R., Greenberg, R., and Geissler, P. (1999) Strike-slip faults on Europa: Global shear patterns driven by tidal stress. *Icarus*, **141**, 287–298.

Hoppa, G.V. et al. (2000) Distribution of strike-slip faults on Europa. *Journal of Geophysical Research-Planets*, **105**(E9), 22617–22627.

Leith, A.C. and McKinnon, W.B. (1996) Is there evidence for polar wander on Europa? *Icarus*, **120**, 387–398.

Ojakangas, G.W. and Stevenson, D.J. (1989) Polar wander of an ice shell on Europa. *Icarus*, **81**, 242–270.

Sarid, A.R., Greenberg, R., Hoppa, G.V., Hurford, T.A., Tufts, B.R., and Geissler, P. (2002) Polar wander and surface convergence on Europa: Evidence from a survey of strike-slip displacement. *Icarus*, **158**, 24–41.

Schenk, P. and McKinnon, W.B. (1989) Fault offsets and lateral crustal movement on Europa: Evidence for a mobile ice shell. *Icarus*, **79**, 75–100.

Tufts, B.R., Greenberg, R., Hoppa, G.V., and Geissler, P. (1999) Astypalaea Linea: A San Andreas-sized strike–slip fault on Europa. *Icarus*, **141**, 53–64.

REFERENCES TO CHAPTER 13

Randy Tufts was at work on a detailed report on the high-resolution images of Astypalaea Linea ("The Fault") at the time of his death in April 2002. He described the major features in several conference presentations, including Tufts et al. (1999). The corrugations in part of the pull-apart zone were described by Prockter and Pappalardo (2000), critiqued in Chapter 17.

Prockter, L.M. and Pappalardo, R.T. (2000) Folds on Europa: Implications for crustal cycling and accommodation of extension. *Science*, **289**, 941–943.

Tufts, B.R., Greenberg, R., Hoppa, G., and Geissler, P. (1999) Strike-slip on Europa: Galileo views of Astypalaea Linea (Extended abstract). *Lunar and Planetary Science Conference*, **30**, #1902.

REFERENCES TO CHAPTER 14

Cycloids and their implications for a liquid water ocean are discussed in Hoppa et al. (1999) and in Wilford (1999). *Galileo* Project Scientist Torrence Johnson cites cycloids as the best evidence for an ocean. Data from the *Galileo* magnetometer also fits a model of induced currents in a layer of salty water still liquid during the *Galileo* mission (Khurana et al., 1998).

Hoppa, G.V., Tufts, B.R., Greenberg, R., and Geissler, P.E. (1999) Formation of cycloidal features on Europa. *Science*, **285**, 1899–1902.
Khurana, K.K. et al. (1998) Induced magnetic fields as evidence for sub-surface oceans in Europa and Callisto. *Nature*, **395**, 777–780.
Wilford, J.N. (1999) Scientists point to new evidence of liquid water on a Jupiter moon. *New York Times*, September 17.

REFERENCES TO CHAPTER 15

The simple earlier interpretation of the azimuthal sequence was first called into question by Hoppa et al. (2001). Nevertheless, it was used as the basis of geological conclusions by Kattenhorn (2002) and Figueredo and Greeley (2000). The survey of azimuthal sequences that forms the basis for this chapter is in Sarid et al. (2004) and unpublished work in progress.

Figueredo, P.H. and Greeley, R. (2000) Geologic mapping of the northern leading hemisphere of Europa. *Journal of Geophysical Research-Planets*, **105**, 22629–22646.
Hoppa, G.V. et al. (2001) Europa's rate of rotation derived from the tectonic sequence in the Astypalaea region. *Icarus*, **153**, 208–213.
Kattenhorn, S.A. (2002) Nonsynchronous rotation evidence and fracture history in the Bright Plains Region, Europa. *Icarus*, **157**, 490–506.
Sarid, A. et al. (2004) Crack azimuths on Europa: Time sequence in the southern leading face. *Icarus*, **168**, 144–157.

REFERENCES TO CHAPTER 16

My research group first laid out our ideas about chaotic terrain in Greenberg et al. (1999), from which much of this chapter derives. This 1999 paper has been a bestseller: Of the top 20 papers in numbers downloaded from the *Icarus* website in 2003, it was third most popular, the only one about Europa, and the only one that was not published in 2003. The Imaging Team's paper on Conamara was originally conceived as reporting evidence of direct oceanic exposure, but, by the time the forces of thick ice were done with the paper, it was much more vague, despite its title (Carr et al., 1998). The claim, based on a failure to account for observational selection, that there is evidence that chaotic terrain is recent is repeated in numerous publications (e.g., in Prockter et al., 1999). We addressed observational selection effects in Hoppa et al. (2001). Failure to take such effects into account contributed to confusion in the literature about the size distribution of small patches of chaotic terrain, as discussed further in Chapter 19. Our survey of chaotic terrain is in Riley et al. (2000). The thermal mechanism for darkening the surface is described by Fagents et al. (2000). The spurious argument against thermal melt-through was made in the team-authorized position paper (Pappalardo et al., 1999). Stevenson's argument that melt-through from below could not occur was presented in Stevenson (2000), but rebutted by O'Brien et al. (2000). The melt-through model was detailed in O'Brien et al. (2002). The concentration of internally-generated tidal heat at spots on the surface of Io is described by Veeder et al. (1994). The possibility of concentration of such surface expression of heat, even through a thick ocean, is discussed by Thomson and Delaney (2001), and the possibility of a thick clay layer was raised by Ghail (1998) and by Ransford et al. (1981). The study of the global dichotomy of the paler leading hemisphere was by Tiscareno and Geissler (2003).

Carr, M.H. et al. (1998) Evidence for a subsurface ocean on Europa. *Nature*, **391**, 363–365.

Fagents, S.A. et al. (2000) Cryomagmatic mechanisms for the formation of Rhadamanthys Linea, triple band margins, and other low-albedo features on Europa. *Icarus*, **144**, 54–88.

Ghail, R.C. (1998) Ice crust thickness and internal composition of Europa. *Lunar and Planetary Science Conference*, **29**, #1766.

Greenberg, R. et al. (1999) Chaos on Europa. *Icarus*, **141**, 263–286.

Hoppa, G.V., Greenberg, R., Riley, J., and Tufts, B.R. (2001) Observational selection effects in Europa image data: Identification of chaotic terrain. *Icarus*, **151**, 181–189.

O'Brien, D.P., Geissler, P., and Greenberg, R. (2000) Tidal heat in Europa: Ice thickness and the plausibility of melt-through. *Bulletin of the American Astronomical Society*, **32**, 1066.

O'Brien, D.P., Geissler, P., and Greenberg, R. (2002) A melt-through model for chaos formation on Europa. *Icarus*, **156**, 152–161.

Pappalardo, R.T. et al. (1999) Does Europa have a subsurface ocean? *Journal of Geophysical Research*, **104**, 24015–24056.

Prockter, L.M. et al. (1999) Europa: Stratigraphy and geological history of the anti-Jovian region. *Journal of Geophysical Research-Planets*, **104**, 16531–16540.

Ransford, G.A., Finnerty, A.A., and Collerson, K.D. (1981) Is Europa's surface cracking due to thermal evolution? *Nature*, **289**, 21–24.

Riley, J. et al. (2000) Distribution of chaos on Europa. *Journal of Geophysical Research-Planets*, **105**(E9), 22599–22615.

Stevenson, D.J. (2000) Limits on the variation of thickness of Europa's ice shell. *Lunar and Planetary Science Conference*, **31**, #1506.

Thomson, R.E. and Delaney, J.R. (2001) Evidence for a weakly stratified Europan ocean sustained by seafloor heat flux. *Journal of Geophysical Research-Planets*, **106**, 12355–12365.

Tiscareno, M.S. and Geissler, P. (2003) Can redistribution of material by sputtering explain the hemispheric dichotomy of Europa? *Icarus*, **161**, 90–101.

Veeder, G.J. et al. (1994) Io's heat flow from infrared radiometry: 1983–1993. *Journal of Geophysical Research-Planets*, **99**, 17095–17162.

REFERENCES TO CHAPTER 17

The details regarding dilation are in Chapter 11, and description of pull-aparts associated with strike–slip are described in Chapters 12 and 13. The corrugations in Astypalaea were discussed in Prockter and Pappalardo (2000). The idea of chaos as a surface area sink goes back to our chaos paper (Greenberg et al., 1999). The idea that Agenor is a convergence features goes back to Schenk and McKinnon (1989), a paper that was ahead of its time and got almost everything right. The imaginative interpretation of Agenor as a strike–slip fault is by Prockter et al. (2000). The Evil Twin of Agenor is described in Greenberg (2004).

Greenberg, R. (2004) The Evil Twin of Agenor: Crustal convergence on Europa. *Icarus*, **167**, 313–319.
Greenberg, R. et al. (1999) Chaos on Europa. *Icarus*, **141**, 263–286.
Prockter, L.M. and Pappalardo, R.T. (2000) Folds on Europa: Implications for crustal cycling and accommodation of extension. *Science*, **289**, 941–943.
Prockter, L.M., Pappalardo, R.T., and Head, J.W. III (2000) Strike-slip duplexing on Jupiter's icy moon Europa. *Journal of Geophysical Research-Planets*, **105**(E4), 9483–9488.
Schenk, P. and McKinnon, W.B. (1989) Fault offsets and lateral crustal movement on Europa: Evidence for a mobile ice shell. *Icarus*, **79**, 75–100.

REFERENCES TO CHAPTER 18

The definitive work on the surface age from the crater record is Zahnle et. al. (2003). The possible implications of the Hilda asteroids are described by Brunini et al. (2003). The authorized *Galileo* Imaging Team survey of craters is reported in Moore et al. (1998, 2001). Numerical simulations of cratering are described in Turtle and Pierazzo (2001) and Turtle and Ivanov (2002). *Science* magazine's position on acceptable models of Europa was made clear by Kerr (2001). Taliesin was identified as a multi-ring crater by Lucchitta and Soderblom (1982). The interpretation of magnetometer evidence for a liquid water layer in Ganymede and Callisto is in Kivelson et al. (2002) and Khurana et al. (1998), respectively.

Brunini, A., Di Sisto, R.P., and Orellana, R.B. (2003) Cratering rate in the jovian system: The contribution from Hilda asteroids. *Icarus*, **165**, 371–378.
Kerr, R.A. (2001) Putting a lid on life on Europa. *Science*, **294**, 1258–1259.
Khurana, K.K. et al. (1998) Induced magnetic fields as evidence for sub-surface oceans in Europa and Callisto. *Nature*, **395**, 777–780.
Kivelson, M.G., Khurana, K.K., and Volwerk, M. (2002) The permanent and inductive magnetic moments of Ganymede. *Icarus*, **157**, 507–522.
Lucchitta, B.K. and Soderblom, L.A. (1982) The geology of Europa. In: D. Morrison (ed.), *The Satellites of Jupiter* (pp. 521–555) University of Arizona Press, Tucson, AZ.

Moore, J.M. et al. (1998) Large impact features on Europa: Results of the Galileo nominal mission. *Icarus*, **135**, 127–145.
Moore, J. M. et al. (2001) Impact features on Europa: Results of the Galileo Europa Mission (GEM). *Icarus*, **151**, 93–111.
Turtle, E.P. and Ivanov, B.A. (2002) Numerical simulations of impact crater excavation and collapse on Europa: Implications for ice thickness. *Lunar and Planetary Science*, **33**, #1431.
Turtle, E.P. and Pierazzo, E. (2001) Thickness of a Europan ice shell from impact crater simulations. *Science*, **294**, 1326–1328.
Schenk, P. (2002) Thickness constraints on the icy shells of the Galilean satellites from a comparison of crater shapes. *Nature*, **417**, 419–421.
Zahnle, K.L., Schenk, P., Levison, H., and Dones, L. (2003) Cratering rates in the outer solar system. *Icarus*, **163**, 263–289.

REFERENCES TO CHAPTER 19

The implications of thick ice for suffocating oceanic life were described by Gaidos et al. (1999), who had fallen for the dogma of PSDs, which was based on Pappalardo et al. (1998). Papers that purported to support the existence of such a class of features, but which used the circular logic of defining them by size and then concluding that they had a certain characteristic size, included Spaun et al. (1999) and Prockter et al. (1999). Note that the author list of Spaun et al. (1999) includes "the *Galileo* SSI Team". Although that paper's title suggests support for the spacing claims about PSDs, the topic is never covered in the paper. The "PSD literature", which propagates the unsupported generalities invented by Pappalardo et al., includes Rathbun et al. (1998), Pappalardo et al. (1999), and Greeley et al. (2000). That repetition and exaggeration has given the PSDs a life of their own. The quoted examples of the misuse of the word *lenticulae* are from Spaun et al. (1999) and Pappalardo and Head (2001). Most PSDs are actually small patches of chaotic terrain, which were surveyed by Riley et al. (2000) Our survey of the pits and uplifts is reported in Greenberg et al. (2003). The opening quotation is from Galileo's 1615 letter to Christine of Lorraine, Grand Duchess of Tuscany (Drake, 1957).

Drake, S. (1957) *Discoveries and Opinions of Galileo* (p. 200). Doubleday Anchor, Garden City, New York.
Gaidos, E.J., Nealson, K.H., and Kirschvink, J.L. (1999) Life in ice-covered oceans. *Science*, **284**, 1631–1633.
Greeley, R. et al. (2000) Geologic mapping of Europa. *Journal of Geophysical Research-Planets*, **105**, 22559–22578.
Greenberg, R., Leake, M.A., Hoppa, G.V., and Tufts, B.R. (2003) Pits and uplifts on Europa. *Icarus*, **161**, 102–126.
Pappalardo, R.T. and Head, J.W. (2001). Thick-shell model of Europa's geology. *Lunar and Planetary Science Conference*, **32**, #1866.
Pappalardo, R. T. et al. (1998) Geological evidence for solid-state convection in Europa's ice shell. *Nature*, **391**, 365–368.

Pappalardo, R.T. et al. (1999) Does Europa have a subsurface ocean? *Journal of Geophysical Research*, **104**, 24015–24056.
Prockter, L.M. et al. (1999) Europa: Stratigraphy and geological history of the anti-Jovian region. *Journal of Geophysical Research-Planets*, **104**, 16531–16540.
Rathbun, J.A., Musser, G.S. and Squyres, S.W. (1998) Ice diapirs on Europa: Implications for liquid water. *Geophysical Research Letters*, **25**, 4157–4160.
Riley, J. et al. (2000) Distribution of chaos on Europa. *Journal of Geophysical Research-Planets*, **105**(E9), 22599–22615.
Spaun, N.A., Prockter, L.M., Pappalardo, R.T., Head, J.W., Collins, G.C., Antman, A., Greeley, R., and the *Galileo* SSI Team (1999) Spatial distribution of lenticulae and chaos on Europa. *Lunar and Planetary Science Conference*, **30** (Abstract).

REFERENCES TO CHAPTER 20

The Galileo quote is from a letter of 1630, quoted from Drake (1978). Other papers discussed in this chapter are listed below.

Collins, G.C., Head, J.W., Pappalardo, R.T., and Spaun, N.A. (2000) Evaluation of models for the formation of chaotic terrain on Europa. *Journal of Geophysical Research-Planets*, **105**, 1709–1716.
Drake, S. (1978) *Galileo at Work* (p. 332). University of Chicago Press, Chicago.
Goodman, J.C., Collins, G.C., Marshall, J., and Pierrehumbert, R.T. (2004) Hydrothermal plume dynamics on Europa: Implications for chaos formation. *Journal of Geophysical Research-Planets*, **109**.
Melosh, H.J., Ekholm, A.G., Showman, A.P., and Lorenz, R.D. (2004) The temperature of Europa's subsurface water ocean. *Icarus*, **168**, 498–502.
Nimmo, F. and Gaidos, E. (2002) Strike-slip motion and double ridge formation on Europa. *Journal of Geophysical Research-Planets*, **107**.
Nimmo, F., Giese, B., and Pappalardo, R.T. (2003) Estimates of Europa's ice shell thickness from elastically-supported topography. *Geophysical Research Letters*, **30**(37), 1–4.
Ruiz, J. and Tejero, R. (2003) Heat flow, lenticulae spacing, and possibility of convection in the ice shell of Europa. *Icarus*, **162**, 362–373.
Showman, A.P. and Han, L. (2003) Numerical simulations of convection in Europa's ice shell: Implications for surface features. *Lunar and Planetary Science*, **34**, #1806.

REFERENCES TO CHAPTER 21

One of the earliest alerts to the possibility of life on (or in) Europa was by Guy Consolmagno, who tells the story in his memoir (Consolmagno, 2000). Pioneering work was described in Reynolds et al. (1983, 1987). The problems with an isolated ocean suffocating life were discussed at the beginning of Chapter 19 and in Gaidos et al. (1999). Clever schemes for supporting life without links to the surface were discussed by Chyba (2000), McCollom (1999), and Schulze-Makuch and Irwin (2002). The idea that terrestrial life may have originated in sub-oceanic volcanic vents was proposed by Baross and Hoffman (1985). Evidence for oxygen compounds at the surface is from Carlson et al. (1998, 1999), Smythe et al. (1998), and Hall et al. (1995). Organic compounds have been seen on Europa's neighbors (McCord et al., 1998). The substances likely to be found in the ocean are discussed in Oró et al. (1992) and Kargel et al. (2000). At the time, Kargel was influenced by the party line of thick ice, with an isolated ocean, but the paper is a valuable resource regarding plausible substances that might be involved. Interpretation of the near-infrared spectra of the dark orangish-brown stuff as frozen brines is in McCord et al. (1998, 2001). Alternative interpretations are in Dalton and Clark (1998) and Carlson et al. (1999). Rates of destruction of organic compounds by radiation at the surface were discussed in Varnes and Jakosky (1999), and burial rates by cratering in Phillips and Chyba (2001). For further discussion of energetic particle bombardment of Europa see Tiscareno and Geissler (2003) and Cooper et al. (2001). Sunlight adequate for photosynthesis was discussed by Reynolds et al. (1983), Chyba (2000), Lorenz and Lunine (1996), and Lunine and Lorenz (1997). Long-term survival of frozen bacteria in Antarctica was discovered by Priscu et al. (1999). The idea that life may have formed preferentially in cold settings is described in Bada et al. (2002), and in tidal settings by Lathe (2004). Transport of organisms among planets has been considered by Mileikowsky et al. (2000). My research group's ideas about the habitable settings allowed by a permeable crust are described in five papers: Greenberg et al. (2000, 2001, 2002), Greenberg (2002), and Greenberg and Geissler (2002). The books by Ward and Brownlee (2000) and by Grinspoon (2003) are recommended as overviews of thinking about astrobiology. The discussion of planetary protection is based on Greenberg and Tufts (2001a, b). The reports cited in this discussion are NRC (2000), COSPAR (1964), and Sagan and Coleman (1964).

Bada, J.L. and Lazcano, A. (2002) Some like it hot, but not the first biomolecules. *Science*, **296**, 1982–1983.
Baross, J.A. and Hoffman, S.E. (1985) Submarine hydrothermal vents and associated gradient environments as sites for the origin and evolution of life. *Origins of Life*, **15**, 327–345.
Carlson, R.W. et al. (1998) Hydrogen peroxide on Europa. *Science*, **283**, 2062–2064.
Carlson, R.W., Johnson, R.E., and Anderson, M.S. (1999) Sulfuric acid on Europa and the radiolytic sulfur cycle. *Science*, **286**, 97–99.
Chyba, C.F. (2000) Energy for microbial life on Europa. *Nature*, **403**, 381–382.
Consolmagno, G.L. (2000) *Brother Astronomer* (pp. 141–143). McGraw-Hill, New York.
Cooper, J.F. et al. (2001) Energetic ion and electron irradiation of the icy galilean satellites. *Icarus*, **149**, 133–159.
COSPAR (1964) *Resolution 26: COSPAR Position with Regard to the Florence Report of Its Consultative Group on Potentially Harmful Effects of Space Experiments* (Article 5, COSPAR Information Bulletin No. 20, p. 26). Committee on Space Research, International Council of Scientific Unions, Paris.
Dalton, J.B. and Clark, R.N. (1998) Laboratory spectra of Europa candidate materials at cryogenic temperatures. *Earth Observations from Space Transactions*, **79**, 541.
Gaidos, E.J., Nealson, K.H., and Kirschvink, J.L. (1999) Life in ice-covered oceans. *Science*, **284**, 1631–1633.
Greenberg, R. (2002) Tides and the biosphere of Europa. *American Scientist*, **90**(1), 48–55.

Greenberg, R. and Geissler, P. (2002) Europa's dynamic icy crust: An invited review. *Meteoritics and Planetary Science*, **37**, 1685–1711.

Greenberg, R. and Tufts, B.R. (2001a) Standards for prevention of biological contamination of Europa. *Earth Observations from Space Transactions*, **82**, 26–28.

Greenberg, R. and Tufts, B.R. (2001b) Infecting other worlds. *American Scientist*, **89**(4), 296–300.

Greenberg, R., Geissler, P., Tufts, B.R., and Hoppa, G.V. (2000) Habitability of Europa's crust. *Journal of Geophysical Research-Planets*, **105**(E7), 17551–17561.

Greenberg, R., Tufts, B.R., Geissler, P., and Hoppa, G.V. (2001) Europa's crust and ocean: How tides create a potentially habitable physical setting. In: *Astrobiology* (a volume of the series Lecture Notes in Physics), pp. 111–124, Springer Verlag, New York.

Greenberg, R. et al. (2002) Tidal tectonic processes and their implications for the character of Europa's icy crust (invited review). *Review of Geophysics*, **40**(2, 1), 1–33.

Grinspoon, D. (2003) *Lonely Planets*. HarperCollins, New York.

Hall, D.T. et al. (1995) Detection of an oxygen atmosphere on Jupiter's moon Europa. *Nature*, **373**, 677–679.

Kargel, J.S. et al. (2000) Europa's crust and ocean: Origin, composition, and the prospects for life. *Icarus*, **148**, 226–265.

Lathe, R. (2004) Fast tidal cycling and the origin of life. *Icarus*, **168**, 18–22.

Lorenz, R.D. and Lunine, J.I. (1996) Light and heat under the snow: Europa habitability revisited. *Abstracts of the Europa Ocean Conference* (p. 48). San Juan Capistrano Institute, San Juan, CA.

Lunine, J.I. and Lorenz, R.D. (1997) Light and heat in cracks on Europa: Implications for prebiotic synthesis. *Lunar and Planetary Science Conference*, **28**, #855–856.

McCollom, T.M. (1999) Methanogenesis as a potential source of chemical energy for primary biomass production by autotrophic organisms in hydrothermal systems on Europa. *Journal of Geophysical Research*, **104**, 30729–30742.

McCord, T.B. et al. (1998) Organics and other molecules in the surfaces of Callisto and Ganymede. *Science*, **278**, 271–275.

McCord, T.B. et al. (1998) Salts on Europa's surface detected by Galileo's near-infrared mapping spectrometer. *Science*, **280**, 1242–1245.

McCord, T.B. et al. (2001) Thermal and radiation stability of the hydrated salt minerals epsomite, mirabilite, and natron under Europa environmental conditions. *Journal of Geophysical Research*, **106**, 3311–3320.

Mileikowsky, C. et al. (2000) Natural transfer of viable organisms in space, Part 1: From Mars to Earth and Earth to Mars. *Icarus*, **145**, 391–427.

NRC (2000) *Preventing the Forward Contamination of Europa* (a report of the NRC Space Studies Board by the Task Group on the Forward Contamination of Europa, June). National Academy Press, Washington, DC.

Oró, J., Squyres, S.W., Reynolds, R.T., and Mills, T.M. (1992) Europa: Prospects for an ocean and exobiological implications. In: *Exobiology in Solar System Exploration* (SP-512, pp. 103–125). NASA, Washington, DC.

Phillips, C.B. and Chyba, C.F. (2001) Impact gardening rates on Europa. *Lunar and Planetary Science Conference*, **32**, #2111 (CD-ROM).

Priscu, J.C. et al. (1999) Geomicrobiology of subglacial ice above Lake Vostok, Antarctica. *Science*, **286**, 2141–2144.

Reynolds, R.T., Squyres, S.W., Colburn, D.S., and McKay, C.P. (1983) On the habitability of Europa. *Icarus*, **56**, 246–254.

Reynolds, R.T., McKay, C.P., and Kasting, J.F. (1987) Europa, tidally heated oceans, and habitable zones around giant planets. *COSPAR Advances in Space Research*, **7**(5), 125–132.

Sagan, C. and Coleman, S. (1965) Spacecraft sterilization standards and contamination of Mars. *Journal of Astronautics and Aeronautics*, **3**(5), 22–27.

Schulze-Makuch, D. and Irwin, L.N. (2002) Energy cycling and hypothetical organisms in Europa's ocean. *Astrobiology*, **2**, 105–121.

Smythe, W. et al. (1998) Absorption bands in the spectrum of Europa detected by the Galileo NIMS instrument. *Lunar and Planetary Science Conference*, **29**, #1532.

Tiscareno, M.S., and Geissler, P. (2003) Can redistribution of material by sputtering explain the hemispheric dichotomy of Europa? *Icarus*, **161**, 90–101.

Varnes, E.S. and Jakosky, B.M. (1999) Lifetime of organic molecules at the surface of Europa. *Lunar and Planetary Science Conference*, **30**, #1082 (CD-ROM).

Ward, P.D. and Brownlee, D. (2000) *Rare Earth*. Copernicus Books, New York.

REFERENCES TO CHAPTER 22

More respectful histories, closer to the official version, of the *Galileo* project are reported in the books by Harland (2000) and by Fischer (2001). Lake Vostok is conventionally cited as an analog to Europa, as reported, for example, by Morton (2000), and Harland (2000). Because Lake Vostok seems to be isolated under a thick ice layer, this analogy helps bolster the canonical, but incorrect, view that Europa's ocean is similarly isolated from the surface. The description of NASA's position regarding the Defense Department's supposed lack of interest is from Space.com (February 11, 2003) quoting Project Prometheus head Alan Newhouse. The elegance and excitement of my research group's discoveries about Europa were highlighted in the article by Benson (2003). The closing quotation is from Galileo's *The Assayer* of 1623, quoted by Drake (1957, p. 239).

Benson, M. (2003) What Galileo saw. *The New Yorker*, September 8. A more complete original version is available online at
http://www.kinetikonpictures.com/books/additionaltexts.htm.

Drake, S. (1957) *Discoveries and Opinions of Galileo* (p. 239). Doubleday Anchor, Garden City, New York.

Fischer, D. (2001) *Mission Jupiter*. Copernicus, New York.

Harland, D. (2000) *Jupiter Odyssey*. Springer-Praxis, Chichester, UK.

Morton, O. (2000) Ice Station Vostok. *Wired*, April, 121–146.

Index

Age of surface 5, 14, 71, 97, 103, 112, 120, 176, 204, 210, 265–268, 325
Agenor Linea 77, 162, 255–62, 348–9
Aksnes, K. 99
Amergin 273–5, 278–9
analogies 33–4, 62, 156–7, 254, 324, 339
Antarctica 107, 338
apocenter 52–3, 55, 64, 80–4, 111, 149–50, 153, 197–8
apojove, *see also* apocenter 56, 92–3
Arctic 3, 24, 30, 69, 117–18, 128, 199, 219, 244, 354
arcuate crack 168, 189, 192, 198, 203, 207, 348
astrobiology 3–4, 2856, 323–36
Astypalaea 77, 145–53, 155–7, 161–3, 166, 178, 181–4, 186–9, 191–3, 195, 201, 207, 209, 211, 213, 217, 252–4, 269, 346
atmosphere 7, 12–13, 48, 79, 326, 339
azimuth sequence 107–8, 110, 145, 151–2, 157, 159, 174, 198, 210–18, 351–2

background plains 22
bands 11, 14, 134–48, 167–73, 176, 183, 184, 186, 189, 193, 202–3, 211, 213, 242, 251–2, 255–63, 273–4, 345–52
Belton, M.J. 32, 41, 123
bias, observational selection 33, 160–1, 227, 231, 287, 290

biosphere 6, 43, 323–36, 339
Bottke, W.F. 353
bright plains 22, 27, 114–15
brittle–elastic 57, 72, 123, 127, 132, 156, 253–4, 314–15, 321
Brunini, A. 268

Callanish 12, 269, 271–3, 281–2
Callisto 36, 57, 58, 85, 95, 280–3
canonical model 32, 34–5, 42, 52, 88, 97–8, 123, 132, 160, 193, 217, 229–33, 236, 248, 250, 253, 269–71, 276–7, 279, 285–7, 308, 317–18, 353
Canup, R. 95
Carr 98, 227–8
central peak craters 278–82
chaos (*see* chaotic terrain)
chaotic terrain 9–29, 57, 71, 97–9, 115, 117–19, 124–5, 128–31, 135, 137, 139, 161–2, 167, 175, 177, 181, 192, 200, 202, 204, 209, 211, 219–51, 254–5, 258, 268, 270–9, 283, 287–98, 300–3, 306–9, 316–17, 325–6, 329–30, 343–7, 349–52, 356
charged particles 7, 234, 326, 328, 332, 340
Cilix Crater 195, 278, 316
clathrate 309
clays 9
Coleman, S. 332–5
Collins, G. 318—20

376 Index

colors 11, 18, 21, 106–7, 121, 130, 168, 234, 258, 271, 273, 326
comets 5, 13, 266–8, 326, 353
Conamara Chaos 10–13, 16, 18–21, 23–6, 98, 125, 128–9, 160, 211, 219–23, 227–30, 232, 234, 236, 238, 241–2, 244, 248–9, 269, 271, 275–6, 287–9, 291–2, 316, 318
conjunction, orbital 51–2, 92–3, 96, 189, 268
contamination, biological 331–6, 339
convection, solid-state, within ice 29–31, 88–9, 97, 123, 128, 228–30, 236, 279, 283, 286–7, 290, 295, 306–8, 315, 318, 320–1
convergence 142, 146, 168–73, 178, 189, 191–3, 197, 251–63, 268, 348
convergence bands 171, 173, 178, 255–63
Coon, M. 117
Copernicus 35–6
core, planetary 7, 48–50, 68, 86
corrugation of crust 188–9, 252–4, 280
crack propagation 23, 131–2, 197–201
craters 12–16, 25, 50, 85, 115–16, 119, 185, 189, 195, 209, 225, 230, 233, 244, 265–9, 273–83, 293, 297, 314–17, 325–6, 343–4
cross-cutting lineaments 11, 19, 21, 25, 107–10, 121, 167, 176, 183, 191, 200, 204, 207–8, 210–16, 350
cryobot 338
cryovolcanism, *see* volcanism
cusps of cycloids 23, 168, 189, 191, 195, 198–200, 207, 348–50
cycloids 23, 84, 147, 165, 168, 170, 183, 185, 188–9, 191–205, 207–9, 213, 217, 258, 260–2, 268, 319, 324–5, 337, 348–50

dark margins 14, 20, 21, 23, 26, 27, 105, 107, 127–8, 130, 214–15, 242, 246
dark material 14, 130, 246, 255, 271, 343
Dark Pool 242, 244
darkening 18, 21, 27, 105, 130–131, 242, 244–5, 247, 271, 292
Darwin, G. 64, 66–7
day on Europa, *see* Europan day
Delaney, J. 250, 318
Delphi Flexus 192, 207–9

densely-ridged terrain 19, 21–2, 26–8, 117, 119, 211, 219, 341
diapirs 123, 125, 127, 295, 307–9, 314
dilational bands 14, 134, 136–42, 145, 148, 166–71, 178, 183–4, 186, 189, 191, 193, 203, 242, 251, 254–7, 261–2, 273–4, 341, 343–52
dilational ridges 130, 133–6
dinosaurs 5, 267
diurnal tides 23, 55, 56–7, 61, 63–8, 72, 80–5, 91, 94, 96, 111–12, 116–17, 120–3, 130–1, 133, 136–7, 146, 150–2, 158, 176, 178, 183, 197–201, 208, 209, 213, 244, 325, 328, 346, 348
Dones, L. 266
downwarping 127–9, 240, 244, 316

E4 orbit 18, 232, 275
E6 orbit 22, 24, 219–20, 222, 277, 287–8
E12 orbit 119, 223
E14 orbit 234, 282, 344
E15 orbit 161–2, 164, 167, 194–5, 210, 212–17, 228–31, 287, 298, 350
E16 orbit 182
E17 orbit 161–2, 165, 167, 182, 195, 210–11, 217–18, 234, 246, 273, 287, 294, 298, 349, 351
E19 orbit 161, 163, 175, 199
Earth, comparison with Europa 3–10, 13, 19, 29, 32–3, 47–9, 51–2, 61, 69, 72–4, 109, 117–18, 131, 142, 145, 191, 219, 227–8, 252, 266, 324–5, 330,
eccentricity, orbital 6, 30, 51–7, 59, 61–8, 80, 85–6, 89, 91, 93–4, 96–7, 268
ecosystem 3, 5, 6, 324, 328–31, 334
ejecta from impacts 12, 17, 19, 25, 189, 244, 269, 273, 330, 333
elastic properties and behavior 49, 57, 62, 71–4, 77–9, 87, 109, 117, 121, 123, 127–8, 132, 152–3, 155–6, 176, 198, 200, 253–4, 314–16
energetic charged particles 7, 234, 326, 328
episodic heating 96–9
Esposito, L. 335
Europa Orbiter 338–9
Europan day 51, 55, 121–2, 150–3, 197–8
Evil Twin of Agenor 258–62

Fagents, S. 131, 242, 247
Figueredo, P. 210–11, 229
Flexi, *see* cycloids
formation of the Galilean satellites 7, 95
Franklin, F. 99
frictional heating 6, 30, 50–1, 54–5, 57–8, 63, 68, 85–6, 94, 131, 314

G1 orbit 12, 106, 109, 112, 213, 215–17, 273, 297
Gaidos, E. 285–6, 314–16, 324, 326, 328
Galilean satellites 7–8, 30, 36, 39, 52, 54, 57–8, 75, 85, 91, 95–7, 106, 120, 233, 267–8, 280
Galileo Europa Mission, *see* GEM
Galileo Galilei 8, 29, 35, 51–2, 99, 285, 313, 353–4
Galileo spacecraft mission 5, 7, 9–11, 14–18, 21–3, 30–42, 63, 74–6, 88–9, 106, 123, 159–60, 182, 209, 217, 227, 229, 248, 297, 323, 328, 330, 337–8, 352, 354
Ganymede 36, 51–2, 54, 57–8, 85, 91–4, 99, 106, 268–9, 280–3, 326
Geissler, P. 87, 106–16, 149, 160, 182, 207, 210, 315, 324, 353
GEM 337–8
geological mapping 97, 160–1, 210, 233–4
Ghail, R. 309
squid, giant 285
Giese, B. 316
Gladman, B. 267
global resurfacing 13–14, 23, 234, 268, 325
Goldreich, P. 176–7
Goodman, J. 318–20
grain size, ice 18, 30, 71, 88, 107, 320–1
Greeley, R. 75–6, 117, 130, 181–2, 210–11, 229, 246

habitable niches 5–6, 28, 323, 329–30, 335
halos, dark, around chaotic terrain 16, 19, 26–7, 273, 292, 295
Hamilton, D. 94
Han, L. 321
Hansen, V. 233
Head, J. 75–6, 98, 123–6, 128, 183, 228–9, 246, 290–1, 314, 318
heat, tidal 6–7, 29–31, 50–1, 54–9, 68–71, 86–9, 94–9, 120, 176–7, 233, 238, 248–50, 269, 280, 286, 309, 318–24, 330
Helfenstein, P. 74–7, 103, 105
high-gain antenna 19, 92, 159
HIIPS 41
Hilda group of asteroids 268
Himalayas 142, 168, 252, 262
Hoppa 76–7, 80, 109–10, 113–15, 149, 151, 157, 159, 175, 177–8, 193, 197, 202, 207–8, 211, 219, 231, 246, 313, 324, 342, 353
hydrated salts 17, 271, 327

ice–grain size, *see* grain size, ice
illumination effects 18, 19–21, 113, 161, 195, 209, 229, 234, 245, 271, 273–4, 287, 292–3, 296, 306
image processing 40–1, 98, 110, 236
Imaging Team 22, 31–2, 34, 37–41, 75–6, 98, 110, 117, 123, 149, 181–2, 197, 227, 229, 233, 238, 246, 248, 275–8, 287, 323–4, 342
impact craters, *see* craters
Inertial Upper Stage 38
Io 9, 36, 39–40, 51–4, 57–8, 68, 76, 85–7, 89, 91–7, 99, 249, 326, 354
isolated ocean paradigm 5, 9, 21, 29–30, 34, 36, 41–2, 89, 167, 178, 251, 253, 270–1, 279–80, 285, 290, 313–14, 318, 324, 326, 332, 338–9, 340, 342, 349–50
isolated tilted rafts 346–7

Jeffreys, H. 64
Johnson, T. 250, 337
Jupiter 3, 5–8, 10–13, 30–1, 34–40, 49–58, 62–9, 73, 76, 78–80, 85–8, 92–7, 106–7, 109–10, 112, 115–17, 151, 157, 159, 177, 182, 201, 204, 207, 209–10, 216, 258, 268, 323, 325–30, 337–8, 352
Jupiter Icy Moons Orbiter (JIMO) 340

Kartchner Caverns 149, 323, 332
Katreus Linea 255–6, 258–9
Kattenhorn, S. 211
Kerr, R. 279–80, 286
Kirschvink, J. 285

Laplace, P.-S. 52, 93
Laplace resonance 51, 54, 57, 59, 68, 85, 87, 91–4, 96
lead ice 120–121, 130
leading hemisphere 115, 162, 174–7, 210–11, 223, 234, 291, 298, 300–1, 303–5, 351–2
Leake, M. 298, 302, 321
Lee, M.H. 96–7
Leith, A. 177
lenticulae 14–16, 26–7, 130, 191, 223, 236, 246, 286–7, 291–3, 295–6, 307, 321
Levison, H. 266
Libya Linea 77, 106, 146, 148, 195
life, *see* astrobiology
lithosphere 57, 72–4, 78, 80, 120, 123, 125, 127–30, 156, 189, 252–4, 307, 309, 314–17
Love, A.E.H. 49
Love number 49, 56, 66, 86

MacDonald, G. 66
MacDonald tide 67
magnetometer 204–5, 280, 325
magnetosphere 7, 234, 326, 328
Malhotra, R. 94
Manannán Crater 119, 273–6, 278, 282, 343–4
mantle (rocky) of Europa 7, 49–50, 68
Mars 160, 268, 331–3, 337, 341
matrix of chaotic terrain 23–5, 27, 219, 221, 223, 225–6, 229–30, 240–2, 244, 251, 270–2, 291–2, 325, 344, 346
McEwen, A. 105–6, 110
McKinnon, W. 88, 134, 145, 177, 251, 256
Melosh, J. 73–4, 122, 319
melting through the crust 4, 6, 9, 19, 24, 26, 28, 42, 98, 128, 130, 219, 223, 227, 231, 238–50, 271, 273, 286, 297, 308–9, 318–320, 325, 327, 346, 348
Mini-Mitten 225, 227, 288
Minos Linea 11, 13–15, 77, 105–12, 116, 207, 210–15
misconceptions 23, 25, 27, 68, 191, 203, 217
Mitten 227, 230, 241, 244–6, 288
mottled terrain (so-called) 16, 22, 27, 115, 119, 256, 296

multi-ring impact features 269, 273, 280, 282

NASA 31–2, 34–7, 41, 76, 120, 123, 145, 158, 160, 162, 182, 232, 266, 286, 308, 323, 330, 332, 334–41
National Research Council (NRC) 286, 331–5
Nealson, K. 285
Near-Infrared Mapping Spectrometer (NIMS) 9, 326–7
Nimmo, F. 314–17
NIMS, *see* Near–Infrared Mapping Spectrometer
Nolan, M. 192, 208
nomenclature 105, 191, 258, 271, 296
non-synchronous 6, 10, 49, 51, 55, 58, 63–9, 72–84, 103–17, 151, 157, 175–6, 178, 202–4, 207–11, 213, 216–18, 253, 325, 329
NRC (*see* National Research Council)

O'Brien, D. 87, 177, 248–50, 319–20
observational selection, *see* bias
ocean floor, *see* seafloor
Ockham, W. 35–6, 42, 71, 107, 205, 231, 246–7, 350–1, 354
Ockham's razor, *see* Ockham
Ojakangas, G. 96, 176–7
orbital evolution 6, 67, 91–9, 265
orbit number, *Galileo* mission, *see* E4, E6, E12, E14, E15, E16, E17, E19, G1
organic compounds 18, 130, 240, 326–9, 334
organisms 3, 6, 324, 328–31, 333–4, 336, 340–2, 348
oxidants 285, 326–9
oxygen 3, 9, 324, 326, 329

Pappalardo, R. 75–6, 98, 123–6, 178, 228–9, 231, 236, 248, 253, 286–8, 290–1, 314, 316, 318–20, 344
Parmentier, M. 74–7, 103, 105
Peale, S. 30, 54, 85, 94–7, 120, 308
pericenter 52–3, 55, 57, 64–7, 80–3, 111, 150, 268
perijove, *see also* pericenter 56, 93

permeable ice 5, 29, 42, 254, 314, 324, 326–7, 332, 334–5, 341–2, 352
Phillips, C. 98, 122
photosynthesis 328
Pierazzo, E. 278–81, 283
PIRL 98
pits 18, 161, 177, 240, 291–303, 306–10, 321, 344, 346
"pits, spots, and domes" 98, 228, 230–1, 279–80, 283, 286–98, 302, 306–8, 310, 314, 321, 344
planetary protection 331–6
polar wander 68, 78–9, 174–8, 202, 204, 218, 309–10, 352
Prime Directive 333–5
prime meridian 10
principal stress 73, 77–9, 103, 150
Prockter, L. 253
Proctor, A. 94
Prometheus 339–40, 343
PSDs, see "pits, spots, and domes"
pull-aparts 148, 163, 168–9, 183–6, 188–9, 195, 251–2, 346
Pwyll Crater 12–13, 16, 185, 270, 273, 275–8, 281–2

Q (tidal parameter) 62–3, 85–8, 293, 321

radiogenic heating 87
rafts in chaotic terrain 19, 23–8, 113, 117, 120–1, 128, 219–21, 223, 225–7, 229, 235, 240–3, 245–6, 251, 270–2, 275–6, 278, 291, 307, 309, 318, 325, 344, 346–7
Rathbun, J. 308–9
reactivation of cracks 214, 218, 329, 348, 350
recognizability of chaotic terrain, pits, and uplifts 25–6, 223, 225, 234, 236–8, 250, 288, 300, 302, 306, 325
regional mapping images 160–5, 167–8, 170, 174–5, 194–5, 199, 210, 212–15, 217–18, 223, 225–31, 234, 236, 269, 273, 287, 294, 298, 300, 305–6, 349–51
RegMap, see regional mapping images
resonance, see Laplace resonance
resonance capture, see orbital evolution
resurfacing, global, see global resurfacing

Reynolds, R. 120, 308, 323–4, 327
Rhadymanthys Linea 292, 296
Riley, J. 232, 234, 270, 288, 290, 298
RNA 330
rotation period 63, 112–13, 115–16, 207–11, 216–18, 325
Ruiz, J. 321
Rummel, J. 331

Sagan, C. 332–5
salts 17, 130, 204, 240, 271, 280, 326–7, 341
San Andreas fault 145
Sarid, A. 158, 162–3, 168, 176, 213, 298
Schenk, P. 134, 145, 251, 256, 278–83, 286, 317
Schubert, G. 88
seafloor 9, 134, 193, 252, 318, 330
selection effects, see bias, observational selection
shear displacement, see strike–slip
Shoemaker, E. 115, 297
Showman, A. 94, 321
Sickle dilational band 135–8, 141, 203, 343, 345–6
Sidon Flexus 192, 207–9
Spaun, N. 125
Spohn, T. 88
Squyres, S. 308
Star Trek 333
Star Wars 37–8
starlings 354
Stevenson, D. 96, 176–7, 248–9
stress 6, 23, 49–51, 55, 57–9, 71–84, 103–11, 116–17, 120, 122, 127–8, 130–2, 149–58, 176–8, 197–204, 208–11, 213–14, 216, 218, 229, 253–4, 262, 315–16, 320, 324–9, 348, 351
strike–slip 138, 141, 145–79, 181, 183, 185, 189, 193, 213, 218, 229–30, 251–2, 255, 257, 261, 287, 298, 302–3, 309, 314–16, 325, 346, 348
sub-jovian 10, 11, 50, 55, 69, 73, 78–80, 157, 176–7, 202–4, 258, 262
sub-Jupiter, see sub-jovian
subduction 142, 192–3, 252, 262
sulfur 9, 17, 326–7
surface area budget 139, 168, 173, 189, 251–2, 254, 261–3
surface convergence, see convergence

Taliesin Crater 282
taxonomy 22, 27, 98, 231, 290–1, 295, 306–8
tectonic terrain 9, 19, 24, 26–7, 115, 119, 133, 168, 176, 209, 211, 219, 221, 223, 225, 232, 234, 238, 242, 273, 298, 308, 346
Tegid 282
Tejero, R. 321
Thera Macula 162, 246, 271, 349
thermal effects, *see* heat and melting
thick–ice paradigm, *see* isolated ocean paradigm
Thomson, R. 250, 318
Thrace Macula 162, 192, 246–7, 271, 349
thrust faults 193, 252, 262
Thynia Linea 77, 106
tidal dissipation, *see* frictional heating and heating, tidal
tidal pumping of water through cracks 133, 137, 230, 251, 325, 328
tidal walking 146, 149–57, 159, 161–2, 174, 176, 189, 314–16, 325
Toomre, A. 176–7
trailing hemisphere 10–11, 64, 115, 157, 161–2, 174, 176–7, 298–310
triple-bands 14–15, 17, 21, 26–7, 105, 107, 109, 116, 123, 128, 167, 214–15, 242, 246, 260, 292, 341, 350
Tufts, B.R. 14, 76, 110, 128, 135–7, 140, 145, 147–9, 160, 178, 181, 193, 197, 246, 308, 316, 332, 335, 346, 349, 352, 353
Turtle, E. 122, 278–81, 283
Tyre Macula 139, 269–71, 273, 281–2

U.S. Geological Survey (USGS) 39, 98, 105, 115, 256, 259, 352
Udaeus 11, 13–15, 77, 105–12, 116, 207, 210–11, 213–15
uplift features 19, 161, 177, 189, 288–99, 302–10, 321, 346
USGS, *see* U.S. Geological Survey

Vening Meinesz, F. 73–5
viscosity 26, 31, 49, 55, 62, 68, 71–2, 79, 87–8, 123, 128, 142, 156, 189, 229, 245, 248–9, 253, 262, 297, 314–15, 321
volcanic vents, *see* volcanism
volcanism 9, 40, 54, 58, 85–6, 97, 134, 181, 192, 229–31, 243–6, 249–50, 285, 309, 324, 330, 309, 331
Vostok, Lake 338–9
Voyager mission 4, 7, 9, 11, 16, 22, 30–1, 39–40, 54, 63, 74, 76, 85, 87–8, 92, 105, 113–15, 117, 120, 123, 145–7, 177–8, 181, 183, 189, 191–2, 201, 207–9, 251, 256, 262, 271, 278, 308, 320, 323, 339, 352
Ward, W. 95

Wedges 11, 13, 134–6, 140, 145, 157–8, 193, 201–4, 234, 246, 256, 258–9, 261–2, 306–7
Wolfe, R. 115

Yoder, C. 94–5
Young's modulus 121

Zahnle, K. 266–8

Printing: Mercedes-Druck, Berlin
Binding: Stein+Lehmann, Berlin